CORPORATE RESPONSES TO
EU EMISSIONS TRADING

Global Environmental Governance

Series Editors: John J. Kirton, Munk Centre for Global Affairs, Trinity College, Canada and Miranda Schreurs, Freie Universität Berlin, Germany

Global Environmental Governance addresses the new generation of twenty-first century environmental problems and the challenges they pose for management and governance at the local, national, and global levels. Centred on the relationships among environmental change, economic forces, and political governance, the series explores the role of international institutions and instruments, national and sub-federal governments, private sector firms, scientists, and civil society, and provides a comprehensive body of progressive analyses on one of the world's most contentious international issues.

Recent titles in the series (full listing continued at the back of the book)

Corporate Responses to EU Emissions Trading
Resistance, Innovation or Responsibility?

Edited by

JON BIRGER SKJÆRSETH and PER OVE EIKELAND
The Fridtjof Nansen Institute, Norway

Routledge
Taylor & Francis Group

LONDON AND NEW YORK

First published 2013 by Ashgate Publishing

Published 2016 by Routledge
2 Park Square, Milton Park, Abingdon, Oxfordshire OX14 4RN
711 Third Avenue, New York, NY 10017, USA

First issued in paperback 2016

Routledge is an imprint of the Taylor & Francis Group, an informa business

British Library Cataloguing in Publication Data
A catalogue record for this book is available from the British Library

The Library of Congress has cataloged the printed edition as follows:
Skjærseth, Jon Birger.
  Corporate responses to EU emissions trading : resistance, innovation or responsibility? / by Jon Birger Skjærseth and Per Ove Eikeland.
     pages cm. -- (Global environmental governance)
  Includes bibliographical references and index.
  ISBN 978-1-4094-6078-7 (hardback)
1. Emissions trading--European Union countries. 2. Greenhouse gas mitigation--Economic aspects--European Union countries. 3. Corporations--Environmental aspects--European Union countries. 4. Industries--Environmental aspects--European Union countries. 5. Environmental policy--European Union countries.
I. Eikeland, Per Ove, 1963- II. Title.
  HC240.9.P55S447 2013
  363.738'76--dc23
                                                                    2012044831

ISBN 13: 978-1-138-27976-6 (pbk)
ISBN 13: 978-1-4094-6078-7 (hbk)

Contents

List of Figures

List of Tables

Notes on Contributors

Anne Raaum Christensen is a Researcher at the Fridtjof Nansen Institute. Her research interests include EU energy and climate policies, with particular focus on transport policies and power companies. She has published in these fields, including *EU Policy on Car Emissions and Fuel Quality: Reducing the Climate Impact From Road Transport* (with Lars H. Gulbrandsen, 2012).

Per Ove Eikeland is Senior Research Fellow at the Fridtjof Nansen Institute. His main research interests include European energy and climate policies, national energy transitions and corporate strategies, with a wide span of publications in these fields of research.

Lars H. Gulbrandsen is Senior Research Fellow and Director of the Global Governance and Sustainable Development programme at the Fridtjof Nansen Institute. He studies environmental politics with a particular focus on market-based policy instruments, non-state governance and corporate social responsibility. He has published a number of journal articles and is the author of *Transnational Environmental Governance: The Emergence and Effects of the Certification of Forests and Fisheries* (Edward Elgar, 2010).

Liv Arntzen Løchen is a Researcher at the Fridtjof Nansen Institute. Her research interests include EU energy and climate policies, with particular focus on the EU ETS and its effect on corporate climate strategies. She has also published in the field of adaptation, including *Adaptation to Climate Change Among Electricity Distribution Companies in Norway and Sweden: Lessons From the Field* (with Tor Håkon Inderberg, 2012).

Jon Birger Skjærseth is Research Professor at the Fridtjof Nansen Institute. His research interests include international climate negotiations, EU climate and energy policies and corporate strategies. He has published numerous articles and books in these fields, including *Climate Change and the Oil Industry* (with Tora Skodvin, 2003) and *EU Emissions Trading* (with Jørgen Wettestad, 2008).

Christian Stenqvist is Master of Science in Engineering and a PhD student at the Division of Environmental and Energy Systems Studies, Lund University, Sweden. He conducts research in the field of market and policy developments of industrial energy efficiency improvement.

Arild Underdal is Professor of Political Science at the University of Oslo and Center for International Climate and Environmental Research – Oslo (CICERO). Most of his research has focused on international environmental governance. Major publications include *Environmental Regime Effectiveness: Confronting Theory with Evidence* (with E.L. Miles and others, 2002), and *Regime Consequences: Methodological Consequences and Research Strategies* (co-edited with O.R. Young, 2004).

Jørgen Wettestad is Research Professor at the Fridtjof Nansen Institute. His main research interest is climate, air and energy policy at the global, EU and national levels. He has published numerous books and articles in this field, including *EU Emissions Trading* (with Jon B. Skjærseth, 2008) and *EU Climate Policy: Industry, Policy Interaction and External Environment* (with Elin L. Boasson, 2012).

Preface and Acknowledgements

The EU aims to put Europe on track toward a low-carbon economy. In this striking challenge, the EU Emissions Trading System (EU ETS) has been singled out as the key climate policy instrument with the ultimate goal to develop this system into an instrument for a global carbon market. In light of such hopes, what can be learned from past experiences with the EU ETS? And how can we best gauge whether and how the system has in fact contributed to starting a process of decarbonization in Europe? These questions have become especially relevant now that continuing the EU ETS as the key climate policy instrument is increasingly contested in Europe, while the rest of the world shows relatively little interest in copying the system.

Previous studies on the consequences of the EU ETS have been dominated by a focus on incentives for change mainly at the aggregate sector or EU level. Studies of single companies in specific industry sectors studies are also growing. This study differs in at least three ways. First, it covers general changes in corporate response at the sector level as well as presenting systematic comparative studies of companies in different sectors. The latter are applied to understanding causal effects and the role played by interaction between various policy instruments and other factors than the EU ETS for corporate climate strategies. Second, we explore a broader set of mechanisms that may link the EU ETS to company climate strategies: how does the system direct attention to opportunities for innovation? And how are corporate norms of responsibility affected by the EU ETS? We have chosen to include norms of responsibility because of the formidable challenges confronting the industry in light of the climate change problem and the EU aim for a low-carbon economy by 2050. The transformation needed in European industry in response to these challenges can hardly be accomplished unless fundamental norms of responsibility can be harnessed to work together with policies for steering company climate strategies. Finally, we explore corporate responses to the EU ETS in all major sectors covered by the system. With this book, then, we undertake a systematic examination of corporate responses to the EU ETS from a broad empirical and analytical social science perspective, covering companies in the main EU ETS sectors: electric power, oil, cement, steel and pulp and paper. The editors have drawn on their formal backgrounds in political science and economics in developing this broad analytical approach to how companies respond to the EU ETS.

Many scholars have provided us with valuable support and suggestions. Arild Underdal has served as a 'mentor' for the project and provided extremely helpful guidance. He has also contributed to Chapter 8. Anne Raaum Christensen has, in addition to writing the chapter on the cement industry, served as research assistant for the project. Special thanks go to Harro van Asselt and Jonatan Pinkse for reviewing parts of the manuscript as well as two anonymous reviewers. David Victor and Tora Skodvin have also provided helpful comments. We are also grateful to Susan Høivik, who has improved the English text considerably, and the Research Council of Norway's Renergi programme for funding the project on which this book is based. Last, but not least, we would like to thank representatives of the companies, industry associations, environmental movement, governmental organizations and the EU for taking the time to talk with us.

Per Ove Eikeland and Jon Birger Skjærseth
Lysaker, Norway

List of Acronyms

ACC	Aker Clean Carbon
ACCF	American Council for Capital Formation
AEII	Alliance of Energy Intensive Industries
BDI	Federation of German Industries
BIR	Business Improvement Review
BOF	Basic Oxygen Furnace
BTU	British Thermal Unit
CCGT	Combined-cycle Natural Gas-based Generation
CCS	Carbon Capture and Storage
CDM	Clean Development Mechanism
CDP	Carbon Disclosure Project
CDU	Christian Democrats
CEPI	Confederation of European Paper Industries
CERs	Certified Emissions Reductions
CH_4	methane
CHP	Combined Heat and Power
CO_2	carbon dioxide
CSI	Cement Sustainability Initiative
CSR	Corporate Social Responsibility
CWT	Complex Weighted tonne
DG Environment	European Commission's Directorate-General for the Environment
EAF	Electric Arc Furnace
ECCP	European Climate Change Programme
ECI	Energy, Climate and Innovation
ECJ	European Court of Justice
ECRA	European Cement Research Academy
ECSC	European Coal and Steel Community
ECX	European Climate Exchange
EEGI	European Electricity Grid Initiative
EEX	European Energy Exchange
EII	European Industrial Initiatives
EMS	Environmental Management Systems
EPTB	Environmental Products Trading Business
ERUs	Emission Reduction Units
ESTEP	EU Steel Technology Platform
ET	Emissions Trading

ETP	European Technology Platform
ETS	Emissions Trading System
EU	European Union
EU ETS	EU Emissions Trading System
EUAs	EU Allowances
EUROFER	European Confederation of Iron and Steel Industries
GCEP	Global Climate and Energy Project
GEMS	Global Energy Management System
GETS	Greenhouse Gas and Energy Trading Simulations
GHG	Greenhouse Gas
GRI	Global Reporting Initiative
HSD®	High Strength and Ductility
HSS	High Strength Steel
HTC	Heidelberg Technology Center
ICE	Intercontinental Exchange
IEA	International Energy Agency
IETA	International Emissions Trading Association
IMO	International Maritime Organization
IPCC	Intergovernmental Panel on Climate Change
JI	Joint Implementation
KP	Kyoto Protocol
MEPs	Members of the European Parliament
MISTRA	Foundation for Strategic Environmental Research
MRV	Monitoring, Reporting and Verification
NAPs	National Allocation Plans
N_2O	nitrous oxide
NER	New Entrants Reserve
Norske Skog	Norske Skogindustrier ASA
OPC	Ordinary Portland Cement
PPI	Pulp and Paper Industry
RDF	Refuse-derived Fuel
RD&D	Research, Development and Demonstration
R&D	Research and Development
RES	Renewable Energy Sources
RGGI	Regional Greenhouse Gas Initiative
RMS	Resource Management System
SCA	Svenska Cellulosa Aktiebolag
SEK	Swedish Crowns (Kroner)
SET plan	Strategic Energy Technology Plan
SPD	Social Democrats
SSAB	Swedish Steel AB
SSE	Scottish and Southern Energy
STEPS	Shell Tradable Emission Permit System
TREC	Tradable Renewable Electricity Certificates

ULCOS	Ultra Low CO_2 Steelmaking
UNFCCC	United Nations Framework Convention on Climate Change
UNIPEDE	International Union of Producers and Distributors of Electrical Energy
VAs	Voluntary Agreements
WBCSD	World Business Council for Sustainable Development
ZEP	Zero Emissions Platform/Zero Emissions Fuel Power Plants/Zero Emission Fossil Fuel Power Plants

Chapter 1

Introduction

Jon Birger Skjærseth and Per Ove Eikeland

The European Union (EU) Emissions Trading System (EU ETS) has been described as a 'grand policy experiment' – the first-ever international system for emissions trading (Kruger and Pizer 2004). Adopted in 2003 and launched in 2005, the EU ETS covers some 11,000 industrial installations in Europe accounting for almost half of EU carbon dioxide (CO_2) emissions. The EU has invested significant political capital in making this innovative cap-and-trade system work as the 'flagship' of its climate policy. The ultimate aim is to create a global carbon market by encouraging other major emitters, not least the USA and China, to follow suit.[1] The learning effect of the EU ETS could thus be tremendous. However, the recent economic recession and slow progress on an international climate agreement for the post-2012 period have put EU climate policy and the EU ETS to the test.[2] Some actors would like to see the EU ETS experiment fail, so as to provide stronger signals for low-carbon investment. Others argue that the system should be strengthened as key driver for decarbonization in Europe (van Renssen 2012). Whether the EU ETS stays on course, keeps afloat or sinks may have ramifications far beyond the EU itself. By exploring how the system actually works on the ground, we seek to shed light on whether the system has helped to start a process towards low-carbon corporate strategies.

With the EU ETS as core instrument, the EU aims to put Europe on a long-term track toward a low-carbon economy. EU leaders have agreed that greenhouse gas (GHG) emissions are to be cut by 20 per cent by 2020 compared to 1990 levels and by 80–95 per cent by 2050 to contribute to limiting the rise in global

---

1   The EU ETS applies to the 27 current EU member states. In addition, Norway, Iceland and Lichtenstein have emissions trading systems linked to the EU ETS. Switzerland, New Zealand and the USA also have emissions trading systems. The USA has one system covering California and one regional system (Regional Greenhouse Gas Initiative, RGGI) covering nine eastern states. A Western Climate Initiative is under development, including California and four Canadian provinces. Emissions trading systems are also being developed in the Republic of Korea, Japan and China (with pilot systems in several cities and provinces).

2   The carbon price is at the time of writing only 6 euros – many times lower than foreseen by EU policymakers.

temperature to 2°C.[3] The long-term task is indeed daunting: to promote almost full decarbonization by 'driving a wide range of low-carbon technologies into the market'.[4] This will entail speeding up the development and diffusion of new breakthrough technologies (Egenhofer et al. 2011).[5] One example is Carbon Capture and Storage (CCS), set to play a key role in meeting EU long-term climate targets.[6]

Studies of the EU ETS abound (e.g. van Asselt 2009, Convery 2009). The story of how the system was initiated, decided and implemented has been analysed in detail (Skjærseth and Wettestad 2008a, b, 2010a, b). We know a great deal about the economic incentives and how they are expected to drive the system toward abatement and innovation. We are also learning more about the actual short-term aggregate effects at sector and EU levels (Ellerman and Buchner 2008, Ellerman et al. 2010) – see below for a more comprehensive review of the literature.

However, less is known about *how* the EU ETS affects *companies* in different sectors. Does it influence their interests, beliefs, norms, strategies, long-term innovation plans and deployment of low-carbon solutions? These questions are of particular salience because private sector companies will be the key transformation agents in dealing with the long-term challenge of climate change. The necessary changes are not only a matter of economic incentives and incremental innovation: ultimately, what is needed is a fundamental transformation of Europe's industries, including changes in company norms of responsibility.

With this book we want to contribute to filling this knowledge gap about how the EU ETS actually works on the ground, and to see whether the system has helped to spur a process towards formulating and implementing long-term low-carbon strategies. This aim led us to embark on a systematic examination of corporate responses to the EU ETS from a broad empirical and analytical social science perspective. The study covers all the main ETS sectors: electric power, oil, cement, steel and pulp and paper industries – which account for most CO_2 emissions covered by the EU ETS. To what extent have companies in these sectors been affected by the EU ETS? How have they responded? Why have

---

3   European Commission (8 March 2011), COM (2011)112 final, Communication to the Council: 'A Roadmap for moving to a low-carbon economy in 2050' (Brussels).

4   European Commission (8 March 2011), COM(2011)112 final, Communication to the Council: 'A Roadmap for moving to a low carbon economy in 2050' (Brussels): 6. See also European Commission (15 December 2011), COM(2011)885/final, Communication to the Council: 'Energy Roadmap 2050 (Brussels).

5   New technologies will reduce the risk of lock-in into high carbon intensive technologies, which could make industries uncompetitive when carbon carries a higher price (Egenhofer et al. 2011).

6   For the role of CCS, see European Commission (15 December 2011), COM(2011)885/final, Communication to the Council: 'Energy Roadmap 2050 (Brussels). A weak carbon price threatens demonstration and deployment of this technology. See ZEP, 2012. Securing the business case for CCS as a key enabler for the decarbonisation of Europe. Zero emissions platform (ZEP) Strategic Review, January 2012 (Brussels).

there been such apparently differing responses to one and the same system? This final question points to the conditions under which different response strategies develop and unfold. The ETS does not operate in a vacuum. What we seek here is a better understanding of how different drivers may co-produce outcomes – do they reinforce or counter the effect of the ETS?

Our study builds on the literature on the relationship between regulation and corporate strategies within a broader institutional approach. Earlier strategy studies may indicate that regulatory pressure is likely to affect corporate climate strategies – but how companies will respond and why they respond in specific ways are still contested questions. The literature has viewed environmental regulation like the EU ETS as either a threat to company profitability and competitiveness, or alternatively as an opportunity to improve profitability and competitiveness through abatement and innovation. Other contributions have taken the social risk of environmental challenges as a point of departure, arguing that corporate norms of responsibility can be strengthened or weakened by new regulation (see Chapter 2).

This chapter begins by introducing the EU ETS, its basic generic features and how it evolved in the EU context. Next, we provide an extended review of earlier studies focusing on the consequences of the EU ETS. We then outline the research approach of this volume, followed by a brief introduction to the chapters that follow.

Evolution of the EU ETS: A Snapshot[7]

The EU ETS is a cap-and-trade system covering about 11,000 installations owned by the electricity-producing and energy-intensive industry companies, representing close to 50 per cent of CO_2 emissions in the EU. Under a cap-and-trade system, each company typically starts the year with a certain number of tons allowed emitted – implying a right to pollute. The company then decides how to use its allowance: it might restrict output, switch to a cleaner fuel, or cut emissions in other ways. Companies with surplus allowances can sell to other companies. With a sufficient volume of allowances traded in the market, a clearing price should emerge that would reflect the marginal costs of CO_2 reductions for the entire system. This price would provide individual facilities with a yardstick for comparing the least costly modes of compliance: investments in abatement, purchases of allowances or acceptance of any penalty set for excess emissions compared to allowances held. The market price for carbon will thus give important signals for changing corporate behaviour. For this price signal to see an upward trend the market should experience scarcity in allowances, with fewer allowances handed out than the projected need.

---

7   This section draws heavily on Skjærseth and Wettestad 2008a, 2008b, 2009, 2010a. 2010b. See also Wettestad 2005 and Skjærseth 2010.

The total cap of the EU ETS became a hotly discussed design element, alongside the method of allocating allowances – for free, for payment (auctioning) or a combination. The allocation method is important as a distribution mechanism for the economic value of the allowances, but auctioning would also impose extra regulatory pressure on the companies as compared to free allowances. Other design elements to evolve since the EU ETS was launched include the degree of harmonization and centralization of the system with regard to allocation of allowances, the coverage of the system in terms of sectors and GHGs, monitoring, verification and enforcement, guidelines on how to use the revenues from auctioning to support 'climate-friendly' technology, and restrictions on the import of external credits for compliance with the EU ETS.

The EU ETS has evolved through various stages, from its inception in the late 1990s to the revisions made in 2008 for the post-2012 period. The system was formally adopted in July 2003 and launched on 1 January 2005, with a pilot phase from 2005 to 2007 as the first step. This was followed by the Kyoto Protocol commitment phase 2008–2012 as the second step. On the basis of experience gained, in December 2008 the EU adopted a revised system for the post-2012 phase (2013–2020). This revised EU ETS formed part of a larger EU climate and energy package which included binding policies on sectors not covered by the ETS, renewables, CCS, fuel quality and vehicle emissions. In the following, we detail how the EU ETS evolved for the two first trading phases and the revision of the system for the third trading phase.

The Making and Implementation of the EU ETS

The 1997 Kyoto Protocol established emissions trading as an optional mechanism for the global deal on reducing GHG emissions, in line with the preferences of the USA.[8] The EU was initially sceptical. And yet, only five years later, the EU agreed on the world's first international emissions trading system for large industrial power-producing and energy-intensive emitters. The reasons for this turnabout were both external and internal. The Kyoto Protocol provided for the optional use of such an instrument for trade between countries, and fixed the target and the timeframe for the EU commitment to reduce emissions by 8 per cent (from 1990 levels) between 2008 and 2012.[9] Although the Kyoto Protocol did not enter into force until 2005, the EU perceived this commitment as mandatory and difficult to achieve without new EU-level climate policy instruments. Economists in the European Commission's Directorate-General for the Environment (DG Environment) acted as policy entrepreneurs – taking the initiative, building up knowledge and crafting support among reluctant stakeholders, including parts of

---

8   The Kyoto Protocol has provisions for emissions trading between countries, not industrial installations as in the EU ETS.

9   The EU commitment under the Kyoto Protocol was based on different reduction targets for each of the then 15 EU member states, through a burden-sharing agreement.

industry, green groups, Members of the European Parliament (MEPs) and most member states, keen to avoid a repetition of the failure in the early 1990s to get support for an EU carbon/energy tax (Skjærseth 1994).

The first hints regarding an EU-wide emissions trading system were given by the European Commission in the late spring of 1998.[10] Based on a June 1998 Communication on the EU's post-2012 strategy, the idea of such a system was taken one step further in a May 1999 Communication on preparing for the implementation of the Kyoto Protocol.[11] This Communication envisaged a system that would initially be fairly narrow in scope (with regard to the coverage of industry sectors and greenhouse gases), targeting only large emitters or a single economic sector, and CO_2 emissions only. Issues of harmonization and central EU control were mentioned only briefly.

The March 2000 ETS Green Paper outlined specific design proposals.[12] These included an implicit recommendation for the total number of allowances to be determined at the EU level, and that the system should cover six sectors, with electricity production by far the largest. In order to sell the idea to reluctant stakeholders, the Commission set up a working group under the European Climate Change Programme (ECCP), where representatives from industry, green groups and the European Commission met regularly in the period March 2000–January 2001. Already at this point then, corporate actors had good reason to take seriously the Commission's plans for an EU-wide trading system.

The framing of the ETS proved effective in reducing resistance and building support. To industry stakeholders, the instrument was framed as a cost-effective tool that could provide economic opportunities for decreasing emitters to sell their allowances. To green groups as well as the European Parliament, it was framed as environmentally effective – a tool that, if appropriately designed, would automatically bring emissions down to the cap set. To governments, these arguments were combined and linked to their commitments for implementation of the burden-sharing agreement and the Kyoto Protocol targets. Initially, only Denmark, Ireland, the Netherlands, Sweden and the UK supported emissions trading within the EU, with the remaining 10 members either opposed or indifferent.

The Commission plans were significantly advanced by the decision of newly elected US President George W. Bush *not* to ratify the Kyoto Protocol. The US exit served to unify the positions within and among the EU actors and institutions in supporting the EU ETS as the key measure for ensuring the entry into force of

---

10   European Commission (June 1998), COM(98)353, Commission Communication to the Council and the Parliament. Climate change – Towards an EU Post-Kyoto Strategy (Brussels).

11   European Commission (19 May 1999), COM(99)230, Commission Communication to the Council and the Parliament. Preparing for Implementation of the Kyoto Protocol (Brussels).

12   European Commission (8 March 2000). COM(2000)87 final, Green Paper on Greenhouse Gas Emissions Trading within the European Union (Brussels).

the Kyoto Protocol.[13] In October 2001, the European Commission adopted the proposal for the Emissions Trading (ET) Directive.[14] The Commission had taken the various member state interests into account, and the final proposal included decentralized allocations of allowances at the member state level. The final ET Directive was agreed in late 2002 and formally adopted by the Council in July 2003.[15] The main shape and content of the proposed directive remained intact through the complex EU decision-making process.[16] By this time, corporate actors knew for certain that a mandatory cap-and-trade system would materialize.

In 2004, the EU adopted the associated Linking Directive connecting the EU ETS to the Kyoto Protocol's flexible project mechanisms, the CDM and JI.[17] This Directive admitted the use of credits stemming from such projects to comply with the EU ETS. These credits would be more affordable than EU ETS allowances, due to the lower abatement costs in developing countries.[18] Together, these two Directives established a three-year pilot phase (2005–2007) of the EU ETS to precede the main commitment period of the Kyoto Protocol (2008–2012).

The central issue concerning implementation of the ET Directive was decentralized cap-setting in the form of National Allocation Plans (NAPs).[19] These

---

13   A direct linkage was made to efforts to save the Kyoto Protocol by accompanying the ET directive proposal with a proposal for a Council Decision to ratify the Kyoto Protocol.

14   European Commission (23 October 2001), COM(2001)581, Proposal for a Directive of the European Parliament and of the Council Establishing a Framework for Greenhouse Gas Emissions Trading within the European Community and Amending Council Directive 96/61/EC (Brussels).

15   European Parliament (25 October 2003), Directive 2003/87/EC of the European Parliament and of the Council of 13 October 2003 Establishing a Scheme for Greenhouse Gas Emission Allowance Trading within the Community and Amending Council Directive 96/61/EC, Official Journal of the European Union L 275, 32–46.

16   The proposal required adoption by a qualified majority in the Council of Ministers and co-decision with the European Parliament. Qualified majority is a weighted voting system based on the principle of giving larger states more votes than smaller ones. The co-decision procedure became instrumental in getting the two biggest EU emitters, the UK and Germany, to accept a mandatory system, as opposed to their initial preference for only a voluntary EU ETS in the first pilot phase (2005–2007). The co-decision procedure also gave the European Parliament, generally seen as the 'greenest' of the EU institutions, the right to propose amendments or veto, but in the end, this had little to say for the result.

17   European Parliament (13 November 2004), Directive 2004/101/EC of the European Parliament and of the Council of 27 October 2004 Amending Directive 2003/87/EC Establishing a Scheme for Greenhouse Gas Emission Allowance Trading within the Community, in Respect of the Kyoto Protocol's Project Mechanisms, Official Journal of the European Union L 338, 18–23.

18   Hence, the use of external credits is generally expected to lead to downward pressure on prices and lower incentives to abatement within the EU itself.

19   Other issues included providing guidance to industry, establishing relevant institutions on monitoring, verification and reporting, and setting up national allowance registers.

NAPs would determine the environmental ambition of the system: set the cap, determine which sectors get how many allowances, and provide the allocation method. According to the ET Directive, at least 95 per cent of the allowances were to be allocated free of charge for the pilot phase, and at least 90 per cent for the second phase starting in 2008. The production of NAPs started in the summer of 2003. The European Commission was given the authority to assess the NAPs and reject those found not in compliance with the relevant provisions of the Directive's Annex III – particularly the need to be on a path towards the Kyoto target in the period 2005–2007 and to be consistent with the Kyoto targets in the period 2008–2012.[20]

When the first verified EU ETS emission figures (for 2005) were announced in mid-May 2006, these indicated strongly that NAP I allocations had in fact distributed large quantities of excess allowances. The figures showed CO_2 emissions at about 80 million tonnes, 4 per cent lower than the number of allowances allocated for that year (Ellerman and Buchner, 2007).[21] This sent shock waves into the allowance-trading market, with the market price for allowances plunging from a top level of around €30 per tonne CO_2 in late April 2006 to around €12 in early May, and further down in the spring of 2007, hitting a low of only €0.5 by the end of April.

To avoid repetition of such a calamity, the Commission in 2006/2007 used the verified 2005 emissions data for its assessments of the NAPs presented for the phase 2008–2012. National emissions from the ETS sectors were set at an average of around 6.5 per cent below 2005 emissions levels, apparently ensuring a higher carbon price. All in all, the decentralized nature of cap-setting resulted in a 'race to the bottom', where each member state had incentives for allocating generously to protect its own industries.

Revision

Already in October 2005, the Commission launched a second phase of the European Climate Change Programme (ECCP II) at a conference attended by over 450 delegates representing all major stakeholders. The purpose was to take stock of experience to date and develop new policies beyond 2012. A review undertaken in 2007 identified various problems and concluded that a more harmonized emissions trading system was imperative.[22] In March 2007, the European Council adopted a new target for climate policy – to achieve at least a 20 per cent reduction

---

20   In 2004, the Commission emphasized that the total quantity of allowances allocated should not be in excess of actual or projected emissions.

21   Some of this gap may have been due to actual reductions and lack of reliable data on historical emissions.

22   See the introduction to the revised ET Directive. 2009. Directive 2009/29/EC of the European Parliament and the Council of April 2009 amending Directive 2003/87/EC

of GHG emissions by 2020 compared to 1990.[23] The Council pointed to the EU ETS as the main climate policy measure to realize this target. In January 2008, the Commission put forward a proposal for a revised EU ETS for a third phase, 2013–2020.[24] This proposal suggested significant changes: a more harmonized and centralized ETS, introducing auctioning as the general principle for allowance allocation, and more restrictive rules on importing credits from third countries. A modified version of the Commission's proposal was finally adopted by the European Council and the Parliament in December 2008, formalized in 2009.[25]

The changes to the existing EU ETS were significant. First, the revised ETS introduced a single EU-wide cap on allowances to be reduced in a linear manner by 1.74 per cent annually from 2013 onwards.[26] Second, allowances were to be allocated on the basis of fully EU-harmonized rules, scrapping the system of NAPs. Third, auctioning was introduced as the main allocation method for the power sector from 2013.[27] Eighty per cent of the revenues from auctioning were to be distributed among the member states according to their ETS emissions in 2005 (or average 2005–2007).[28]

For other ETS industries, the new system established a gradual phase-in of auctioning (minimum 20 per cent of allowances in 2013, increasing to 70 per cent by 2020, with a view to reaching 100 per cent by 2027). Further, under the revised ETS, industrial sectors or sub-sectors particularly exposed to global competition, and hence in danger of 'carbon leakage', would get allowances

---

so as to improve and extend the greenhouse gas emission allowance trading scheme of the Community. Official Journal of the European Union, L140/63.

23   European Council. 2007. Presidency Conclusions 8/9 March 2007. 7224/07-Brussels: European Council,. 2 May.

24   Proposal for a Directive of the European Parliament and of the Council amending Directive 2003/87/EC so as to Improve and Extend the Greenhouse Gas Emission Allowance Trading System of the Community. Brussels: COM(2008) 16 provisional, 23 January.

25   Directive 2009/29/EC of the European Parliament and the Council of April 2009 amending Directive 2003/87/EC so as to improve and extend the greenhouse gas emission allowance trading scheme of the Community. Official Journal of the European Union, L140/63

26   This would lead to a reduction of ETS emissions by 21 per cent by 2020 relative to 2005. This aimed at producing scarcity of allowances and higher allowance prices compared to the first two phases.

27   There was a differentiation within the power sector. Installations poorly integrated into the EU electricity grid that were operational or under construction no later than the end of 2008, or which individually provided more than 30 per cent of national electricity in countries with relatively low GDP (such as Poland and other Central and Eastern European countries), could get up to 70 per cent of all allowances free of charge in 2013, decreasing gradually to zero by 2020.

28   Ten per cent of these revenues in turn would be redistributed from high to low per capita income member states, to strengthen the capacity of the latter to invest in climate-friendly technologies. An additional 2 per cent was to be distributed to countries whose GH emissions in 2005 were at least 20 per cent below their Kyoto base-year emissions.

for free in the period 2013–2020. Transitional free allocation to installations, however, would be based on harmonized benchmark rules taking the most energy-efficient techniques, substitutes and alternative production processes into account. Allocation of allowances based on benchmarking would provide a competitive advantage to the most energy-efficient installations. Another mandatory element was the use of 300 million allowances from the new entrants' reserve to support up to 12 carbon capture and storage (CCS) demonstration projects and projects demonstrating renewable energy technologies.

The rules on the use of CDM and JI credits are now included in the ET Directive itself. Transfer of CDM and JI credits is permitted from the second to the third trading period, giving companies roughly similar principal access to such credits as before. The scope of the system has also been expanded to new sectors (petrochemicals and aluminium) and greenhouse gases. In 2012, international aviation was included. This expansion has been postponed, opposition coming particularly from the USA, China and India.

The EU ETS revision formed part of a broader EU post-2012 energy and climate strategy. In addition to the 20 per cent climate target, the European Council stressed the need to increase energy efficiency by 20 per cent relative to projections for 2020; the Council adopted a binding target of a 20 per cent share of renewable energy in total EU energy consumption by 2020, including a 10 per cent binding minimum target for renewables in total EU transport fuel consumption.

The ETS reform was the centrepiece of a package of climate and energy policy instruments, presented by the Commission in January 2008, that also included: (1) a decision on effort-sharing among member states in the form of differentiated targets for sectors not covered by the EU ETS such as transport and agriculture, (2) promotion of renewable energy sources, and (3) frameworks for CCS. New legislation was also underway for reducing CO_2 emissions from new cars and for fuel quality – the latter including a requirement to reduce emissions from 'well' to 'wheel'.[29] Seen from the perspective of industry, the interaction among these various policies was to become increasingly important.

In January 2009, the EU sought to consolidate the EU ETS by facilitating an agreement, intended for the upcoming Copenhagen Summit, of an OECD-wide carbon market by 2015. This ambition seemed quite realistic, given the signals from President Obama that a federal cap-and-trade system would be created in the USA as well. Then those plans stalled in the US Congress, and an OECD-wide carbon market became totally unrealistic for the foreseeable future, also threatening future support of the unilateral EU ETS. Additional challenges to the EU ETS came from the financial crisis, estimated to account for an 11 per cent

---

29 The European Council's success in adopting a more stringent ETS in December 2008 on the basis of unanimity in the Council despite the global financial crisis can be explained partly by the EU's package approach. The ETS revision was negotiated in parallel with other legislative proposals, providing ample room for issue linkages, fairness in burden sharing and mutually reinforcing climate and energy policy goals.

drop in EU ETS emissions in 2009 relative to 2008, and thus for a plunge in demand for allowances (with average EU allowances prices falling from €24/t for 2008 to €14/t in 2009).[30] The ensuing surplus of allowances available in the second phase could, as noted, be saved for use until 2020, lessening the need for abatement in the third trading phase. These challenges, and others, are discussed in the concluding chapter.

Consequences of the ETS: Review of the Literature

Several studies have analysed the consequences of the EU ETS with particular emphasis on abatement and innovation. In this section, we present some of this literature, so as to take stock of the development of state-of-the-art knowledge about the EU ETS, while also identifying research gaps.[31]

The first generation of studies focused on expected (*ex-ante*) effects of the ETS. These studies made an effort to predict whether and how the system would influence the sectors covered by the system, generally asserting that the initial incentives for innovation under the system were rather weak. These studies argued that lenient emission targets, allocation of allowances for free through 'grandfathering', generous opportunities to use of CDM/JI credits for ETS compliance purposes, and future carbon price uncertainty would result in incentives for innovation that were at best modest (e.g. Brewer 2005, Gagelmann and Frondel 2005, Egenhofer et al. 2005).

The next generation of studies centred on *ex-post* evaluations after the first 2005–2007 trading phase. These studies concluded that while the EU ETS had entered the boardrooms and affected the corporate agenda (through, for instance, requirements as to monitoring, reporting and verification), the system had not induced any fundamental shifts in corporate strategy and corporate investments decisions (Egenhofer 2007, Hoffmann 2007, Ikkatai et al. 2007, Kenber et al. 2009). Ellerman, Convery and de Perthuis (2010) provide the most extensive account of actual short-term CO_2 abatement in the sectors covered by the ETS, and conclude that there has been abatement in both the energy-intensive and power sectors since the introduction of the EU ETS, despite 'over-allocation' of allowances and the generally low CO_2 price. One example mentioned is the switch from coal-powered stations to gas-powered ones. Other scholars have contested such findings, holding that changes in emissions are difficult to attribute to the ETS as such (McIlveen and Helm 2010).

Since 2009, several studies have focused on the long-term effects of the EU ETS, with greater emphasis on technological innovation and long-term strategy

---

30   Bloomberg New Energy Finance Press Release 18 January 2010, 'The global carbon market increases by 5 per cent in 2009'.

31   We would like to thank Anne Raaum Christensen for preparing the basis for this literature overview.

changes. Some general lessons can be drawn from these studies: the EU ETS has been exerting an impact on innovation in firms (especially in the power sector), but, due to design flaws and teething problems, the effects have been limited and varying (see e.g. Cames 2010, Rogge and Hoffmann 2010, Egenhofer et al. 2011, Martin, Muûls and Wagner 2011, and Rogge, Schmidt and Schneider 2011). The limited impact is due to the scheme's initial lack of stringency and predictability, as well as the importance of contextual factors. Egenhofer et al. (2011) see a positive impact of the EU ETS on innovation in the power sector, but as an integral part of the wider policy mix, including feed-in tariffs for renewables.

Most studies agree that this picture can (and probably will) change in the next phase (2013–2020) of the EU ETS: auctioning in the power sector, allocations based on benchmarks in energy-intensive industries significantly exposed to international competition, a tighter cap, greater predictability and a higher carbon price will provide more positive incentives for innovation (see e.g. De Bruyn et al. 2010, Johnstone et al. 2010, Martin, Muûls and Wagner 2011, and Rogge, Schneider and Hoffmann 2011). Lack of 'stringency' is the factor most often mentioned when scholars seek to explain why the EU ETS induced relatively little innovation in the first phases.

These valuable studies show that our understanding of the consequences of the EU ETS is expanding, but also that there is a need for broader empirical and analytical studies to learn more. For one thing, previous studies have been dominated by a focus on economic incentives as the only mechanism. Incentives are clearly important for changing company behaviour, but a broader analytical approach is also needed in order to capture additional mechanisms and processes. In the present study, we seek to achieve this by systematically applying different perspectives or models on corporate response to regulation, based on various assumptions of how companies act and the factors affecting their behaviour. A second, related, point involves gaining a better understanding of how individual companies actually respond to the EU ETS and the wider policy context, particularly in their long-term innovation plans. There have been few systematic company studies in previous research, and such analyses can strengthen our understanding of how – through which mechanisms – the ETS actually works. Third, empirical studies to date have focused mainly on the power sector in Germany. We need to learn more about the system particularly in the energy-intensive industries: steel, cement, pulp and paper and oil sectors.

Research Approach

The principal aim of this study is to understand how companies respond to the EU ETS. As in any study of institutional consequences, this entails the risk of overstating the effects of the selected institutional arrangement. Researchers may, like companies, become myopic in their analysis. To reduce this problem and keep an open mind for a richer understanding of company responses to regulation, we have selected *corporate climate strategies* as our main focus for explanation. The aim of the study can thus be reformulated as '*to understand the consequences of the EU ETS for corporate climate strategies*'. As climate strategies are shaped also by other factors than anticipated or actual regulation, this point of departure implies that we do not assume that the EU ETS works in isolation from other drivers. It is neither feasible nor necessarily most relevant to try to 'isolate' the effect of the EU ETS. However, we will pay greater attention to the relationship between the EU ETS and climate strategies than other explanatory factors. We hope to reduce the risk of overstating the effect of the ETS by focusing on climate strategies, but it cannot be eliminated. Readers should bear this particular caveat in mind throughout this book.

The selection of corporate climate strategies as our focus for explanation gives rise to several methodological challenges and pitfalls. The first point to note is that we seek to understand how and why different strategies change and vary, rather than ranking companies in some sort of green 'beauty' pageant. Conceptualizing and measuring climate strategies is not an easy task. One challenge is to distinguish between general corporate (business) strategies and climate strategies; another is to separate the public profile of a company from its actual behaviour. We seek to move beyond simple rhetoric by exploring what companies actually do, particularly in terms of significant investments and activities with potential long-term implications for future operations. These methodological challenges are further discussed in Chapter 2.

Second, how can we then establish a causal relationship between the EU ETS and climate strategies? We should first note that some companies had a climate strategy in place before the EU ETS was adopted; others did not. This means that in some cases the key question is how the EU ETS has affected pre-existing strategies, whereas in other cases the EU ETS may have spurred new strategies. One implication is that we need to study corporate strategies prior to 2003, the year the EU ETS was formally adopted by the EU before trading started in 2005. A related challenge is that companies may respond to expectations about planned changes in the regulatory environments. The first vague signals of emissions trading in Europe can be traced back to 1998. The implication is the same as above, but distinguishing between pre-2003 climate strategies affected by ETS anticipations or other drivers is extremely difficult. This challenge will be made explicit in the case studies.

The research design aims to capture how the ETS interacts with other factors to co-produce climate strategies. The case studies will first analyse changes and differences in corporate climate strategies. The next step involves exploring the link between change in strategies, the EU ETS and interaction with other relevant climate and energy policies at the EU level. The final step is then to analyse how other company-external factors (such as relevant national climate and energy policy) and company-internal factors affect ETS response and climate strategies. The structure of the case studies is the same throughout the book, but each author has singled out the most pertinent tracks that can contribute to a richer understanding of the topic under scrutiny. This in turn implies that some cases are more technical than others.

In this endeavour, we apply three perspectives or models of corporate response to regulation, on the assumption that ETS companies will respond in one way or another. These are not models in a formal sense, but are used as three different analytical 'lenses' that indicate different behavioural assumptions and mechanisms linking regulation to response strategies. The essence of our first model is that companies will adapt reluctantly to new, mandatory, regulation. This model sees companies as profit-maximizing actors that respond by minimizing the short-term costs of compliance. The second model predicts that companies will do something more than merely adapt in terms of cost minimization. Companies are here seen as myopic, needing appropriately designed regulation that can enable them to recognize new innovative low-carbon opportunities that may benefit the company in the short- and the long-term. The final model depicts companies as socially responsible actors. Managers have mixed motivations – including norms of responsibility that may be 'crowded out' or 'crowded in' by new regulation.

These models, based largely on a perspective that treats companies as unitary actors, are subsequently extended. We include factors from the 'inside' and more widely from the 'outside' of the company, so as to capture how climate strategies are affected by factors beyond the EU ETS and other relevant EU policies. This extension to internal factors such as company leadership, and external factors like national policies, should alert us to other factors affecting climate strategies than the ETS, and help us to identify the conditions under which different responses and strategies occur.

The 'models' will be assessed by various methodological techniques, including pattern matching, explanation building and heuristic value. Pattern matching aims to compare expectations with observations. The better the match, the more confidence will we have in the model's explanatory power. Explanation building takes into account that our knowledge base may restrict our ability to extract empirically assessable expectations, and is used when the final explanation is not fully stipulated at the outset of the study. This approach is based on 'narratives'; it is particularly useful for capturing complex relationships between companies and the context within which they operate (see Skjærseth and Skodvin 2003).[32] Finally, we will assess the heuristic value of our three models or perspectives. The heuristic value is specifically concerned with the mechanisms that link regulation to corporate response. As pointed out by Underdal and Hanf (2000), different ways of assessing models can add separate insights into causal relationships, with each way providing insights that the others cannot.

We have selected five EU ETS sectors – electric power, oil, steel, cement, and pulp and paper – and two companies within each sector. The purpose of the

---

32 Causal complexity refers to situations where the direction of influence between two variables depends on the value of a third variable, making the direction of influence difficult to determine (Ragin 1987).

sector studies, based on analysis of the relevant European industry association and the major EU ETS companies, is to understand change in aggregate company strategies. As to the comparative company studies within each sector, the aim is to understand how and why strategies change and vary. The companies within each sector have been selected according to two main criteria: expected differences in climate strategies based on pilot studies, and access to information. The first criterion is applied to shed light on the conditions under which different strategies emerge, as differences in climate strategies between ETS companies in the same sector are likely to result from different combinations of explanatory factors. The second criterion is practical: companies differ widely in their dissemination of information and openness to researchers. The two criteria are somewhat conflicting: we could have found significantly more variation in climate strategies if practical considerations were unimportant. The result of weighing these criteria in each case is that the book covers mainly North European companies that display limited variation in climate strategies. Greater variation in strategies could have been achieved by including, for instance, companies from southern and eastern Europe. The case-study authors have also applied additional case-specific criteria, such as previous experience and knowledge.

Data collection has been based on multiple sources. Interviews with representatives from the companies, industry associations, observers and policy-makers constitute a major source. Another important source has been company self-reporting through the Carbon Disclosure Project.[33] Analyses of corporate annual reports and sustainability reports, position papers and secondary studies have been relevant as well. Finally, several databases have been utilized, among them, Sandbag and the EU Industrial and R&D Investment Scoreboard. Despite our efforts to use a range of data sources to enable triangulation of data, much of the information has come from the industry associations and the companies themselves, and thus reflects how they want to display their activities to the outside world. Here we must simply acknowledge that there will always be an element of subjective judgment in assessing companies and their climate strategies.

Outline of this Volume

The five empirical chapters of this book analyse the electric power, oil, pulp and paper, cement and steel industries. Observations from these individual chapters are in turn compared and assessed in terms of the analytical framework. This framework, presented in Chapter 2, will guide our analysis of corporate climate strategies, responses to the EU ETS and other company-internal and wider external factors.

Chapters 3 to 7 present the empirical case studies. First comes Chapter 3, on the electric power industry – the most important sector covered by the EU ETS in

---

33   https://www.cdproject.net/en-US/Pages/HomePage.aspx

terms of allowances and emissions. The German RWE and the Swedish company Vattenfall have been singled out for more in-depth analysis here. Chapter 4 deals with another segment of the energy industries: oil companies. These companies are affected directly by the EU ETS because of their emissions from production and processing activities, and more strategically by carbon pricing that may spread beyond Europe. The multinational oil majors Shell (Dutch/British) and ExxonMobil (US) have been selected as representatives from this sector. The remaining empirical chapters deal with energy-intensive manufacturing industries. Chapter 5 focuses on the pulp and paper industry, with the Swedish company Svenska Cellulosa Aktiebolag (SCA) and the Norwegian Norske Skog selected for in-depth analysis. Chapter 6 studies the cement industry, and includes a focus on the Swiss company Holcim and the German Heidelberg Cement. Finally, Chapter 7 explores the German company Thyssenkrupp and the Swedish company SSAB as agents of the steel industry. Each of these empirical chapters presents its own analysis of the role that the EU ETS has played alongside other factors for sector and company climate strategies.

In Chapter 8, we search for comparative patterns between sectors and companies in light of the analytical framework. Chapter 9 concludes the analysis by summarizing the empirical findings, with reflections on the strengths and weaknesses of the research approach. This chapter includes an epilogue on the further road for the ETS and EU climate policy since the time of data collection for this study.

References

Asselt, H. van 2009. *Study on the Effectiveness of the EU ETS: The EU ETS in the European Climate Policy Mix: Past Present and Future*. Report for the ADAM project, 8 July 2009. Amsterdam: IVM.

Brewer, T.L. 2005. Business perspectives on the EU emissions trading scheme. *Climate Policy*, 5(1), 137–44.

Bruyn, S. de, Markowaska, A. and Neilssen, D. 2010, *Will Energy-Intensive Industry Profit from ETS under Phase 3?* Delft: CE Delft, October.

Cames, M. 2010. *Emissions trading and innovation in the German electricity industry*. Dissertation. Berlin: Technishen Universität Berlin

Convery, F. 2009. Reflections – The emerging literature on emissions trading in Europe. *Review of Environmental Economics and Policy*, 3(1), 121–37.

Egenhofer, C. 2007. The making of EU Emissions Trading Scheme: Status, prospects and implications for business. *European Management Journal*, 25(6), 453–63.

Egenhofer, C., Alessi, M., Georgiev, A. and Fujiwara, N. 2011. *The EU Emissions Trading System and Climate Policy Towards 2050 – Real Incentives to Reduce Emissions and Drive Innovation?* Brussels: CEPS Special Report.

Egenhofer, C., Fujiwara, N. and Gialoglou, K. 2005. *Business Consequences of the EU Emissions Trading Scheme.* CEPS Task Force Report No. 53. Brussels: Centre for European Policy Studies.

Ellerman, A.D. and Buchner, B.K. 2007. The European Union Emissions Trading Scheme: origins, allocation, and early results. *Review of Environmental Economics*, 1(1), 66–87.

Ellerman, A.D. and Buchner, B.K. 2008. Over-allocation or abatement? A preliminary analysis of the EU ETS based on the 2005–06 emissions data. *Environmental Resource Economics*, 41(2), 267–87.

Ellerman, A.D, Convery, F. and de Perthuis, C. 2010. *Pricing Carbon: The European Union Emissions Trading Scheme.* Cambridge: Cambridge University Press

Gagelmann, F. and Frondel, M. 2005. The impact of emission trading on innovation – Science fiction or reality? *European Environment*, 15(4), 203–11.

Hoffmann, V. H. 2007. EU ETS and investment decisions: The case of the German electricity industry. *European Management Journal*, 25(6), 464–74.

Ikkatai, S., Ishikawa, D. and Orhori, S. 2007. *The Effects of the EU ETS on Companies: Research by Conducting Interviews in European Companies.* Discussion Paper No. 627, Kier Discussion Paper, Kyoto: Kyoto Institute of Economic Research.

Johnstone, N., Hascic, I. and Kalamova, M. 2010. Environmental policy design characteristics and technological innovation: Evidence from patent data. *OECD Environment Working Papers*, No. 16. Paris: OECD.

Kenber, M., Haugen, O. and Cobb, M. 2009. *The Effects of EU Climate Legislation on Business Competitiveness: A Survey and Analysis.* Climate and Energy Series 09. Washington, DC: The German Marshall Fund of the United States.

Kruger, J. and Pizer, W.A. 2004. The EU Emissions Trading Directive: Opportunities and

Potential Pitfalls. Washington, DC: Resources for the Future.

Martin, R., Muûls, M. and Wagner, U. 2011. *Climate Change, Investment and Carbon Markets and Prices – Evidence from Manager Interviews.* Carbon Pricing for Low-Carbon Investment Project, Climate Strategies. Berlin: Climate Policy Initiative.

McIlveen, R. and Helm, D. 2010. Greener, Cheaper. London: *Policy Exchange.*

Ragin, C. 1987. *The Comparative Method.* Berkeley: University of California Press.

Renssen, S. van 2012. The fate of the EU carbon market hangs in the balance. *Europeann Energy Review*, 12 April.

Rogge, K. S. and Hoffmann, V.H. 2010. The impact of the EU ETS on the sectoral innovation system for power generation technologies – findings for Germany. *Energy Policy*, 38, 7639–52.

Rogge, K.S., Schmidt, T.S. and Schneider, M. 2011. *Relative Importance of Different Climate Policy Elements for Corporate Climate Innovation Activities: Findings for the Power Sector. Carbon Pricing for Low-Carbon Investment Project, Climate Strategies.* Berlin: Climate Policy Initiative.

Rogge, K.S., Schneider, M. and Hoffmann, V.H. 2011. *The Innovation Impact of EU Emission Trading – Findings of Company Case Studies in the German Power Sector*, Working Paper Sustainability and Innovation No. S 2/2010. Karlsruhe: Fraunhofer ISI.

Skjærseth, J.B. 1994. Climate policy of the EC: Too hot to handle? *Journal of Common Market Studies*, 32(1), 25–45.

Skjærseth, J.B. 2010. EU Emissions trading: Legitimacy and stringency. *Environmental Policy and Governance*, 20, 295–308.

Skjærseth, J.B. and Skodvin,T. 2003. *Climate Change and the Oil Industry: Common Problem, Varying Strategies*. Manchester: Manchester University Press.

Skjærseth, J.B. and Wettestad, J. 2008a. *EU Emissions Trading: Initiation, Decision-making and Implementation*. Aldershot: Ashgate.

____, 2008b. Implementing EU Emissions Trading: success or failure? *International Environmental Agreements*, 8(3), 275–90.

____, 2009. The origin, evolution and consequences of the EU Emissions Trading System. *Global Environmental Politics*, 9(2), 101–22.

____, 2010a. Fixing the EU Emissions Trading System? Understanding the post-2012 changes. *Global Environmental Politics*, 10(4), 101–23.

____, 2010b. Making the EU Emissions Trading System: The European Commission as an entrepreneurial epistemic leader. *Global Environmental Change*, 20: 314–21.

Underdal, A. and Hanf, K. (eds) 2000. *International Environmental Agreements and Domestic Politics: The Case of Acid Rain*. Aldershot: Ashgate.

Wettestad, J. 2005. The making of the 2003 EU Emissions Trading Directive: Ultra-quick process due to entrepreneurial proficiency? *Global Environmental Politics*, 5(2), 1–24.

Chapter 2
Analytical Framework

Per Ove Eikeland and Jon Birger Skjærseth

We need systematic examination of corporate responses to the EU ETS in several different industrial sectors to learn more about how the system works on the ground. Learning from previous experience is vital, since the EU has also identified the ETS as its key climate policy instrument for achieving its future goal of decarbonization, and because the system might come to serve as a governance model for other countries. This chapter outlines a broad analytical framework that will guide our empirical research into whether, how and why companies in different sectors have responded to the EU ETS. Approaching the questions posed in the first chapter takes us to various strands of the social science literature, including economics and political science.

Our point of departure is that companies' responses to regulation can be understood and observed in their *climate strategies*. The EU ETS stands out as the first and most prominent EU-level approach to climate regulation that may serve to promote or inhibit pre-existing climate strategies or shape new strategies. The first part of this chapter elaborates on the concept 'corporate climate strategies'. The next part links such strategies to explanations, so that we may assess how and why the ETS has caused observed variation in climate strategies. Explanations are framed in terms of three general perspectives or 'models' of corporate responses to regulation. All models treat companies largely as unitary actors; but they are based on different behavioural assumptions and point to different mechanisms linking regulation to responses.

In the final part, we extend these unitary models to include company-internal and wider external factors. After all, the EU ETS did not evolve in a vacuum: also other company-internal and -external factors need to be accounted for as independent or interacting drivers of corporate climate strategies. That said, the empirical case studies focus most on the relationship between the ETS and climate strategies, since the main purpose of this book is to explain how and why this system has worked on the ground.

Corporate Climate Strategies

The climate strategy of a company is revealed through the company's objectives concerning climate change mitigation and the actual mobilization of resources (or failure thereof) to achieve those objectives. The motives and processes behind the development of objectives vary (Bourgeois and Brodwin 1984, Mintzberg 1993). Strategy formulation may be rooted in expectations from external stakeholders or from internal processes.[1] In the former case, a climate strategy may merely reflect the company's external public relations position – 'as a facade to impress outsiders' (Mintzberg 1993:41).[2]

The literature on corporate climate strategies has evolved in line with the broader environmental management and strategy literature that sees company strategies according to either a continuum model or typologies (Kolk and Pinkse 2004: 305). Continuum-based classifications are frequently framed as linear stage-based, purporting to identify change in strategies over time. They often assume that firms will improve their environmental performance and move in stages toward the higher goal of environmental excellence (Lee 2012). The continuum approach has been criticized for determinism and for being one-dimensional, less suited for handling ambiguities in strategies (Kolk and Mauser 2002, Kolk and Pinkse 2004, Lee 2012).[3] Typology-based classifications, by contrast, identify a number of ideal or actual types of strategies, without implying improvement processes; typology-based studies tend to be static, however, focusing on inter-company differences at a given point in time (Lee 2012: 38).[4]

Our main aim is to explain company response to the EU ETS, a regulatory system that has developed and changed over time. We opt for a very simple dynamic classification scheme of reactive and proactive strategies, inspired by continuum models (e.g. Sharma and Vredenburg 1998, Aragon-Correa and Sharma 2003, Sharma and Henriques 2005). We do not assume any deterministic, linear movement from one to another strategy class. In studies of environmental

---

1   Another distinction is whether internal processes involve many planners or only a few. Objectives may reflect the preferences of a few senior executives rather than a broad-based consensual process.

2   Such a strategy of 'greenwashing' may reflect reputation risks faced by companies that fail to live up to their stated aspirations and standards, and be associated with failure in mobilizing resources for real GHG mitigation. Strong language and mission statements will create expectations not only among the public and civil society organizations, but also among employees, shareholders and investors (Kolk 2000).

3   Determinism means that once undertaken, strategies will automatically move along a continuum from a reactive to a proactive strategy, and thus criticized, for lacking the perspective of strategic choice in explaining inter-company variation (Lee 2012: 38). Being one-dimensional, continuums are seen as too narrow tools for handling the empirical observation that firms may be ambiguously placed in two or more stages along different environmental strategy dimensions (Kolk and Pinkse 2004: 305, Lee 2012: 38).

4   Typical of the typology approach is the work of Kolk and Pinkse (2004a, 2004b) where climate strategies are classified along two dimensions: the company's aim (strategic intent) and the extent of cooperation. This classification was empirically derived from a large survey of company reports to the Carbon Disclosure Project.

strategies, reactive responses generally feature compliance with legal regulations.[5] Opposed to this, proactive strategies demonstrate practices that go beyond simply obeying the law.[6] We also account for the fact that strategic responses tend to be multi-dimensional, by including positions on the climate-change problem 'diagnosis'. This would involve accepting/rejecting the climate problem as such, and responsibility for contributing to mitigation. Yet, it would also include market responses.[7] The indicators below and the analytical models presented in the next section are specified to deal with such apparent ambiguities.

To fit the analytical models applied, we focus on innovation as an element in a proactive strategy, frequently applying the concept of 'innovative strategy'. An *innovative strategy* points towards responses beyond what is required by regulation, something more than business-as-usual activities to contribute to short- and long-term climate problem solving. The innovative strategy concept would take into account the above definition of a proactive strategy, in line also with Joseph Schumpeter's definition of innovation as the discovery and commercial or industrial application of something new – a new product, process or method of production; a new market or source of supply; a new form of commercial, business or financial organization (Schumpeter 1934). By contrast, a *reactive strategy* is defined as companies opposing climate regulation, and when/if regulation is adopted, will either not comply or comply only short-sightedly and not with activities beyond business-as-usual. As noted, responses might go either way – from reactive to innovative, or from innovative to reactive.

For this study of the effects of the EU ETS within the framework of specific analytical models of company behaviour, the main point is to assess the direction and content of changes, not to place companies in specific categories (see Chapter 1). Empirical assessment of changes in climate strategies for companies in different sectors poses great challenges and requires more general indicators than comparing companies within the same sector. We therefore start out with four sets of fairly general climate strategy indicators, later adapted to the specific case studies:

---

5   Reactive responses may also involve non-compliant behaviour and obstruction of environmental regulation (Meznar and Nigh 1995, Newton and Harte 1997, Cho, Patten and Roberts 2006, Lepoutre 2008), as well as market-responses that do not question the business-as-usual of the firm (Hunt and Auster 1990, Russo and Fouts 1997, Sharma and Vredenburg 1998).

6   Even completely redesigning products and processes (Hart 1995), redefining the business model (Sharma and Henriques 2005) or engaging in active roles in the industry or society to change behavioural patterns (Hunt and Auster 1990, Hart 1995).

7   As noted in the typology applied by Kolk and Pinkse 2004, Lee 2012 or e.g. Steger 1993 in studying general environmental strategies, tailored to study the effect of market risks and opportunities.

- acknowledgement of the climate problem
- position on climate regulation
- short-term abatement measures
- measures to promote long-term low-carbon solutions.

The first set of indicators is based on the extent to which the companies acknowledge the problem of climate change and take responsibility for contributing to mitigation. The question is whether companies acknowledge the main conclusions of the Intergovernmental Panel on Climate Change (IPCC) and take responsibility for problem-solving in the form of *voluntary* individual or cooperative action. Grossly simplified, the main conclusions from the IPCC state that the problem is real and that there is sufficient scientific evidence for acting accordingly.[8] Companies can respond to the EU ETS by changing their voluntary contribution to dealing with the problem, individually or jointly.

The second set of indicators is simply a question of whether companies accept, support or oppose new regulation. Opposition tends to materialize in the form of lobbying activities aimed at weakening or abolishing the regulation in question. In the ETS context, a company's position on regulation will be related to its statements and activities related the development of the system from prior to 2003 – the year the ETS was adopted. A central question here is whether the companies accepted or supported the emerging ETS or not. Also, whether they supported design elements that would make the system mandatory and stringent, such as cap-and-trade instead of baseline-and-credit (the latter not mandating absolute reductions in emissions). Other important design elements in connection with company positions include allowance allocation methods and import restrictions for external credits (see Chapter 1).

The third set is based on short-term responses in the market in the form of abatement and activities that will compensate for abatement. Short-term abatement includes targets; measures; monitoring, reporting, verification and actual behavioural change in terms of energy efficiency improvement and reductions in GHG emissions. Activities compensating for abatement include external trading in EU Allowances (EUAs), Certified Emissions Reductions (CERs) and Emission Reduction Units (ERUs).[9] Other response options are non-compliance

---

8   Ideally, we should apply more specific indicators within three broad dimensions: whether the Earth is warming (scientific evidence overwhelmingly says yes); whether this warming is caused by natural or anthropogenic causes (scientific evidence says 'very likely') and whether something should be done about it (costs of action vs. costs of inaction). However, many companies do not express nuanced positions on this issue. Our thanks to Harro van Asselt for making this point.

9   EUAs are EU ETS carbon credits which can be traded within the EU only. CERs are carbon credits from CDM projects and ERUs are carbon credits from JI projects. External credits can be used for EU ETS compliance with limitation. CDM and JI are the flexible project mechanisms under the Kyoto Protocol (see Chapter 1).

or relocating production to other countries with less stringent climate policies. In our context, short-term responses cover the two first trading phases (2005–2007; 2008–2012) and implementation of already decided measures directed at the third phase (2013–2020).

The fourth set of indicators includes responses aimed at a future low-carbon society.[10] These could be company long-term visions, or more committing long-term quantitative goals and timetables for emissions reductions. More importantly, such responses could be scaled up investments in research, development and demonstration of new or still immature low- or zero-emission process technologies and products – both within and beyond business-as-usual, such as carbon capture and storage (CCS), highly energy-efficient combustion technology and technologies enabling fuel switching, load management and more effective transmission of electricity.[11] Also included would be engagement in cooperative large-scale technological problem-solving and commercialization projects, jointly with the company's main competitors, governments or civil society actors. Engagement in joint activities involving the sharing and pooling of human and financial resources would be a particularly strong indicator of a proactive strategy because of the presumably massive efforts needed to make available not only incremental but system-transforming solutions for Europe, in line with the 80–95 per cent cuts in emissions by 2050 aimed at by EU leaders.

Based on these empirical indicators, we can offer a tentative portrait of the extreme *ideal-type* 'reactive' and 'proactive' or innovative company strategies. A typical 'reactive' company will fail to acknowledge the climate problem and its own responsibility for contributing to problem-solving. It will lobby against mandatory climate regulation, including the EU ETS, be cautious about commitment to quantitative goals and timetables where the company could later be held accountable, and will implement only short-term least-cost mitigation options – whether incremental business-as-usual measures, relocation of production and emissions to non-regulated areas or outsourcing of mitigation by merely purchasing low-cost CERs or ERUs. By contrast, a typical 'innovative' company will signal a firm sense of social responsibility for problem-solving, will support mandatory climate regulatory action, including the EU ETS as the most comprehensive binding EU-level measure, will search and implement short-term abatement options within and outside business-as-usual for the company, will put up visions and goals, and engage in cooperative long-term development

---

10　Applied to the EU ETS, the distinction between decisions directed at short-term abatement and long-term low-carbon solutions is not always clear-cut. However, long-term low-carbon solution criteria aim to cover present visions, goals, activities and investments directed towards radical or major change in products, production process or the quality of the outputs. One example is CCS, which is not expected to be commercialized within the current time-frame of the ETS trading phases.

11　Load management involves balancing the supply of electricity on the network with the electrical load by adjusting or controlling the load rather than the power station output.

of low-carbon solutions.[12] The latter type of responses can be seen as necessary to realizing a European low-carbon economy in line with the ambitions of the EU. As noted, however, we cannot expect real-life companies to fit perfectly with such archetype ideal strategies. Companies engage in a wide range of political and market responses that may change in different directions along the basic reactive/ proactive continuum. These ideal types function solely as devices for assessing company changes so that we may offer conclusions about the direction in strategies from before the EU ETS was adopted till the end-date of this study (see below).

In order to assess companies according to the above criteria, we must deal with at least two methodological challenges. First, the public profile of a company may diverge significantly from its actual behaviour, for strategic and/ or practical reasons. Thus we need to move beyond simple rhetoric, and explore what companies actually do, particularly in terms of significant investments and activities with potential short- and long-term implications for future operations. A further problem here is that the availability and reliability of 'hard data' differ significantly between companies and sectors.

Second, climate strategies should be assessed according to a certain baseline and end-date. Since companies may take action based on anticipation of upcoming regulation, in the case studies we will have to pay attention to changes before the 2003 baseline year when the ETS was formally adopted. The cut-off point is 2010/2011, reflecting that case-study authors have conducted data collection at differing points in time.

Explanatory Perspectives

Regulatory pressure has been singled out as a key cause of corporate climate strategies (see Kolk and Pinkse 2004, Hoffman 2006, Okereke 2010). The way companies are likely to respond to regulation is contested, however, depending on different underlying perspectives on corporate behaviour. The business strategy literature presents a range of models and propositions prescribing how companies will respond to environmental regulation in general and emissions trading in particular. Below, we introduce three models together with derived expectations about corporate climate-strategy responses to regulation, the EU ETS in particular. Next, expectations are modified on the basis of a cluster of additional company-internal and company-external factors.

---

12   There is a likely relationship between the scores on these criteria, but these scores are not necessarily additive. A company might invest significantly in low-carbon solutions for purely commercial reasons without paying any attention to the challenge of climate change as such.

Model I: Companies as Reluctant Adapters

This model of regulatory response is grounded in the mainstream economic view of the firm as a unitary rational profit-maximizing agent that adopts strategies on the basis of full information on the relative costs of various alternatives (Gravelle and Rees 1981, Ambec et al. 2011). Company operations prior to the adoption of regulation will be at a level that minimizes input costs. The model builds on the following basic assumptions: (1) environmental regulation will impose new costs on the company; (2) managers will re-evaluate the options in terms of these new costs to find the new short-term optimal levels; (3) the wider economic, social and political context is not relevant for explaining (change in) strategic decisions. Everything except regulation is held constant.

This behavioural model is not taken to represent an accurate description of how companies will behave in the real world, but as a parsimonious theory based on simplified assumptions, taking into account that profit-seeking is a central goal of any company.[13]

The main assumption of this model is that environmental regulation adds costs that will erode a company's profits and competitiveness unless similar companies are also subjected to such regulation. Profits will be reduced because the company is forced to allocate labour and capital to abatement, which is not productive from a business perspective. For the EU ETS specifically, capping emissions and putting a price on CO_2 will charge a company for a by-product of its production that was free before.[14] Administrative costs from compliance activities like monitoring, verification and reporting of emissions data will add further costs. Such costs will necessarily divert capital away from productive investments, so profits will go down. A company's options for short-term cost reductions will be restricted because a full-information profit-maximizing company would already have discovered all the 'low-hanging fruits' and taken advantage of those opportunities before regulation was put into place (Ambec et al. 2011). Or, put differently, you never find ten-dollar bills lying on the ground because someone else will have picked them up already (Porter and van der Linde 1995b: 98). This notion of profit maximization leads to a trade-off between compliance to regulation and competitiveness; between ecology and economy where the social benefits that arise from strict environmental standards conflict with the industry's private costs of abatement (Porter and van der Linde 1995a, b).

This model basically predicts that a company will resist any mandatory environmental regulation, including the EU ETS. If such regulation is implemented, the company will respond with only low-cost incremental business-

---

13   According to Porter and van der Linde (1995a, b), this way of thinking within economics has contributed to shape a mind-set that actually affects how companies respond to regulation.

14   This may lead to opportunity costs – the costs of the alternative foregone by choosing a particular activity.

as-usual abatement options. The EU ETS will thus be seen as a regulatory threat to company profitability and competitiveness – a threat likely to make it start lobbying to defeat or weaken the regulation, although after a comparison of costs of lobbying against those of complying. Secondly, the model predicts that if and when the ETS is adopted, the choice of compliance response will depend on what pays off in the short term, based on a ranking of various options according to their costs. If non-compliance can be detected and penalties for non-compliance are in place (and are higher than the market price of allowances),[15] the response will take the form of reluctant compliance with the ETS, in the form of phasing in only the lowest-cost solutions at first.[16] We would not expect to find abatement beyond business-as-usual activities in the short term, or activities aimed at longer-term low-carbon innovation. In fact, activities beyond those conforming to the least-cost option would be deemed paradoxical within this model, similar to the 'Harrington paradox' from the ecological economics literature (Harrington, 1988).[17]

Table 2.1   Essence of Model I: Companies as reluctant adapters

Key external explanatory factor	Behavioural assumption	Wider economic, social and political context	Expected ETS response and climate strategies
Mandatory regulation.	Rationality and profit maximization: minimization of new regulatory costs.	Not relevant for explaining (change in) strategic decisions .	Reluctant compliance.* Towards reactive climate strategy.

Note: * If non-compliance is more costly than compliance.

The model does not predict that all companies will respond identically, however. Regulatory costs and pressure may vary from company to company and among sectors even though everything else is assumed to be constant, including carbon prices, technology, consumer needs and products. In the context of the EU ETS, regulatory pressure is a product mainly of the shortage/surplus of allowances

---

15   The penalty is EUR100 for each tonne of CO_2 equivalent emitted by the installation for which the operator has not surrendered allowances. Payment of the penalty will not release the operator from the obligation to surrender the allowances required. The penalty for the first trading phase (2005–2007) was EUR40 per tonne.

16   Including incremental abatement, trading or relocation of production to countries with more lenient climate policies.

17   Harrington (1988) describes 'compliance' with regulation as a paradox within the model of the profit-maximizing firm in cases where regulation lacked control and punishment.

compared to actual emissions (emissions-cap ratio) and the method for allocating allowances. Companies short in allowances that must be bought in the market will face high regulatory pressure. Such companies can be expected to have more extensive activities in trading or abatement than companies that have a surplus of allowances or get their allowances for free. However, what they do more of will depend on the marginal costs of various compliance options.

In the empirical analysis in Chapter 8, we examine this model in terms of pattern-matching. Expectations derived from the model will be compared to observed changes and differences in climate strategies. The better match between expected and observed strategies, the more confidence will we have in the model's ability to explain. We also explore the heuristic value of this perspective by assessing how well the main mechanisms of profit maximization and incentives explain the shaping of climate strategies.[18]

Model II: Companies as Innovators

The traditional perspective on the relationship between regulation and competitiveness has been notably challenged by the work of Michael Porter and Claas van der Linde (Porter and van der Linde 1995a, b). These authors turn the above line of reasoning on its head: regulation, they hold, can strengthen international competitiveness and make companies more profitable. More stringent but properly designed regulation will not only facilitate short-term compliance, but also stimulate exploration, experimenting, learning and innovation in firms. Versions of this argument, now known as the Porter Hypothesis, have received significant theoretical and empirical attention (e.g. Ambec et al. 2011).

The innovation 'model' recognizes that companies strive for profits, but companies tend to be short-sighted and unable to make optimal choices for various reasons, including boundedly rational managers, market failure and organizational inertia. 'Bounded rationality' implies that firms intend to be rational, but have limited success in achieving this (Simon, 1976). An early explicit bounded-rationality perspective was 'the behavioural theory of the firm' propounded by Cyert and March (1963). They claimed that the major observations underpinning the bounded rationality scheme were that rational actors experience significant constraints as to information and cognitive capacity for calculation, making explicit and timely calculations of optimality costly or impossible.[19]

---

18 See Underdal (2000) and Skjærseth and Skodvin (2003) for examples of comparative analysis along these lines at national and company levels respectively.

19 As a result, firms will simplify the decision-process and problem-solving in various ways. They set targets and look for alternatives that satisfy those targets, rather than trying to find the best imaginable solution. They allocate attention by monitoring performance with respect to targets. They attend to goals sequentially and not simultaneously. They follow rules-of thumb and standard operating procedures (Cyert and March 1992[1963]: 214–15).

More specifically, this perspective assumes that: (1) managers evaluate regulation in terms of costs *and* benefits to their companies; (2) their climate strategy response is shaped by limitations in information and calculation, sequential attention, habits and routines; (3) appropriate regulation will generate new attention to earlier non-apprehended opportunities, spur innovation and make companies more competitive; (4) companies operate in a world of dynamic competition, constantly finding innovative solutions to pressures from competitors, customers and regulators.

The main implication is that companies *need* regulation in order to recognize new and innovative opportunities that may benefit the company (Porter and van der Linde 1995b). Using the earlier metaphor, ten-dollar bills are often left on the ground because people simply do not look in the right direction. Regulation can raise corporate awareness; it can spur new learning about resource inefficiencies and potential technological improvements, reduce uncertainty about investment for the future, create pressures to motivate innovation and offset compliance costs. Innovation may lead to early-mover advantages – the new technologies, processes or products can be sold in the marketplace at a later stage.

Porter and van der Linde add various conditions under which environmental regulation can be considered 'appropriate' for spurring innovation (Porter and van der Linde 1995a, 1995b).[20] The following features characterize appropriate regulation. First, regulation should focus on outcomes, not specific technologies. Best-available-technology regulation will hinder innovation, as there will be scant incentives for companies to progress beyond the technology required. Second, regulation must be 'strict' in order to spur real or 'radical' innovation, since companies can handle lax regulation incrementally by short-term adjustment. Regulation should be developed in line with other countries or slightly ahead, to reduce competitive disadvantages and stimulate early-mover advantages. Third, continuous improvement should be encouraged and be based on market incentives, including taxes and emissions trading. Finally, uncertainty should be reduced by coordination of regulation. Uncertainty will also be reduced by making the regulatory process predictable and based on phase-in periods to avoid the over-hasty implementation of expensive solutions.[21]

The Porter Hypothesis has been tested in major survey-based studies and case studies. Several large-scale statistical studies have variously supported or rejected the Porter Hypothesis (see e.g. Jaffe and Palmer 1997, Popp 2006, Arimura, Hibiki and Johnstone 2007, Lanoie et al. 2007). In this book, we operate with

---

20   There is, however, a large literature on the pros and cons of various types of environmental policy instruments and how they work in different situations (e.g. Vedung 1997).

21   Governments can aim to reduce regulatory uncertainty, but never remove it since technologies and economies tend to change.

the 'weak' version: properly designed regulation will tend to spur innovation. [22] The generality of this hypothesis is still poorly tested, and studies have emerged with varying conclusions (see e.g. Burtraw 2000, Stavins 2001, Johnstone and Labonne 2006). Whether the ETS qualifies as a 'properly designed instrument' is an empirical question that will form part of our analysis.

Concerning the ETS response, we thus first expect that companies will accept or support the system as new business opportunities are discovered and emphasized. Once the ETS is adopted, companies will comply and start searching for new opportunities beyond business-as-usual in order to create early-mover advantages. Incremental innovation (short-term abatement beyond what is needed to comply) and long-term acitivities directed at innovation can be expected, since company management is concerned with new opportunities in addition to costs and risks.

Table 2.2   Essence of Model II: Companies as innovators

Key external explanatory factors	Behavioural assumption	Wider economic, social and political context	Expected ETS response and climate strategies
Mandatory regulation. Type of regulation: stringent and 'appropriate' in spurring innovation. Coordination between relevant regulations.	Bounded rationality: myopic attention and search for new market opportunities.	Dynamic competition relevant for explaining (change in) strategic decisions.	Compliance and search for innovative business opportunities. Towards innovative climate strategy.

Variation in responses will mainly be the result of how the ETS was designed and how other regulations are coordinated and interact with the ETS. Regulatory pressure will vary among sectors and companies, as depicted by Model I. In contrast to Model I, we expect less opposition to more stringent regulation, since new search activities enable companies to see new short- and long-term opportunities from up-scaled activities in R&D and innovation. Second, the degree of coordination and harmonization with other relevant regulations, such

22   This version is 'weak' in the sense that it does not say anything about whether the innovative activities are good or bad for company competitiveness. The 'strong' version argues that innovation will tend to offset any additional regulatory costs, making companies more competitive. There is also a 'narrow' version that holds that flexible regulatory policies give firms greater incentives for innovating and are thus better than a prescriptive form of regulation (Ambec et al. 2011).

as EU renewable policies, will vary among companies and sectors. We expect that a high degree of harmonization between the ETS and other EU climate and energy policies will lead to stronger pressure and more innovative activities.[23] Finally, regulation should be in line with or slightly ahead of other countries, so as to facilitate early-mover advantages and minimize competitive disadvantages. Thus, innovation is likely to increase when companies are exposed to appropriate regulation that is stringent, harmonized and in line with or slightly ahead of other countries.

In the empirical analysis, we explore Model II in terms of pattern matching and heuristic value. The assessment will be based on an analysis of changes and differences in climate strategies linked to the EU ETS. In the analysis of the heuristic value of the model we will pay particular attention to a wider set of mechanisms, including how regulation may set in motion a process whereby companies are stimulated to explore, experiment and learn how they can benefit from abatement and innovation.

Model III: Companies as Socially Responsible

Models I and II do not include the possibility that firms are concerned with the greater *social* risk of environmental challenges. They focus not on the social benefits of environmental regulation, but solely on private costs and benefits (Porter and van der Linde 1995b). The third perspective assumes that social responsibility may be part of company behaviour. In fact, it could also be argued that environmental values is needed to make individuals and companies contribute sufficiently to EU climate targets – which leaves open the question of whether the incentives provided by today's market-based policy instruments are appropriate.

Model III assumes that company managers can have mixed motivations that may include social norms of responsibility. This perspective builds tentatively on the following assumptions: (1) managers evaluate options broadly in subjective terms of social, economic and political aspects, and their response to regulation is affected by social norms of responsibility; (2) regulation can affect such norms of responsibility; (3) companies operate in a complex political and social environment where consumers and civil society organizations play an important role.

Norm-guided behaviour has increasingly been incorporated into economic studies of responses to governmental regulation (Frey 1997, Nyborg and Rege 2003).[24] Norm-guided behaviour is central to the growing literature on motivation

---

23   The EU ETS may interact with other relevant regulation at both the EU and the national levels. We have chosen to narrow in on regulation at the EU level in the 'pure' version of this model, as our main focus is on corporate responses to policies adopted by the EU. Relevant national policies are included below as an extension of Model II.

24   Above, we mentioned Harrington's Paradox discussed in ecological economics – that companies may comply with environmental regulation despite lack of sanctions. This paradox has been explained by various forms of norm-guided behaviour.

crowding theory (Frey and Jegen 2001, Frey and Stutzer 2008) and is discussed in the vast literature on corporate social responsibility (CSR) (e.g. Barth and Wolff 2009).[25] Companies can contribute to provide public environmental goods, for instance through voluntary CSR principles and measures.[26] However, since voluntary contributions are rarely deemed sufficient to provide important public goods, like keeping the average global temperature below critical levels, additional state regulation is normally viewed as necessary.[27] Motivation crowding theory stipulates that external intervention can 'crowd in' or 'crowd out' norms of responsibility (Frey and Stutzer 2008).[28] A crowd-in effect will occur when regulation is perceived as supportive and based on mutual acknowledgement of obligations and responsibilities, communication between principals and agents, and participation in decision-making. A crowd-out effect will occur when regulation is perceived as controlling, which may reduce actors' motivation to act. The mechanisms are a shift in the locus of control which may reduce self-determination and a feeling of responsibility. External intervention may also violate the norm of reciprocity based on mutual acknowledgement.

With regard to the EU ETS, Frey and Stutzer (2008) argue that tradable emission rights in particular will tend to crowd out environmental 'morale' and can be counterproductive in promoting voluntary climate mitigation. The basic idea underlying the EU ETS is that the use of the air to discharge pollutants is no longer free of charge and must be paid for. Emissions trading in fact authorizes pollution, turning it into a commodity that can be bought and sold through allowances in the market (Ott and Sachs 2000). Payment for the right to engage in an undesired activity is then seen as fostering a cynical attitude: sinning is OK as long as you

---

25   Some strands in the literature make a distinction between social norms and moral norms: the first depicting social approval and control as motivation, and the latter referring to deeply ingrained self-imposed moral restrictions based on intrinsic motivation (Frey and Jegen 2001, Nyborg and Rege 2003).

26   There is a growing literature that questions the profit-maximization assumption by including also intrinsic norm-oriented behaviour to explain why some companies can be described as norm-entrepreneurs in global governance (see Flohr et al. 2010). Alternatively, CSR or entrepreneurship can be explained, partly or fully, as efforts to reduce reputational costs due to environmental values among consumers and civil society organizations, or to avoid state intervention. These alternatives are compatible with corporate profit-seeking in a broad sense.

27   The relationship between environmental policy and motivation for voluntary action thus becomes important for the choice of policy instruments. In the CSR context, responsibility is often understood as relational and shared between the state, business and civil society (Barth and Wolff 2009). Governments may stimulate, regulate or even overtake corporate CSR activities.

28   The theory is based on motivation for individual behaviour from the perspective of economics and psychology, the argument being that company behaviour can and should be analysed using crowding theory to understand more about the conditions that lead managers to make environmentally-friendly decisions.

pay for it. Some sinners may even make a substantial profit from the system.[29] The trade system will shift attention from abatement towards the profits to be made from pollution, and refraining from polluting activity on moral grounds becomes irrelevant or even seen as naive or irrational (Frey and Stutzer 2008).

The crowding-out and price-incentive effects of emissions trading will work in opposite directions because a price signal will tend to reduce emissions. The outcomes may thus be the same, but the mechanisms that produce them are markedly different. To be sure, the crowding-out effect may be counterbalanced by a crowding-in effect. As noted above, that effect would be expected if the process of making the EU ETS is characterized by mutual acknowledgement of responsibilities, communication and participation. We cannot rule out the possibility that a strategic response to a cap-and-trade system conforming to specific criteria can, in the long term, induce a corporate sense of responsibility. This means it is extremely difficult to extract expectations of outcomes on the basis of the relationship between the ETS and social norms of responsibility. However, the presence of social responsibility norms can be expected to lead to a search for innovative opportunities even when regulation is 'inappropriate'.

Table 2.3   Essence of Model III: Companies as socially responsible

Key external explanatory factors	Behavioural assumption	Wider economic, social and political context	Expected ETS response and climate strategies
Mandatory regulation. Type of regulation: Emissions trading may 'crowd out' norms of responsibility. Regulatory process: Industry participation may counteract 'crowd-out' effect and may possibly 'crowd in' norms of responsibility.	Mixed motivation: Profit and social norms of responsibility.	Social risk of environmental challenges relevant for explaining (change in) strategic decisions.	ETS response and climate strategies depends on norms of responsibility. If norms of responsibility are present, search for innovative opportunities may take place despite 'inappropriate' regulation.

---

29   It has been documented that power companies in particular have reaped 'windfall profits' from the trading system. See Chapter 3.

Empirical assessment of this perspective is a challenging task, since corporate norm-guided behaviour is extremely difficult to distinguish from other motivations.[30] Moreover, the norm-based perspective is as yet less developed and less empirically scrutinized than the previous models. A weaker knowledge base requires a more explorative approach to expectations and empirical assessment. We will begin assessing this model in terms of explanation-building by focusing on change and differences in stated responsibility, CSR and cooperation as part of company climate strategies. Next, we analyse the relationship between the EU ETS and any variation in (stated) responsibility. This analysis must take into account that: (1) determining the norm element of climate-change responsibility is extremely difficult; (2) the 'crowding-out' effect can take place only if a significant amount of environmental social responsibility norms were internalized before the ETS; (3) changing norms of corporate behaviour either way may take a long time, possibly extending beyond the time-frame of the empirical analysis.

Company-internal and Wider External Factors

The three models above have been developed for the specific purpose of analysing how the EU ETS affects company climate strategies. However, we know that the EU ETS was not implemented in a vacuum, so the observed climate strategies could well have been affected also by other factors external to the firms. Moreover, the three models treat companies more or less as ideal types, with no room for variation in internal managerial factors impacting on company strategies.

The environmental management literature has indicated a range of company-specific factors that should be accounted for in analysing climate strategies – including the specific history of each company, size of the resource base (e.g. capital and human resources), management capabilities and leadership, ownership, regulatory risk inherent in overall company business strategy, environmental reputation and stakeholder influence (see Levy and Kolk 2002, Skjærseth and Skodvin 2003, Kolk and Pinkse 2004, Pinkse and Kolk 2007).

In this section we explore how the three models can be extended by the inclusion of factors from the 'inside' and more widely from the 'outside' of the company. A broader perspective will enhance our exploration of the extent of ETS impact by bringing out factors that affect climate strategies other than the ETS.

30　A criterion for climate strategies based on an element of intrinsic social responsibility is that they would include joint innovative efforts based on cooperation with competitors, governments and civil society organizations to devise comprehensive solutions to collective problems. Cooperation may be purely strategic (reduce R&D costs, provide early-mover advantage within a broader cooperative framework), but genuine social responsibility for collective problem-solving is unlikely to be compatible with inward-looking individual corporate behaviour. Some degree of joint effort may thus constitute a necessary but not sufficient condition for social responsibility. That does not mean that individual corporate behaviour cannot be 'good' for society.

Further, a broader perspective will help us in identifying the conditions under which different responses and strategies emerge.

Company-internal Factors

This study focuses specifically on stability or movements between reactive and innovative climate strategies. In this particular context, two factors appear as especially relevant. The first is the inherited general business strategy: the structure of input factors and technologies in the production process, the choice of products to make and sell, and the markets to serve. The choice of general business strategy is a managerial task, but once decisions have been taken and investments made, sunk costs and organizational inertia within the firm will contribute to lock-in of main business strategies. Alongside any configuration of input factors, the products and market follow a specific *carbon intensity*, in turn creating variation in risks to any regulation that places a price on carbon emissions.

The 'inherited product market' would add yet another source of variation in regulatory risk. Some companies offer their products in an international market, where regulatory risks are generated if the competitors are not bound by climate regulations that impose additional costs. Other companies may sell their products in the more restricted EU market, where all competitors will be targeted by the same regulatory instrument. Companies will have higher risks from regulation in Europe if their products are sold to countries with no carbon regulations or if they are subject to competition from products imported to Europe from these countries. The more vulnerable a company is to *international competition*, the higher will be its risks and the lower its possibilities to pass on carbon costs to the customers. With the EU ETS, particular concern has been expressed with regard to international competition and the risks of possible 'carbon leakage', including the possibility that high carbon costs may lead companies to decide to relocate to areas subject to less strict climate policies. Carbon costs may be direct or indirect – the former associated with the costs of fulfilling allowance obligations, and the latter resulting from the ETS impact on electricity tariffs with additional costs for electricity-intensive industries.

Such company-specific variation in sources of regulatory risks would have implications for the expectations derived from our main models. For Model I the implications appear clear-cut. High carbon intensity will increase the costs of regulation and make companies (even) more reluctant to adopt an environmental policy. Compliance will require more allocation of capital and labour, and will probably increase administrative costs. Concerning Model II, high carbon intensity will increase the regulatory 'pressure'; incremental adjustment will become less likely and radical innovation more likely. The implications for Model III are less clear-cut, but it seems reasonable to assume that higher regulatory risk combined with regulatory pressure will be seen as more controlling than supporting, thus stimulating a crowd-out of social responsibility norms. On the other hand, such companies are vulnerable to societal pressure and may engage in CSR activities for purely strategic reasons to avoid regulation or reduce regulatory pressure.

Concerning the risks from international competition, the implications for Model I are again straightforward. The more a company is exposed to international competition, the higher the risk and (potential) costs of regulation, and the more reluctant will compliance be. Model II holds that stringent regulation will increase competitiveness. However, regulation should be developed in line with other countries or slightly ahead, so as to stimulate early-mover advantages and minimize competitive disadvantages relative to foreign companies that are not subject to the same standards. If standards are too far ahead (or too different in character), industries may innovate in the wrong direction, according to Porter and van der Linde (1995b: 124). In our context, the 'wrong direction' means that early movers may invest in low-carbon solutions that will not be commercialized. With Model III, exposure to competition combined with regulation is likely to be perceived as more controlling and may thus stimulate a crowding-out effect. The combination of high carbon intensity and exposure to international competition will pull largely in the same direction as regards Model I and Model III.[31] In Model II, they may pull in different directions depending on how far ahead regulations are compared to other countries.

A second factor in focus in this study, termed 'dynamic capabilities', refers to the leeway available to managers for doing more than merely serving as administrators of given resources within a given locked-in business strategy and structure. An early contribution to such a dynamic, change-oriented view of the firm came from Penrose (1959, second edition 1995). Penrose made a sharp distinction between company resources, and management's ability to master these. To Penrose, the managers of the firm were central for detecting alternative services or productive opportunities to be provided by existing resources, for bringing in new resources, and for re-combining the resources into new business activities.

A large 'capability of the firm' literature followed in the wake of Penrose, with one approach focusing specifically on managerial capabilities for entrepreneurship and innovation – the 'dynamic capability' perspective (see Eisenhardt and Martin 2000, Helfat and Raubitschek 2000, Winter 2000). Based on a summary of this literature, Eisenhardt and Martin (2000: 1107) define dynamic capabilities as: 'The firm's processes that use resources – specifically the processes to integrate, reconfigure, gain and release resources – to match and even create market change. Dynamic capabilities thus are the organizational and strategic *routines* by which firms achieve new resource configurations as markets emerge, collide, split, evolve, and die.' In the context of climate strategies, 'strong' dynamic capabilities can be understood as suitable routines for discovering new market opportunities and utilizing resources for low-carbon solutions, such as R&D and investment in CCS or renewables. 'Weak' dynamic capabilities means unsuitable routines

---

31 Competition can be seen as an external factor, but is included here to capture its relationship to carbon intensity.

for identifying and exploiting new opportunities: companies stick to their core business, as with oil companies that focus exclusively on oil and gas exploration and production.

Once settled as management routines, weak dynamic capabilities can be hard to change unless strong external or internal signals are provided, for instance in the form of stringent new governmental regulations or replacing central company management or top leadership. Variation across companies in dynamic capabilities will interplay differently with our three models of ETS response. For Model I, the concept of dynamic capabilities or change in leadership is simply irrelevant. The model does not include variation in internal management variables. Any manager will maximize profits and optimize the organization. Concerning Model II, the Porter Hypothesis holds that companies facing an external intervention in the form of stringent regulation may actually lead management to look in new directions for previously unheeded opportunities. However, the most innovative strategic response would be expected from companies already equipped with strong dynamic capabilities. Moreover, proactive and innovative climate strategies may be expected from companies with high dynamic capabilities even if the regulation is non-stringent, whereas those with weaker dynamic capabilities would tend towards more reactive strategies. Regarding Model III, we expect synergies between dynamic capabilities and social responsibility norms. Strong dynamic capabilities will foster social responsibility norms more readily than weak dynamic capabilities will. Hence, any ETS crowding-out of voluntary action will be less likely for companies with strong dynamic capabilities. This combined effect leads us to expect strong support for regulation and further strengthening of innovative climate strategies that were already in place before regulation.

Company-external Factors

Research has shown how various aspects of 'nationality' are important for company attitudes, culture and strategies (see Gleckman 1995, Rowlands 2000, Skjærseth and Skodvin 2003). One type of 'nationality' influence is likely to come from a company's home-base country, where the company will face particularly strong inherited resource-dependency relationships with market agents and governments. The greatest impact of 'nationality' is expected when a company's history, headquarters and activities are located in the same country. Climate strategies can be affected by social demand for environmental protection, governmental supply of climate or R&D policies and the political institutions linking demand and supply through company–state relationships. High social demand, governmental supply and cooperative state–society relationships constitute an enabling national context that will promote social responsibility and innovative corporate strategies. As to ETS impacts, a specific 'nationality' twist could be expected, because the ETS started out as a decentralized system based on National Allocation Plans that provided room for national governments to affect the regulatory stringency for

**Table 2.4   Selected company-internal and wider external factors:
Implications for Models I, II and III**

External and Internal factors	Implications for Model I	Implications for Model II	Implications for Model III
High regulatory pressure. High carbon intensity. High exposure to international competition.	Companies strongly reactive.	Innovative strategy more likely, depending on how far ahead regulation is, compared to other countries.	Engagement in CSR for strategic reasons.
Strong dynamic capabilities.	Not relevant.	Innovative strategy more likely, even when regulation is weak.	Promote social norms of responsibility and strengthen innovative strategy.
Enabling national context.	Not relevant beyond location of regulated installations.	Innovation more likely, depending on coordination between EU and national policies.	Promote social norms of responsibility and strengthen innovative strategy.
Ambitious international regimes levelling the playing field.	Ease resistance.	Innovation more likely.	Depends on participation and legitimacy.

specific companies and sectors (see Chapter 1). Over time, the system has evolved towards greater harmonization in allocation rules across EU member states.

National contextual factors will moderate differently the effect of the EU ETS expected from our three basic models. Such factors can also affect climate strategies independently of the ETS. With regard to Model I, national context is relevant only for the national location of installations which can affect regulatory costs. Model II is concerned with the consequences of coordination and harmonization of regulation and policies. Emissions trading is frequently implemented alongside existing national policy instruments, like carbon taxes and voluntary agreements (see Sorrell and Sijm 2003). In addition to horizontal links between the ETS and other relevant EU policies, vertical links between the ETS and national policy may be important for the risks and opportunities of climate strategies (Skjærseth and Wettestad 2008). With Model III, an enabling national context, particularly as regards societal demands for more ambitious climate policies, may also help to explain why some companies have incorporated stronger social norms of responsibility than others.[32]

International regimes such as the UNFCCC and the Kyoto Protocol (KP) can affect climate strategies (Skjærseth and Skodvin 2003). Such influences are particularly important for multinational companies with operations all over the world. International developments can affect company expectations about the future and national policies. The KP's provision for international emissions trading and the EU obligations under the Protocol were pivotal to the initiation of the ETS in the first place, as noted in Chapter 1. International cooperation and harmonization in coping with climate change can reduce competitive disadvantages, level the playing field and reduce uncertainty for early movers. This is likely to ease resistance concerning Model I and increase the likelihood of innovation according to Model II. The implications for Model III will depend on participation in making the regime, and the perceived legitimacy of the regime.

Finally, climate strategies and response to the ETS may be affected by factors exogenous to all our explanatory perspectives – specific circumstances at company, national or international levels. Here we should note the financial crisis that started in 2008, which has affected the carbon price and made capital less available for companies. While such occurrences may show up in the risks and opportunities of regulation, they should be given separate consideration.

Company-internal and -external factors beyond the ETS will be analysed on an *ad hoc* basis in the case studies. In each instance, these factors will be taken into consideration in explaining change and differences in corporate climate strategies.

Conclusions

This chapter has presented the analytical framework for explaining variation in climate strategies, with special emphasis on corporate response to the EU ETS. We began by defining and conceptualizing corporate climate strategies in terms of objectives and mobilization of resources on a continuum from reactive to innovative strategies. This particular continuum was constructed because we seek to assess changes in strategies, and to evaluate only broadly any differences observed across companies and sectors. We have pointed out the methodology challenges and pitfalls involved in assessing individual company strategies, such as separating the public profile of a company from its actual behaviour.

We went on to develop several models and noted how companies could be expected to respond to a regulation like the ETS. The first and most parsimonious model – companies as reluctant adapters – expects companies to oppose any environmental regulation, adapt at the lowest cost possible and comply only to the extent that compliance entails lower costs than non-compliance. According to this model, we would expect responses in the direction of reactive strategies. The

---

32   This can be referred to as 'home-state socialization' (e.g. Flohr et al. 2010).

second model – companies as innovators – expects that appropriately designed regulation will stimulate exploration, experimentation and innovation in firms. According to this model, we would expect responses in the direction of innovative strategies. The final model portrays companies as socially responsible actors. This model is less developed than the first two, but underscores the importance of corporate social responsibility and how regulation may 'crowd in' or 'crowd out' environmental norms. It is extremely difficult to extract empirically assessable implications from this model, but one main expectation from crowding theory is that cap-and-trade may reduce social responsibility norms, here operationalized as stated responsibility and voluntary (CSR) action on climate change. This may in turn serve to counteract an innovative low-carbon strategy.

As climate strategies are likely to result from a wider set of factors than those incorporated in the models above, our final step in this chapter was to identify internal and external drivers beyond the EU ETS. We wanted to explore how the three models could be extended by including a wider range of explanatory factors that might affect variation in climate strategies. Regulatory risks in terms of carbon intensity and exposure to international competition, dynamic capabilities and national and international policies may affect changes in climate strategies (to the extent that these factors change), but are particularly useful for explaining why climate strategies differ.

Table 2.5   Summary of main expectations

Perspectives on regulatory response	EU ETS response	Significance of company internal and wider external factors for climate strategy	Climate strategy
Model I: Companies as reluctant adapters.	Resistance/reluctant compliance.	Low	Reactive strategy.
Model II: Companies as innovators.	Search for new innovative market opportunities if system is 'appropriate'.	High	Towards innovative strategy.
Model III: Companies as socially responsible.	'Crowd-in' or 'crowd-out' of social responsibility norms .	High	Norms of social responsibility will reinforce innovative strategy.

References

Ambec, S., Cohen, M.A., Elgie, S. and Lanoie, P. 2011. *The Porter Hypothesis at 20: Can Environmental Regulation Enhance Innovation and Competitiveness?* Washington, DC: Resources for the Future.

Aragón-Correa, J.A. and Sharma, S. 2003. A contingent resource-based view of proactive corporate environmental strategy. *Academy of Management Review*, 28(1), 71–88.

Arimura, T., Hibiki, A. and Johnstone, N. 2007. An empirical study of environmental R&D: What encourages facilities to be environmentally innovative? In *Corporate Behaviour and Environmental Policy*, edited by N. Johnstone. Cheltenham: Edward Elgar (in association with the OECD).

Barth, R. and Wolff, F. 2009. *Corporate Social Responsibility in Europe: Rhetoric and Realities*. Cheltenham: Edward Elgar.

Bourgeois, L. and Brodwin, D. 1984. Strategic implementation: Five approaches to an elusive phenomenon. *Strategic Management Journal*, 5, 241–64.

Burtraw, D. 2000. *Innovation Under the Tradable Sulphur Dioxide Emission Permits Program in the U.S. Electricity Sector.* Discussion Paper 00-38, Washington, DC: Resources for the Future.

Cho, C., Patten, D. and Roberts, R.W. 2006. Corporate political strategy: An examination of the relation between political expenditures, environmental performance, and environmental disclosure. *Journal of Business Ethics*, 67(2), 139–54.

Cyert, R.M. and March, J.G. 1992[1963]. *A Behavioral Theory of the Firm*. Englewood Cliffs, NJ: Prentice Hall.

Eisenhardt, K.M. and Martin, J.A. 2000. Dynamic capabilities: What are they? *Strategic Management Journal*, 21, 1105–21.

Flohr, A., Rieth, L., Schwindenhammer, S. and Wolf, K.D. 2010. *The Role of Business in Global Governance: Corporations as Norm-entrepreneurs*. New York: Palgrave Macmillan.

Frey, B.S. 1997. *Not Just for the Money. An Economic Theory of Personal Motivation*. Cheltenham: Edward Elgar.

Frey, B.S. and Jegen, R. 2001. Motivation crowding theory. *Journal of Economic Surveys*, 15(5), 589–611.

Frey, B.S. and Stutzer, A. 2008. Environmental morale and motivation. In *The Cambridge Handbook of Psychology and Economic Behaviour*, edited by Alan Lewis. Cambridge: Cambridge University Press, 406–28.

Gleckman, H. 1995. Transnational corporations' strategic responses to 'sustainable development', *Green Globe Yearbook*. Oxford: Fridtjof Nansen Institute, Oxford University Press.

Gravelle, H. and Rees, R. 1981. *Microeconomics*. London: Longman.

Harrington, W. 1988. Enforcement leverage when penalties are restricted. *Journal of Public Economics*, 37, 29–53.

Hart, S.L. 1995. A natural-resource based view of the firm. *Academy of Management Review*, 20(4), 986–1014.

Helfat, C.E. and Raubitschek, R. 2000. Product sequencing: Co-evolution of knowledge, capabilities and products. *Strategic Management Journal*, 21, 961–79.

Hoffmann, A.J. 2006. *Getting Ahead of the Curve: Corporate Strategies That Address Climate Change*. Prepared for the Pew Centre on Global Climate Change.

Hunt, C.B. and Auster, E.R. 1990. Proactive environmental management: Avoiding the toxic trap. *MIT Sloan Management. Review*, 31(2), 7–18.

Jaffe, A.B. and Palmer, K. 1997. Environmental regulation and innovation: A panel data study. *Review of Economics and Statistics*, 79(4), 610–19.

Johnstone, N. and Labonne, J. 2006. *Environmental policy, management and research and development*. Paris: OECD Economics Department Working Paper No. 457 – ECO/WKP (2005) 44.

Kolk, A. 2000. *Economics of Environmental Management*. Harlow: Financial Times Prentice Hall.

Kolk, A. and Mauser, A. 2002. The evolution of environmental management: From stage models to performance evaluation. *Business Strategy and the Environment*, 11, 14–31.

Kolk, A. and Pinkse, J. 2004. Market strategies for climate change. *European Management Journal*, 22(3), 304–14.

Lanoie, P., Laurent-Lucchetti, J., Johnstone, N. and Ambec, S. 2007. Environmental policy, innovation and performance: New insights on the Porter hypothesis. Centre Interuniversitaire de Recherché en Analyse des Organisations (CIRANO), *Série Scientifique*, 2007, Quebec: CIRANO.

Lee, Su-Yol. 2012. Corporate carbon strategies in responding to climate change. *Business Strategy and the Environment*, 21, 33–48.

Lepoutre, J. 2008. *Proactive Environmental Strategies in Small Businesses: Resources, Institutions and Dynamic Capabilities*, Dissertation submitted to the Faculty of Economics and Business Administration, Department of Management, Innovation and Entrepreneurship, Universiteit Gent, Ghent, Belgium.

Levy, D. and Kolk, A. 2002. Strategic responses to global climate change: Conflicting pressures on multinationals in the oil industry. *Business and Politics*, 4(3), 275–300.

Meznar, M. and Nigh, D. 1995. Buffer or bridge? Environmental and organizational determinants of public affairs activities in American firms. *Academy of Management Journal*, 38, 975–96.

Mintzberg, H. 1993. The pitfalls of strategic planning. *California Management Review*, 36(1), 32–47.

Newton, T. and Harte, G. 1997. Green business: Technicist Kitsch? *Journal of Management Studies*, 34(1), 75–98.

Nyborg, K. and Rege, M. 2003. Does public policy crowd out private contributions to public goods? *Public Choice*, 115(3), 397–418.

Okereke, C. 2010. Regulatory pressure and competitive dynamics: Carbon management strategies of UK energy-intensive companies. *California Management Review*, 52(4), 100–124.

Ott, H. and Sachs, W. 2000. *Ethical Aspects of Emissions Trading.* Contribution to the World Council of Churches Consultation on 'Equity and Emissions Trading–Ethical and Theological Dimentions', Saskatoon, Canada 9–14 May. Wuppertal: Wuppertal Papers 110, September.

Penrose, E. 1959. *The Theory of the Growth of the Firm.* New York: Wiley (2nd edn: Oxford University Press, 1995).

Pinkse, J. and Kolk, A. 2007. Multinational corporations and emissions trading: Strategic responses to new institutional constraints. *European Management Journal*, 25(6), 441–52.

Popp, D. 2006. International innovation and diffusion of air pollution control technologies: The effects of NO_x and SO_2 regulation in the US, Japan, and Germany. *Journal of Environmental Economics and Management*, 51(1), 46–71.

Porter, M. and van der Linde, C. 1995a. Toward a new conception of the environment–competitiveness relationship. *Journal of Economic Perspectives*, 9, 97–118.

Porter, M. and van der Linde. 1995b. Green and competitive: Ending the stalemate. *Harvard Business Review*, September/October, 120–34.

Rowlands, I.H. 2000. Beauty and the beast? BP's and Exxon's positions on global climate change. *Environment & Planning C: Government and Policy*, 18(3), 339–54.

Russo, M.V. and Fouts, P.A. 1997. A resource-based perspective on corporate environmental performance and profitability. *Academy of Management Journal*, 40(3), 534–59.

Schumpeter, J.A. 1934. *The Theory of Economic Development.* London: Oxford University Press.

Sharma, S. and Henriques, I. 2005. Stakeholder influences on sustainability practices in the Canadian forest products industry. *Strategic Management Journal*, 26, 159–80.

Sharma, S. and Vredenburg, H. 1998. Proactive corporate environmental strategy and the development of competitively valuable organizational capabilities. *Strategic Management Journal*, 19(8), 729–53.

Simon, H.A. 1976. *Administrative Behavior. A Study of Decision-making Processes in Administrative Organization*, 3rd edn. London: The Free Press, Collier Macmillan.

Skjærseth, J.B. and Skodvin, T. 2003. *Climate Change and the Oil Industry: Common Problems, Varying Strategies.* Manchester: Manchester University Press.

Skjærseth, J.B. and Wettestad, J. 2008. Implementing EU Emissions Trading: Success or failure? *International Environmental Agreements*, 8(3), 275–90.

Sorrel, S. and Sijm, J. 2003. Carbon trading in the policy mix. *Oxford Review of Economic Policy*, 19(3), 420–37.

Stavins, R.M. 2001. *Experience with market-based environmental policy instruments*. Resources for the Future Discussion Paper dp-01-58. Washington, DC: Resources for the Future.

Steger, U. 1993. The greening of the board room: How German companies are dealing with environmental issues. In *Environmental Strategies for Industry. International Perspectives on Research Needs and Policy Implications* (eds) K. Fischer and J. Schot. Washington, DC: Island Press, 147–67.

Underdal, A. 2000. Comparative analysis: Accounting for variance in actor behaviour. In *International Environmental Agreements and Domestic Politics*, edited by A. Underdal and K. Hanf. Aldershot: Ashgate.

Vedung, E. 1997. *Policy Instruments: Typologies and Theories*. Uppsala: Uppsala University, Department of Government.

Winter, S.G. 2000. The satisficing principle in capability learning, *Strategic Management Journal*, 21, 981–96.

Chapter 3
Electric Power Industry

Per Ove Eikeland

1. Introduction

This chapter analyses the effects of the EU ETS on climate strategies observed in the European electric power industry. In contrast to other industries, this industry saw a substantial deficit of allowances under the evolving EU ETS, making it particularly suitable as object of study. Also its share dominance within the EU ETS system, with heat and electric power combustion installations accounting for 2/3 of total allowances allocated, makes this industry worth particular attention.[1]

Our data indicate that European electric power companies cautiously implemented more proactive and innovative short- and long-term climate strategies during the first decade of the 2000s. Our main conclusion is that these strategic changes were causally linked to the EU ETS through various mechanisms.

Earlier studies have reached similar but non-equivocal conclusions. Cames' survey study of German power companies concluded that emissions trading had been one important factor behind 'soft' institutional innovations such as the establishment of routines and tools for monitoring emissions, the integration of CO_2 into existing trading floors, and the adoption of risk-hedging strategies (Cames 2010: 174–5). The study documented that expected carbon prices were important for investment decisions, but also that the EU ETS had played no major role in advancing 'hard' technological innovations, except for demonstrating carbon capture and storage (CCS) (Cames 2010: 175).

In another study of German power companies, Rogge, Schneider and Hoffmann (2010) concluded, from multiple case-study data from 2008 and 2009, that the EU ETS had only had limited impact on engagement in research, development and demonstration (RD&D) of new technology, because of the scheme's initial lack of stringency and predictability. Also this study found the strongest impacts in the areas of CCS and clean-coal technologies. Another finding was organizational/

---

1   Heat and power installations were allocated 1.47 billion allowances of a total 2.18 billion in the first trading period and 1.26 billion of a total 1.86 billion in the second trading period. The top 12 installations under the system in 2008 were all tied to combustion of fuels (Carbon Market Data, press release 6 April 2009, http://www.carbonmarketdata.com/cmd/publications/Press%20release%20EU%20ETS%202008%20-%206%20April%20 2009.doc).

vision change in the companies in preparation for investments in renewable energy, driven by the long-term climate targets that had been adopted.

A Swedish survey study (Sandoff, Schaad and Williamsson 2011) noted widespread adoption of carbon mitigation strategies in 115 Swedish heat and power companies, but found scant impacts from the EU ETS.[2] This study concluded that companies appeared to have limited opportunities to make rational choices based on the EU ETS because of risk aversion, complex operating conditions, and lack of knowledge about cost structures. Observed impacts included predominantly short-term learning in the operational management of emissions allowances. This led the authors to conclude that the EU ETS had primarily facilitated short-run operating flexibility, not long-run cost reallocation (Sandoff, Schaad and Williamsson 2011: 24).

Our study differs from these earlier works in design and range of data collected and triangulated in the analysis. The design has enabled knowledge about a wider set of mechanisms by which the EU ETS has affected the industry. Our study data covers the top 20 EU ETS allowance holders in Europe, information about strategic responses of the European industry association Eurelectric, and in-depth studies of two major companies in the sector. Additionally, our study is based on more recent data, enabling a stronger focus on the effects emerging after the EU ETS revisions in 2009.

Despite the general shifts observed in climate strategies, our study points, as do earlier ones, to substantial inter-company differences. This indicates that a full explanation of company climate-strategy dynamics requires attention to additional factors internal or external to the firm.

Vattenfall AB and RWE AG were selected as cases for in-depth study, on the basis of several selection criteria. These two companies appeared with differences in the dependent variable that were sufficiently significant to lead us to expect to find that company-specific factors would have explanatory power. This said, at the end of the decade both companies undertook a substantial shift towards more proactive and innovative responses. By selecting cases among the top five in Europe in terms of power generation and revenues, we expected both companies to have superior financial resources that could be mobilized for GHG mitigation. Selecting companies with history and headquarters in different national contexts (Sweden, Germany) was in line with our aim of investigating such national contextual variables. Like most cases selected for deeper scrutiny in this book, both companies have their origin in a Northern European context. This could be viewed as a clear limitation in that fewer contextual differences would be expected than if cases had been selected more evenly across Europe. As noted in the introduction to this book, this was a deliberate choice, intended to reduce the costs of data collection.

---

2   Data were collected in 2006 and then in 2009, to capture changes after the Swedish heat and power industry received no free allowances in the second trading period.

Section 2 provides background information on the technical possibilities for the power industry to deal with the climate change problem in the short and long term, as well as the factors that make companies differ in implementation costs. Section 3 maps and analyses changes in climate strategies in the electric power industry over the past decade – with the situation at the start of the decade taken as baseline. We substantiate more thoroughly the changes observed for Vattenfall and RWE. Section 4 analyses our general observations indicating that the industry has increasingly moved towards more proactive and innovative strategies, and asks whether the EU ETS and other EU-level regulations are plausible explanatory factors. Section 5 analyses the role played by company-specific internal and external factors. Section 6 provides a comprehensive discussion of co-evolving factors impacting on climate strategies, and Section 7 rounds up with key conclusions.

2. Sector Characteristics

The electric power industry constitutes an important pillar for economic growth and welfare, generating vital inputs to all other sectors of the economy. Modern high-tech societies are based on reliable supplies of electric power. On the other hand, individual countries differ significantly in the share of electricity in their energy supply mix. For instance, in countries where electric-powered heating is common, electricity constitutes a high share of total energy supply. Other countries have built up extensive district heating systems where electricity constitutes a far lower share.

The electric power industry is the single major emitter of greenhouse gases (GHGs) in Europe and the key industry under the EU ETS system, with a share of around 60 per cent of total emissions covered by the system. The major GHG emitted is CO_2 originating from combustion of fossil fuels in heat and electricity generation. The power industry differs from other industries in the overall higher deficit of allowances allocated as compared to actual emissions throughout all trading periods. It differs also because the carbon leakage problem associated with the EU ETS – relocation of production to non-ETS areas – has been seen as less relevant because of the limited opportunities for trade with non-ETS areas due to limitations in existing transmission capacities.[3]

Fossil fuel-based electricity accounted for more than half of total supply in Europe in 2005. Gas and hard coal accounted for nearly 20 per cent each, lignite for 11 per cent and oil for 4 per cent (European Commission Directorate for Environment et al. 2006: 17). The remaining non-fossil electricity was generated from nuclear power (nearly one third) and renewables, including hydropower

---

3 Concerns over carbon leakage have been raised in the Baltic countries, where transmission capacity to neighbouring Russia and Belarus is substantial.

(17 per cent). There are, however, major variations in the primary energy mix used for electricity production across Europe.

The carbon content of various fossil fuels differs greatly, as does carbon intensity (tonnes of CO_2 per MWh) of power generated by different conversion technologies with different efficiency characteristics. This is shown in Table 3.1.

Table 3.1   Estimated carbon intensity in power generation (tonnes CO_2 per MWh)

	Direct CO_2 emissions
Lignite, old	1.11
Lignite, new	0.91
Hard coal, old	0.91
Hard coal, new	0.75
Gas standard tech, old	0.63
CCGT, new	0.35

Source: European Commission Directorate for Environment et al. (2006).

This fact leaves substitution among various fossil fuel-based technologies one option for abatement of carbon emissions.[4] Shifting to co-generation units (simultaneous generation of usable electricity and heat) would reduce emissions even more, since combined heat and power plants could achieve fuel efficiency of 89 per cent, as against 55 per cent for the most efficient conventional plants today. Major challenges in co-generation are finding stable demand for the heat produced, and the fact that necessary investments in heat distribution networks will increase the costs of co-generation projects.

Another abatement option is shifting to carbon-neutral primary energy sources (nuclear power or renewables). Also here, complementary infrastructure investments could be needed, as the most favourable sites for generating wind or solar power are not necessarily located close to where power consumers are located. A further problem with intermittent renewable electricity sources like wind and solar power is the need for back-up capacity to balance supply and demand when the wind is not blowing or the sun is not shining. Several options are possible: back-up non-intermittent generation (fossil fuels or biomass-based capacity), various storage options, or load management options, such as smart-grid solutions.

---

4   Notably shifts from old lignite-based power technology to new integrated lignite gasification combined-cycle generation or new combined-cycle natural gas-based generation (CCGT).

A broader carbon emissions abatement strategy could include efforts at reducing the demand for heat and electricity (assisting customers in energy savings) or the opposite option: expanding electricity supply to sectors currently applying less carbon-efficient energy solutions. A notable example would be electrification of the transport sector, where highly decentralized combustion of fuels (in individual vehicles) makes substantial emissions reductions difficult. This would be potentially far easier if the combustion process were moved to a central plant. Yet another option would involve investments to reduce losses in the transmission and distribution of power. Not all abatement options will be relevant in direct response to commitments under the EU ETS, where the focus is on abatement of CO_2 from existing *production* installations. A company cannot offset emissions under the EU ETS by facilitating emissions reduction at the premises of a customer. An option for the longer term would be RD&D of new climate-friendly solutions, such as avoiding CO_2 emissions from the combustion of fossil fuels through CCS. Currently, three basic CCS options are under development for application in large-scale power plants: pre- and post-combustion capture as well as oxy-fuel combustion.[5]

The costs of various CO_2 abatement technologies have been significantly reduced over the past decade, and some of them are now being implemented on a massive scale, aided by political support mechanisms at the national level. Through RD&D, biomass-, solar-, and wind power technologies have reached cost levels that make them commercially interesting at current prices.

Implementation of new technological solutions comes at a commercial risk, however, accentuated by the free market regulation of the sector. Companies can no longer automatically pass on any investment costs to their customers in the form of higher tariffs (if competitors offer supply at far lower market tariffs). Nor is the pre-liberal market practice of nationalization of costs over the state budget for state-controlled companies fully open any longer. State aid is under the European free-market regulations basically viewed as illegal as it could distort competition in the internal market. On top of this, the new equity organization of the sector entails that running a company at low profits would not be possible, as investors require annual dividends. In this respect, the room for manoeuvre of the electric power companies appears narrower today than 30 years ago.

The costs of large-scale emissions abatement investments would vary from company to company. Abatement costs would be greater for a company whose existing plants were still not amortized, than for a company with fully depreciated plants that would need replacement anyway. The lifetime of a power plant is

---

5   Post-combustion capture removes the CO_2 from the flue gases at power stations. In pre-combustion capture, the fossil fuel is partially oxidized through gasification into CO_2 and H2 with the former captured and the latter used as fuel. In oxy-fuel combustion the fuel burned is oxygen instead of air. The flue gas consists mainly of pure CO_2 that can be transported to the sequestration site and stored. The process relies, however, on the highly energy-consuming initial separation of air to make pure oxygen.

typically 45 years. It is usually depreciated after 20 years and then bears no more capital costs (European Commission Directorate for Environment et al. 2006: 19). Many of Europe's power plants are ageing and are likely to need replacement within the next 15 years; indeed, almost half the capacity is more than 30 years old (KPMG 2011).

3. Changes in Climate Strategies

Our conclusions on a trend-shift towards more proactive and innovative climate strategies in the European electric power industry after the year 2000 are based on a combination of indicators and data sources presented in this section. At the aggregate level, the changes are first indicated by actual reductions in CO_2 emissions by the 20 main EU ETS allowance-holders in the electric power industry. On average, these companies cut their emissions by 18.3 per cent for the period 2005 to 2009 and 6.7 per cent from 2005 to 2008, the latter controlling better for impacts from the financial crisis.[6] In 2010, EU economic growth was positive, with GDP increasing by about 2 per cent compared to 2009. According to Eurostat's 2010 energy balances, this caused a strong increase in primary energy consumption, but at a rate lower than the increase in final energy available to the end-use sectors, explained by a strong increase in renewables and higher transformation efficiency in conventional thermal power (European Environment Agency 2012: 11). In 2011, the electric power industry again saw emissions down, dropping 3.1 per cent from 2010.[7]

Other aggregate industry data pointing in a similar direction include the more active role played by the European industry federation Eurelectric[8] in coordinating and spreading information about best-practice emissions reduction opportunities (section 3.1) and a shift in Eurelectric's positions on climate change policies generally and the EU ETS specifically (section 3.2). Company-level information includes R&D data for the major companies indicating new long-term innovative strategies (section 3.3). Section 3.4 sums up these diverse data sources which indicate significant industrial strategic change. Beyond the general tendencies, variation in climate-strategy responses can still be observed among specific companies. Section 3.5 examines some of this variation in greater detail.

---

6   The figures are computed from data presented in Appendix Table 3.1.

7   Platts, 3 April 2012, 'Verified EU ETS CO_2 Emissions Fall 2.4 per cent in 2011: Partial Data'.

8   Full members of Eurelectric include the national electric power industry federations of EU and EFTA countries as well as those of Croatia, Macedonia and Turkey. Affiliate members include national power industry federations of other Eastern European countries (Russia included) and some African and Asian nations.

3.1 Eurelectric – Accepting the Problem and Responsibility for Solutions

Already in the early 1990s the European electric power industry association UNIPEDE, sister organization of Eurelectric,[9] largely accepted the conclusions of the UN climate scientific panel, IPCC, that climate change represented a serious problem, and accepted responsibility for contributing to a solution, preferably through the joint implementation of a worldwide climate protection strategy, as well as *voluntary self-commitments* by the industry (Eurelectric, December 1994).

Eurelectric backed up this position by initiating an Energy Wisdom Programme that regularly collected and spread information about best practice voluntary emissions reductions in the European power industry. The first results, covering company initiatives taken in the 1990s, were published in 2002 in the Energy Wisdom Programme Global Report 1990–2000.[10] The fifth and most recent 2011 report presented experiences from 19 major European power companies and 500 projects estimated to have saved 630 Mt CO_2 (Eurelectric 2010a: 5). These projects used a wide range of abatement solutions, corresponding with the opportunities as outlined in section 2 of this chapter.

In 2005, Eurelectric initiated a 'Role of Electricity' scenario project that in 2007 reported how the industry could halve CO_2 emissions by 2050, formulated as an industry commitment. After the fourth IPCC Assessment Report indicated the need for even more drastic cuts to prevent dangerous climate change, Eurelectric initiated a new study, *Power Choices – Pathways to carbon-neutral Electricity in Europe by 2050.* In 2009, this study outlined how this ambitious goal could be achieved, relying heavily on massive energy-efficiency efforts combined with electrification and extensive fuel conversion for transport and other sectors of the economy. As such, the report emphasized new business expansion opportunities for the industry. The study also indicated that the costs of achieving carbon-neutral electricity supply would be nearly offset by saved allowance payments and revenues from higher electricity tariffs, the latter seen as an inevitable effect of an EU ETS with a continuously shrinking amount of allowances. Recognizing these new opportunities, 61 CEOs, representing nearly all the major power companies in Europe, in 2009 signed a declaration committing their companies to achieving carbon-neutral power supply in Europe by 2050. The recent fifth Energy Wisdom Programme report referred to this commitment and stated the inventory of the Energy Wisdom Programme as an ideal place to demonstrate the wide range of existing abatement opportunities.

---

9   Eurelectric was established in 1992 and merged in the late 1990s with the leading European electricity industry think-tank UNIPEDE (International Union of Producers and Distributors of Electrical Energy). Both groups had formed climate-change working groups in the early 1990s to monitor the work of the IPCC and the political processes prior to the establishment of the UNFCCC in 1992.

10   Here, 11 companies shared their experiences in emissions reductions from some 200 projects.

3.2 Eurelectric's Position on Climate Policies and the EU ETS

We note a parallel development in Eurelectric's position on climate change policies. In the mid-1990s, while endorsing the flexible mechanisms of the Kyoto Protocol – joint implementation (JI), the clean development mechanism (CDM) and an international emissions trading system – Eurelectric still insisted that climate regulation should be voluntary. Several power companies became active implementers of JI projects under the pilot phase set by the UNFCCC in the late 1990s.

Eurelectric also warned against any unilateral EU regulatory action,[11] but changed this position later in the same decade and took a generally overall positive stand on the EU pilot emissions trading system. Then head of the association's Climate Change Working Group, representative of the French group EDF, strongly supported the EU initiative, as did British member companies. Some members, notably German companies, still opposed unilateral EU regulation and opted for only voluntary measures, in line with the regulation established in their home market.[12]

Preparing for what might come, Eurelectric in 1999 wrote essential rules for how such an EU ETS could work and, together with the Paris Bourse, established a trading desk for simulation of trade in energy and carbon (Greenhouse Gas and Energy Trading Simulations – GETS). Nineteen companies participated in a first round of trading. This number had increased to 34 companies in a second round.[13] This time, most German companies chose not to participate, except for HEW, at that point partly under control of Swedish Vattenfall.

Eurelectric's largely pro-EU ETS position was now challenged by the anti-EU ETS campaigns of German companies. However, the association managed to convince them there would be little to fear, after the GETS simulations showed higher electricity tariffs and profits for the industry.[14] Eurelectric went public with the GETS results in a 2004 EU ETS impact assessment, declaring that the system was superior for achieving cost-effective emissions reductions (Eurelectric June 2004). The assessment went on to state that, as with any economic instrument, the EU ETS would cause electricity tariffs to rise, with subsequent distributional effects, and warned that counteracting measures would break with the principle of cost effectiveness (Eurelectric June 2004).

---

11 This is exemplified by its opposition to the EU carbon/energy tax proposed by the Commission in the early 1990s; it was claimed to be harmful to EU competitiveness. See Eurelectric (March 1996).

12 Interview with John Scowcroft, Head of Environment and Sustainable Development, Eurelectric, 17 November 2010.

13 Interview with John Scowcroft, Head of Environment and Sustainable Development, Eurelectric, 17 November 2010.

14 Interview with John Scowcroft, Head of Environment and Sustainable Development, Eurelectric, 17 November 2010.

One factor threatened to cap the higher industry profits foreseen, however: the Commission's plans for mandatory auctioning of emissions allowances. Eurelectric lobbied consistently for free allocation of allowances (based on the principle of 'grandfathering': the idea that liable entities should be allocated free allowances in some relation to past emissions) when responding to the Commission's Green Paper in 2000 and the proposal for a directive in 2001. It also lobbied for the system to be voluntary (at least for the period 2005–2007, before a full-fledged international system could start according to the Kyoto Protocol);[15] for broad coverage of all GHGs (accepting CO_2 only in a voluntary trial period 2005–2007); and for no restrictions on the use of credits from CDM and JI projects (including credits from domestic projects in sectors not covered by the EU ETS scheme).

The Commission proposal made no reference to making the system 'voluntary'. It promised free allocations only for the first trading period (no mention of the second period), and said nothing about how to deal with the CDM and JI mechanisms. It further proposed excess emissions penalties to be set at EUR 100 per tonne of carbon dioxide, which was considerably higher than the EUR 20/ tonne proposed by Eurelectric (Eurelectric 15 May 2002).

When the EU ETS Directive was finally adopted by the Parliament and the Council in 2003, it marked a compromise. The system was made mandatory for member states but would allow an opt-out of installations in the first trading period, subject to specific annexed criteria; allocations were to be free but with the option for member states to auction 5 per cent of the allowances in Phase I and 10 per cent in Phase II. The penalty level remained at EUR 100. The decision on CDM and JI credits was postponed to a Linking Directive, agreed on in 2004. Against the wishes of the power industry, this Linking Directive set out several restrictions on the use of the flexible mechanisms.[16]

When in 2007 the EU leaders decided to revise the EU ETS to accomplish the new 20 per cent GHG reduction goal set for 2020, Eurelectric had further shifted its stance and now mainly supported the Commission proposal. Eurelectric welcomed cap-setting at the EU level and accepted full auctioning as the method of allocation (Eurelectric July 2007). After much public exposure of the windfall profits earned by the industry because of the EU ETS system, Eurelectric acknowledged that

---

15 Eurelectric here made concessions to member companies already regulated under voluntary environmental agreements, arguing that voluntary emissions reductions should be made compatible with the EU ETS.

16 The Commission suggested vaguely that the maximum share of total emissions that could be offset by such credits should be 8 per cent, but the final decision was left to be specified by the member states in their national allocation plans. Excluded were nuclear power and emissions reductions from land use, land-use changes and forestry activities, in line with the guidelines adopted under the Kyoto Protocol. Hydropower projects above 20 MW would need compatibility with the guidelines of the World Commission on Dams.

continued resistance to auctioning would be a dead case.[17] It still warned against any allocation method that would discriminate specifically against the electricity sector, and dismissed the proposal for more favourable allocation rules for co-generation than for electricity. This notwithstanding, Eurelectric accepted more beneficial levels of auctioning for sectors vulnerable to global competition (Eurelectric April 2008). Earlier demands for no limitations on the use of CDM and JI credits were repeated, however.

3.3 Long-term Strategy – Changes in Patterns of Industrial R&D

The 2009 Eurelectric umbrella agreement on zero-emissions electricity supply by 2050 indicates that the industry at the aggregate level moved beyond a purely short-term climate strategy response. Achieving such a long-term goal would be contingent on RD&D of new technological solutions, and indeed, our data indicate a cautious shift in the industry toward the end of the decade in R&D expenditure on carbon-neutral solutions.

A first indicator is that after 2005 most major electric power companies in Europe seemed to reverse or at least halt the trend of declining intensity in R&D, documented in Appendix Table 3.2. Up to this point, R&D intensities had been on a long-term declining trend starting well before the 2000s, as documented in R&D expenditure surveys by UNIPEDE/Eurelectric in 1998 and 2002. In 1998, Eurelectric concluded that the downward trend in long-term R&D activities, set to decrease further, had brought the average normal R&D/turnover rate down from a normal level of around 1 to 0.6 per cent (UNIPEDE December 1998).

This trend was described as eliciting no surprise, reflecting downsizing to remain competitive in the new liberalized market context where internal R&D structures were increasingly seen as a luxury and thus among the first targets of budgetary cuts to keep costs as low as possible to retain customers and stable profit margins (UNIPEDE December 1998). The 2002 survey revealed a further decrease in budgets and human resources applied for R&D and added: 'The free exchange of information is hampered by competitive pursuits and R&D structures have had to adapt to these new conditions' (Eurelectric January 2003).

Table 3.2 shows R&D investments in absolute figures for those electric supply companies that consistently figured on the EU Industrial R&D Investment scoreboard list throughout the whole period 2004–2009. Also this table indicates a downward trend in R&D activities at the start of the decade and a later reversal.

---

17   Interview with John Scowcroft, Head of Environment and Sustainable Development, Eurelectric, 17 November 2010.

Table 3.2 **R&D investments of selected European electricity companies, million €/year**

	2004	2006	2008	2009
EDF	425	389	421	438
RWE	128	135	210	214
Vattenfall	59	96	155	148
GDF Suez	175	170	127	218
E.ON	55	27	102	81
Enel	20	20	68	88
Energie Baden	8	21	29	35
Fortum	26	17	27	30
Total	**896**	**875**	**1139**	**1222**

Source: EU Industrial R&D Investment Scoreboards 2004–2009.

These data alone would, however, not suffice as evidence of more long-term innovative *climate* strategies, as they do not specify R&D expenditure devoted specifically to low-carbon solutions. Yet, such a re-direction is surely indicated in the Eurelectric R&D expenditure surveys. In its 2002 survey, Eurelectric noted a new awareness of environmental impacts after the Kyoto Conference (Eurelectric January 2003). Moreover, the 2010 R&D survey data more explicitly document a re-direction during the decade, with typical climate change mitigation solutions now figuring at the top of research priorities listed by the industry: 73 per cent of the companies ranked the development of renewable electricity technologies and smart grid solutions as the priority issue, followed by electrical vehicles (58 per cent), safety and improvement of power plants (50 per cent), CCS technologies (50 per cent) and energy efficiency (38 per cent) (Eurelectric 2010b).

Yet another recent trend shift appears to be greater involvement in joint long-term co-operative R&D in low- and zero-carbon solutions. Electric power companies have been among the founding fathers of several European Industrial Initiatives (EII) aimed at further development of low-carbon technologies (wind power, solar power, electricity grids, CCS, and bioenergy). The rationales cited for these joint initiatives are risk-sharing, fostering of public–private partnerships, pooling of public and private funding, and leveraging additional resources. Also this trend becomes more evident if benchmarked against the situation recorded by UNIPEDE in 1998: 'Joint projects are contradictory to a competitive market at first glance and mostly take too much time' (UNIPEDE December 1998). To be sure, UNIPEDE still concluded that: 'while utilities appear hesitant to be involved in joint projects on a national scale, a more open attitude is noted towards cooperation on an EU level' (UNIPEDE December 1998).

3.4 Summary of Observed Changes in Strategy

The combination of data provides evidence of a European power industry moving towards more proactive short-term and innovative long-term strategies during the first decade of the 2000s. A series of collective actions coordinated by Eurelectric at the sector level aimed to stimulate cross-company learning about best-practice voluntary emissions reductions and how an EU ETS would affect the industry. Acknowledging that the EU ETS could work well for the industry, Eurelectric became a supporter, even as regards full auctioning of allowances for the power industry that was decided in 2009. Cross-company learning about the possibilities for long-term full decarbonization of power supply in Europe was stimulated by scenario work conducted by Eurelectric, resulting in the 2009 voluntary agreement between 61 power company CEOs to work toward achieving such a goal by 2050. Through these activities, the strategic focus of the industry shifted from a narrow preoccupation with the burden of unilateral EU climate policies, toward the industrial opportunities that proper regulation could provide. Company-level information complements the picture: emission cuts among the major EU ETS allowance holders in the industry,[18] company adoption of individual emissions reduction goals and greater focus on low-carbon solutions in their investments. At the end of the decade, the data indicate upscaled efforts in long-term R&D to produce new low-carbon solutions in many companies.

3.5 Variation at Company Level

Despite the general trend, the data presented above also show significant inter-company differences. Companies varied considerably as regards developments in CO_2 emissions (Appendix Table 3.1),[19] and R&D intensities (Appendix Table 3.2).[20] Other data pointing to differences are listed in Appendix Table 3.3. Some companies had made quantitative and verifiable goals part of their strategy, while others merely reported on past achievements. Some companies had developed elaborate implementation schemes, specifying quantitative short- and mid-term targets for various low-carbon technologies, whereas others had only formulated

---

18   In 2005, these 20 companies held a 60 per cent share of total emissions from the combustion sector in the EU.

19   Significant above-average emissions reductions were recorded for the South European companies Enel (Italy), Endesa and Iberdrola (Spain), EDP (Portugal) as well as Scottish and Southern Energy (UK) and EnBW (Germany). An increase in emissions was reported by five of the companies in the period 2005–2008, among them Edison (Italy), Drax Power (UK) and RWE nPower (UK).

20   Some companies appeared with stabilization at a higher level, including RWE, GDF Suez, and Fortum. Others appeared with reduced R&D intensities, such as EDF and British Energy (part of the EDF group).

more general commitments. Most but not all of the companies had signed the 2009 industry agreement to work for carbon-neutral electricity supply in Europe by 2050.

In the following, we scrutinize two of the top five European electric power companies: Swedish Vattenfall AB and German RWE AG. The companies shared many features concerning activity structure and geographical market exposure. Both were vertically integrated in the electricity supply chain. Both were engaged in the gas and district heat supply chains and in lignite production. RWE AG employed around 70,000 people; Vattenfall had 38,000 employees. Revenues were approximately €53 billion and €24 billion, respectively. Internationalization after 2000 made the UK a major target for RWE asset acquisition (e.g. the acquisition of nPower in 2002), and later the Netherlands (acquisition of Essent in 2009). Aside from its role as major power producer in Germany, it ranked as second largest in the Netherlands, and third largest in Britain, with substantial activities in other European countries as well. Similar internationalization after 2000 brought Vattenfall to its current position as one of the top generators in Germany and the Netherlands, with major assets also in Finland, Denmark and Poland adding to its dominant position in its home-base, Sweden. The ownership structure of the companies differs: Vattenfall AB is 100 per cent owned by the Swedish state; RWE AG is semi-private, with around 30 per cent of the shares owned by municipalities of the German state of North Rhine-Westphalia.

3.5.1 *Vattenfall AB*

Short-term climate strategy. Vattenfall AB signalled strong climate leadership ambitions at the start of the 2000s, backing also the EU leadership in developing an emissions trading system that would eventually become global in scale. The company vision was to be 'number one for the customer, the environment and the economy' (Vattenfall, Sustainability Report 2002). Then-CEO of Vattenfall, Lars Josefsson, became a renowned advocate of a global emissions trading system, which led *Time Magazine* in 2005 to name him 'a European Hero' for his consistent and ambitious environmental engagement.[21]

However, beyond these visions, Vattenfall failed early in the decade to develop any forceful strategy to bring down company GHG emissions. No verifiable quantitative emissions reduction goals were formulated, and in its annual reporting the company focused on past reductions in specific emissions, largely the credit of earlier owners of the coal plants acquired after 2000 (Vattenfall, Annual Report 2001). Future measures were formulated very vaguely as 'expectations' for company reductions in GHGs (Vattenfall, CSR Report 2003).

---

21 In 2006, Lars Josefsson became part of Chancellor Angela Merkel's advisory group on international climate change policies and in 2007, he initiated the 3C – a business initiative to Combat Climate Change with more than 50 major companies as signatories, working on a road map for new global agreement on reducing greenhouse gases by 2013.

As to specific short-term investments, Vattenfall flagged energy-efficient coal plants and mixing coal with biomass in its CHP plants as planned solutions.[22] No quantitative targets for future renewable energy capacity had been set up (Vattenfall, CSR Report 2003), and actual investments were in fact quite modest until the middle of the decade.[23] In 2006, acquisitions in Denmark increased Vattenfall's wind power capacity by a factor of twenty (Vattenfall, CSR Report 2005, published 2006). Investments in coal-based capacity continued, however, with the aim of replacing older capacity with more efficient generation.[24]

The year 2006 also saw the company's first quantitative emissions reduction goals, with 50 per cent reduction in *specific* CO_2 emissions (g/kWh) to be accomplished by 2030 against 1990 figures (Vattenfall, CSR Report 2006). Since 30 per cent had already been achieved, according to the company, the additional 20 per cent would be within reach, given new generations of more efficient technologies. In 2007, the company started setting short-term annual goals for reductions in *absolute* emissions.[25] However, Vattenfall did not achieve these short-term goals; by 2010, emissions had increased substantially.

In 2010 came a strategic shift that set the far more ambitious goal of cutting annual emissions from the Group by 20–24 million tonnes of CO_2 from 2010 to 2020,[26] an estimated reduction in corresponding carbon intensities from 465g CO_2e/kWh to 350g CO_2e/kWh (Vattenfall, CSR Report 2010: 24). This shift promised a new pace in low-carbon investments; according to the company, it implied that no new coal-fired plants without CCS would be built, and the company would evaluate divestment of fossil fuel–based generation associated with 12–14 million tonnes of emissions before 2014. From this point onwards, the company would aim for 50 per cent biomass co-firing in its coal plants and fuel switching from

---

22   This had been carried out on a massive scale in Sweden during the 1990s; Vattenfall covered 50 per cent of fuel needs for its district heat and CHP plants by biofuels (Vattenfall, Annual Report 1999).

23   For example, Vattenfall recorded only 5 GWh higher wind power output in 2002 than in 2004 – up from 55 GWh to 60 GWh (Vattenfall, Environmental Report 2002 and CSR Report 2004).

24   24 By 2010, Vattenfall had nearly finished the replacement of two coal-fired power plants in Germany, Boxberg and Moorburg, with up-scaled generation capacity that would increase company CO_2 emissions by about 10 Mt per year (Vattenfall, CSR Report 2010).

25   The 2007 CSR Report promised a cut in emissions by 3 million tonnes for 2008–2010 (from the level recorded in 2007 of 84.5 million tonnes). The 2008 CSR Report goal had been reduced to 2 per cent (2 million tonnes) in the period 2009–2011. This goal was retained in the 2009 CSR Report, adding also the interim goal of 1.3 per cent reductions in 2010 (1.2 million tonnes). In 2010, Vattenfall reported emissions of about 91.5 million tonnes.

26   The goal set was to reduce emissions from 91.5 million tonnes in 2010 to 65 million tonnes by 2020.

coal to gas, to achieve the overall emissions reduction goals set for 2020.[27] Future expansion plans presented for non-hydro renewable energy included increasing the output from 3.9 TWh in 2010 to 8 TWh by 2020.[28]

Long-term climate strategy. As noted, the quantitative long-term goal set in 2006 concerned only specific emissions and was admittedly achievable through investments in more efficient coal generation technology. In 2009, Vattenfall committed to achieve zero-emissions supply in Europe by 2050 under the umbrella agreement coordinated by Eurelectric.

Supporting the achievement of such a goal came signs of stronger Vattenfall commitments to increase R&D efforts, indicated by changes in R&D intensities and absolute R&D expenditure. Vattenfall had entered the 2000s as a company with a relatively high R&D intensity, kept stable in the range 1.5 per cent to 1.9 per cent in the 1990s (Eikeland 2013, forthcoming). Table 3.3 shows how R&D intensity plunged at the start of the decade to a level the company termed 'on par with comparable companies' (Vattenfall, Annual Report 2002). A significantly higher level is recorded from 2007 onwards.

Table 3.3   R&D expenditure (SEK million) and R&D intensity (R&D/sales in Vattenfall 2000–2010)

	2000	2001	2002	2003	2004	2005	2006	2007	2008	2009	2010
R&D (SEK mill)	481	564	481	478	529	650	761	1,015	1,529	1,322	1,545
R&D (euro)*	61	66	52	51	60	69	81	108	163	141	165
R&D/sales (%)	1.81	0.89	0.48	0.43	0.51	0.53	0.56	0.71	0.7	0.6	0.7

Note: * Data for 2000–2007 from Sterlacchini (2010), where figures are converted by a fixed exchange rate of 0.1065 euro per SEK. Figures for 2008–2010 from Vattenfall annual reports.
Source: Vattenfall annual reports 2000–2010.

---

27   At this point Vattenfall's generation portfolio included 1,448 MW wind power capacity and 448 MW biomass-based heat and power generation capacity (Vattenfall, CSR Report 2010). Two per cent of total heat and power output came from wind power/biomass, up from 1 per cent in 2002. (Vattenfall, Environmental Report 2002 and CSR Report 2010.)

28   This included completion of and further development of offshore wind-farm development in the UK in partnership with Scottish Power Renewables; construction started of offshore Germany and expansion of wind power capacity in the Netherlands and Belgium. Short-term plans for expansion in biomass-based capacities were presented for Germany, the Netherlands and Poland (Vattenfall, 'Biomass', http://www.vattenfall.com/en/biomass-energy.htm).

Not all was spent on developing and demonstrating new climate solutions, however. In 2006 Vattenfall spent roughly half its R&D budget on finding secure long-term storage for spent nuclear fuel; less than 10 per cent was used for renewable energy R&D, and approximately 20 per cent for other ways of reducing CO_2 emissions. CCS RD&D constituted an important part, with most resources covering the costs of constructing/operating the demonstration facilities in Germany (NyTeknik 25 April 2007). Vattenfall had already in 2001 become the first power company in Europe to launch plans for testing CCS technology on a power generation plant. Throughout the decade, Vattenfall engaged in a series of test rigs, EU projects and joint efforts at various pilots and research stations, to gain know-how and experience of all technology types (pre-combustion, oxy-fuel and post-combustion). Construction of the world's first pilot industrial-scale oxy-fuel pilot plant (30 MWth) started in 2006 at the lignite-fired Schwarze Pumpe plant in Germany, and was inaugurated in 2008. The next step involved plans for commissioning a 300 MWe demonstration plant at the Jänschwalde site in Germany in 2015 and, after the acquisition of Nuon in 2009, for the pre-combustion Magnum II plant (1.2 GW multi-fuel plant) in Eemshaven in the Netherlands.[29]

Vattenfall's long-term R&D efforts show more active long-term cooperation with traditional competitors. The company was active in initiating and taking the lead in several European Industrial Initiatives under the EU Strategic Energy Technology Plan adopted in 2007 to strengthen and align R&D resources in specific low-carbon technological solutions. It featured on the list of founders of the European Industrial Initiative on CCS and the European Smart Grid Industrial Initiative, a sub-project under the European Electricity Grid Initiative (EEGI).[30]

3.5.2 RWE AG
Short-term strategy.   In 2000, RWE AG stated:

> It has still not been scientifically proven beyond doubt that and to what extent global warming is anthropogenic. For precautionary reasons, the international political arena has taken far-reaching decisions to reduce worldwide GHG emissions. Germany and Europe are taking the lead here. Being Germany's largest and Europe's third-largest energy utility, RWE regards climate protection as a central task which will remain extremely important in the future in terms of environmental and economic policy. (RWE AG, Environmental Report 2001: 8)

---

29   Through Nuon, Vattenfall operates another Dutch pilot pre-combustion plant at the Willem Alexander power plant in Buggenum (operations started February 2011). In the UK, Vattenfall runs the post-combustion project 'CCpilot100+' at the Ferrybridge Power Station in cooperation with Scottish and Southern Energy (SSE) and Doosan Babcock.

30   Vattenfall reports participation also in another EEGI R&D project, Active Distribution Network with Full Integration of Demand and Distributed Energy Resources (Vattenfall 2010b).

This statement suggests that RWE, although it had indeed accepted the climate problem at the start of the decade, still emphasized that scientific uncertainties existed as to the role of human influences. Despite this, the company clearly opposed the current EU leadership in developing emissions trading as a core climate policy instrument. In its 2001 Environmental Report, RWE acknowledged emissions trading at the global level as a cost-effective measure, but dismissed EU efforts at making such a system part of EU climate policy alone.

Then-CEO of RWE, Dietmar Kuhnt, openly opposed the EU ETS and warned that unless reshaped, the system would force RWE to cancel planned investments in new and far more carbon-efficient lignite power plants in Germany (Power Engineering International, 14 August 2002). A more moderate position was taken by the new CEO from 2003, Harry Roels, who declared himself in favour of the EU ETS on the condition that it would not harm the German economy.[31] Support remained lukewarm, however, and RWE still claimed that the system would hamper clean-coal investments when the German Parliament passed tougher national allocation plans for the second trading period (Reuters 22 June 2007).

RWE formulated no quantitative short-term CO_2-reduction targets in this period, and retained in 2005 investments in more efficient lignite and hard-coal power plants as the main measure to be taken by the company.[32] Other pillars announced for its carbon strategy included investments in onshore and offshore wind power, and involvement in JI/CDM projects (RWE AG, CSR Report 2005).[33]

In 2008, RWE announced a shift in climate strategy, formulating for the first time short-term quantitative targets for CO_2 reductions (30 million tonnes to be cut by 2012 and 60 million by 2015, 20 per cent and 30 per cent compared to emissions in 2008.[34] Key elements of the new strategy were stepping up participation in CDM/JI projects, increasing the share of CHP, constructing the world's first large-scale coal-fired power plant featuring CCS, maintaining the existing nuclear power capacity, and increasing the use of renewable energy (RWE, CSR Report 2008).

---

31   Presentation 'Powering Up – From Consolidation to Delivery' held by Harry Roels at Analyst and Investor Conference, Essen, 26 February 2004, Available at: http://www.rwe.com/web/cms/mediablob/de/213642/data/213106/1/rwe/investor-relations/events-praesentationen/archiv/archiv-2004/blob.pdf

32   In the 2005 CSR Report, RWE noted: 'We expect lignite and hard coal to remain the corner-stones of our electricity generating industry for some decades to come.'

33   RWE had started cautious investments in renewable energy when the merger between RWE and VEW in 2000 made the subsidiary Harpen AG the competence centre of renewable energy for the company. By 2001, RWE had started investments in wind power and reported it was the largest manufacturer of solar panels in Germany through its subsidiary RWE Solar GmbH (RWE AG, Annual Report 2001). By 2003, RWE reported installed renewable energy capacity of 1,085 MW, of which 300 MW of wind power included offshore capacity in the UK (60 MW) through its British subsidiary RWE npower. Plans were presented for further investments in the UK and Spain. In Germany, biomass-based CHP was given priority (RWE AG, Annual Report 2004).

34   Emissions in 2008 were 172 million tonnes.

An organizational change in early 2008 united the management of all renewable energy projects in one company, RWE Innogy, staffed by 600 people and with an annual investment budget of at least €1bn. Presented plans promised expansion mainly in onshore and offshore wind power, vigorous expansion in biomass and hydropower, and the pursuit of opportunities in solar and geothermal energy, wave and tidal power.

The strategic shift also came to include revision of plans for expanding coal-based power. Early 2009, Johannes Lambertz, CEO of RWE Power, announced: 'RWE will suspend large-scale coal or lignite power plant projects in western European countries such as Germany and the UK' (Power Engineering International 21 January 2009). He added that plants under construction would be finalized, and the search for investments in Eastern European countries would continue. In September 2010, however, RWE announced the suspension also of a planned 800 MW plant in Poland (Industrial Info Resources 14 September 2010). Also, in mid-June 2011, RWE announced that, from 2014, the company would move away from conventional power generation and step up its focus on renewable energy sources (Deutsche Welle 14 July 2011). Two coal-fired power plants under construction in Germany would be finalized (adding 3,600 MW capacity) but no new coal or gas power plants would be built after that.

By June 2011, RWE Innogy reported clear achievements in renewable energy, with wind power capacity nearly doubling from 2008.[35] New short-term plans included 2400 MW additional capacity already under development or with construction starting in 2011.[36] Medium-term plans included massive development of offshore wind power in the UK,[37] and becoming a European market player in biomass-based CHP.[38]

Long-term strategy.   The CEO of RWE was signatory to the 2009 Eurelectric agreement to achieve carbon-neutral electricity supply in Europe by 2050. In its 2010 CSR report, RWE outlined very ambitious intermediate targets: a 75 per cent cut in emissions by 2025, and reducing specific emissions to the industrial average

---

35   Of a total 1,706 MW installed, 150 MW was offshore in the UK and 232 MW added after the acquisition of Dutch Essent in 2009, information available at: http://www.rwe.com/web/cms/en/87264/rwe-innogy/renewable-energies/wind/.

36   These plans included new offshore capacity in the UK, Belgium and Germany.

37   These plans reflected that RWE had been chosen as development partner in the UK Zone Development Agreements tenure process for another 1,500 MW capacity in the Bristol Channel Zone and the potential 9000 MW capacity in the Dogger Bank Zone (in a consortium with Scottish & Southern Energy, Statoil and Statkraft), information available at: http://www.rwe.com/web/cms/en/86676/rwe-innogy/renewable-energies/wind/wind-offshore/.

38   These plans included increasing biomass-based CHP capacities from 115 MWe in 2011 to 600 MWe by 2012, information available at: http://www.rwe.com/web/cms/en/87098/rwe-innogy/renewable-energies/biomass/.

by 2020 (to around 0.45g CO_2/kWh, down from 0.732g CO_2/kWh in 2010) (RWE AG, CSR Report 2010: 21). Like Vattenfall, RWE mentioned upscaled R&D intensities and expenditure as part of its new long-term carbon strategy.

Table 3.4   R&D expenditure (million euro) and R&D intensity (R&D/sales) in RWE 2000–2010

	2000	2001	2002	2003	2004	2005	2006	2007	2008	2009	2010
R&D (euro)	101	108	87	78	78	55	73	74	105	110	149
R&D/sales	0.25	0.21	0.19	0.18	0.19	0.14	0.17	0.17	0.21	0.23	0.28

Source: Data for 2000–2007 from Starlacchini (2010). Data for 2008–2010 from RWE AG annual reports.

RWE had, like Vattenfall, seen a major drop in R&D intensity in the early 2000s, from a ratio of well above 1 per cent recorded in the 1990s.[39] The company's main focus in energy technology R&D had historically targeted the efficiency of coal combustion. From mid-decade onwards, RWE made use of these competencies to evolve as one of Europe's main developers of CCS technology applied to coal- and lignite-fired power plants. A major project concerned the plans presented in 2006 for a new Integrated Gasification Combined Cycle plant with CCS in Hürth near Cologne, with construction start planned for 2012 and inauguration in 2014. But CCS development work included other technologies and national locations as well.[40] In 2007, RWE started preparations for testing CO_2 scrubbers for retrofitting existing plants with CCS. A pilot plant was commissioned at the Niederaussem 1000-MW lignite power plant in 2009, also adding a prototype lignite-drying

---

39   In 1995, the RWE group employed 2000 people in its R&D division; the R&D/sales ratio was 1.2 per cent. Most resources were concentrated in the printing machine business, where RWE AG held majority ownership in Heidelberger Druckmaschinen. When RWE AG sold off the printing business in 2004, R&D expenditures fell considerably (Johansen and Damm 2004: 35–6).

40   RWE's UK subsidiary became the first company in the UK to demonstrate CO_2 removal from the flue gases. RWE became broadly involved in various projects evaluating sites for storing and CO_2 transport infrastructure, exemplified by the Snøhvit cooperative project in the Barents Sea and the Ketzin project in Brandenburg outside Berlin. The acquisition of Dutch Essent in 2009 brought RWE into the plans for applying CCS at the 1.6GW coal- and biomass-fired Eemscentrale plant, applying for EU funding from the New Entrant Reserve 300 scheme (see ICIS 15 February 2011).

facility to test CO_2 scrubbing on such processed fuel.[41] Also added was a plant that fed algae production with flue gases, to investigate possibilities for the production of biofuels. The strategic shift in 2008 included more resources allocated to in-house R&D at the Niederaussem Coal Innovation Centre, including also new R&D staff.[42]

Like Vattenfall, RWE reported extended collaboration in long-term climate technology R&D with traditional competitors at the end of the decade. This marked a shift from the late 1990s, when RWE had reported to the UNIPEDE R&D survey: 'Joint projects are contradictory to a competitive market at first glance and mostly take too much time.'[43]

RWE was a founding member of the European Industrial Initiative on joint R&D in CCS technology under the EU Strategic Energy Technology plan adopted in 2007 for strengthening and aligning R&D resources in specific low-carbon technological solutions. Together with its competitor *e.on*, it was also among the founding members of the Desertec Industrial Initiative aimed at further development and imports of solar power from Africa to Europe.

3.5.3 Comparing Climate Strategy Dynamics in Vattenfall and RWE – Similarities and Differences

The recent strategic moves of RWE AG and Vattenfall AG strengthen the conclusions above, that the European power industry at large had begun moving towards more proactive and innovative climate strategies by the end of first decade of the new millennium. Observed similarities include moving from non-verifiable commitments to quantitative short- and long-term reduction goals; halting the further development of coal-based power beyond finalizing the projects already under construction; staging future investments to involve mainly renewable energy; and a shift towards higher R&D intensities after a substantial decline early in the decade.

Clear differences in climate strategies were recorded for the baseline situation at the start of the decade when Vattenfall signaled acceptance of climate science and commitments to solving the problem, whereas RWE was more hesitant, emphasizing scientific uncertainty and opposing EU ETS regulation. As the decade progressed, the two companies evolved in a more similar direction in acceptance of the problem and their stated commitments to solving it.

Differences remained, however, in the timing of company implementation of short- and long-term strategies. For Vattenfall this brought a dramatic rise in

---

41   The coal pre-drying technique developed by RWE would enable more fuel-efficient combustion processes, even without the combination with CCS technology (increasing fuel efficiency from 43 per cent for conventional coal to 47 per cent with pre-dried coal).

42   R&D staff in the RWE Group increased from 233 in 2005 to 360 by 2010 (RWE AG, CSR Report 2010).

43   RWE was responsible for reporting on the situation in Germany to the UNIPEDE 1998 R&D Survey.

its carbon intensity throughout the decade, whereas RWE experienced a gradual declining trend. At the end of the 1990s, Vattenfall's electricity and heat operations had been nearly carbon-neutral (mainly hydropower and nuclear power generation assets).[44] Acquisitions of coal-fired heat and power generation assets outside Sweden changed this picture dramatically after 2000.[45] From this point, total emissions increased steadily,[46] as did specific CO_2 emissions.[47] RWE had quite stable CO_2 emissions in the period 2003–2009,[48] with specific emissions showing a slightly declining trend.[49] The two companies came to settle on a converging trend in carbon intensity: from a very low level, Vattenfall saw a dramatic increase during the decade; from a very high level, RWE saw a declining trend.

As to specific low-carbon investments, Vattenfall appeared early in the decade to lose pace in renewable energy investments compared to the 1990s. On the other hand, the company became frontrunner in the electric power industry in Europe as regards investments in CCS technology for its coal-based power plants. By comparison, RWE saw investments in renewables on the rise already early in the decade, from an insignificant level in the 1990s. Both companies undertook strategic shifts in the second half of the decade that markedly increased their investments in renewables and demonstration of CCS technologies. As to long-term R&D, Vattenfall remained a far more R&D-intensive company than RWE throughout the decade, but both companies experienced the similar dynamics of R&D intensity falling sharply at the start of the decade and increasing again towards the end of the decade.

---

44   Emissions in Sweden amounted to less than a million tonnes of CO_2 in 2001 (Vattenfall, Sustainability Report 2001).

45   The acquisitions entailed that the company in 2001 recorded total emissions of about 80 million tonnes of CO_2 (Vattenfall, Sustainability Report 2001).

46   Total CO_2 emissions reached 91.5 million tonnes in 2010 (Vattenfall, CSR Report 2010).

47   Specific CO_2 emissions saw corresponding growth, recorded at less than 400g CO_2/kWh in 2005, 442 in 2007, and 452 in 2008 (Vattenfall, CSR Report 2010).

48   RWE annual CO_2 emissions were around 150 million tonnes/year in the period 2003–2009, with a deviation in 2007 when emissions amounted to nearly 190 million tonnes. In 2010, emissions rose to 167 million tonnes after the purchase of Dutch Essent.

49   RWE reported a carbon intensity of 0.81g CO_2/kWh for 2005 (RWE AG, CSR Report 2005). This peaked again in 2007 at 0.866g CO_2/kWh, and thereafter fell to 0.768 in 2008, up to 0.796 in 2009, and down again to 0.732 in 2010 (RWE, Our Responsibility Report 2010).

4. Explaining the General Trend Towards More Proactive and Innovative Climate Strategies

4.1 The Role of the EU ETS

According to Eurelectric, the adoption of the EU ETS immediately generated a wide range span of impacts on the industry:

> The emerging carbon market is causing significant changes in electricity companies' strategic decisions and daily operations. It has affected electricity companies in the following areas: climate strategy and risk management systems; financial and accounting arrangements; taxation requirements (e.g. corporate, capital gains, VAT); legal, permitting issues; investment planning; production planning; organization and administration (e.g. monitoring, reporting, verification, allowance recording, trading); IT systems; communication with shareholders and stakeholders. Furthermore, the marginal cost of generating electricity has now to take into account the price of CO_2. The valuation of assets in the long and short term, including dispatch decisions and investments in new capacity, is impacted by the value of CO_2 ... By making carbon-intensive generation more expensive, emissions-trading affects dispatch orders and thus encourages the gradual increase of low-carbon intensive generation. This is the intent of the emissions trading scheme. (Eurelectric September 2005: 6–7)

Company-level data similarly show that the EU ETS was increasingly regarded as the primary regulatory risk factor that could impact on the companies.[50] Most of the major electric power companies regulated by the system appeared short in allowances and evolved with higher and higher deficits, as shown by Table 3.5 for the period 2005–2009.[51] This applies also for our two case companies: Vattenfall went from a long position in 2005–2007 to a very short position thereafter; RWE went from a comfortable short position to a very short one.[52]

---

50   From 2003 onwards, the EU ETS gained leverage as the most important regulatory factor reported to affect RWE and Vattenfall (see RWE AG and Vattenfall annual reports, sustainability reports, reporting for the Carbon Disclosure project in the period 2003–2010).

51   Czech CEZ and Polish PKE had on average a surplus of allowances. British Drax Power saw on average the shortest position and CEZ the longest. Other companies with significantly short positions were Belgian Electrabel, German RWE, the two UK companies RWE npower and ScottishPower, the two Spanish companies Endesa and Iberdrola, UK SSE (Scottish and Southern Energy), Italian Edison, as well as French GDF Suez. In general, we see long or near break-even positions for companies from Poland, the Czech Republic and Greece.

52   Drax Power ended up with an extremely short position in 2008. Scottish and Southern Energy seems to have experienced the opposite development, entering a more comfortable short position in Phase II than in Phase I.

Table 3.5   **Development in deficit/surplus of allowances for the major 20 electric power companies under the EU ETS (in percentage of allowances allocated)[53]**

Company	2005	2006	2007	2008	2009	Average 2005–2009
RWE	6.6	7.0	7.7	82.2	64.1	33.5
E.ON	9.4	9.9	11.3	38.5	27.9	19.4
Vattenfall	-4.4	-3.1	-2.9	47.6	46.7	16.8
Enel	17.5	26.2	31.9	22.8	1.0	19.8
EDF	7.0	4.8	6.3	28.5	14.5	12.2
PGE	-2.7	-2.3	-5.7	14.0	8.3	2.3
PPC	0.9	-3.2	1.9	17.6	12.7	6.0
CEZ	-11.3	-7.4	2.2	-4.4	-9.0	-6.0
Endesa	21.2	28.9	41.3	36.7	11.1	27.8
GDF SUEZ	17.8	14.4	13.0	47.7	40.9	26.7
Iberdrola	10.0	30.4	26.9	38.9	34.2	28.1
EDP	15.3	3.5	3.1	29.1	32.3	16.7
Electrabel	34.1	30.0	30.2	68.8	63.5	45.3
Edison	4.9	23.2	32.7	41.6	13.3	23.1
PKE	-9.1	-4.6	2.0	3.9	-2.4	-2.0
Scottish and Southern Energy	32.4	41.7	26.3	2.4	13.7	23.3
RWE npower	10.0	50.3	26.8	65.3	10.9	32.7
Drax Power	40.1	56.4	52.3	134.7	108.9	78.5
EnBW	6.7	6.5	4.2	24.9	7.8	10.0
ScottishPower	11.4	32.4	15.9	50.7	48.5	31.8

However, the recorded deficit was not overly coercive, as long as total carbon costs under the EU ETS Phase I and Phase II (cost of covering the deficit of allowances) was generally offset by revenues generated from higher electricity tariffs seen when the carbon price was factored in for the marginal tariff-setting plant (Sijm et al. 2005, Hermann et al. 2010). Producers of zero- and low-carbon electric power became particular beneficiaries, enjoying the same higher market price for electricity but no additional carbon costs, like the producers of fossil fuels-based power. Hermann et al. (2010: 3) estimated total additional revenues specifically for major German power companies from this combined passing-on of costs to electricity tariffs/free allocation to be around €39 billion for the entire

---

53   The figures have been computed by subtracting the allocated allowances from actual emissions and making these a percentage share of allocated allowances for each year. A negative figure indicates surplus of allowances and a positive figure a deficit.

period 2005–2012. The shares estimated for RWE and Vattenfall were €11.25 billion and €7.23 billion (Hermann et al. 2010: 19).

Still, a tightening of the EU ETS from the first to the second trading period took place, and in response to this, companies started to reconsider their climate strategies, as illustrated by the case of RWE AG. Germany cut the number of allowances from 499 Mt in NAP I to 453 Mt in NAP II and required RWE to acquire around 40 per cent of its needed allowances through auctioning (Hermann et al. 2010). In response, the company announced it would suspend its clean-coal investment plans already adopted to cut short-term emissions. When the EU ETS revision was adopted in late 2008, regulation became substantially more stringent for the electricity industry from 2013 (no free allocation of allowances except for plants in Eastern Europe). In response, both RWE AG and Vattenfall AB adopted more far-reaching strategic shifts framed precisely as adaptation to the expected higher future carbon costs. RWE Power's CEO, Johannes Lambertz, explained the new decision to suspend further plans for coal-based capacity in Western Europe as the 'result of full auctioning of CO_2 rights from 2013' (Power Engineering International 21 January 2009). He added that investments would be resumed if power prices reached a level that would offset the costs of CO_2 auctioning, and continued in Eastern European countries because of free allocation of allowances there. When RWE in September 2010 suspended plans for a major new coal plant also in Poland, the global economic downturn in combination with concerns about the future price of carbon emissions in Europe were indicated as explanations (Industrial Info Resources 14 September 2010).

Another more indirect effect of the EU ETS was seen towards the end of the decade. Global credit rating institutions now started to include carbon costs as part of their evaluation of company risk profiles and credit worthiness. In 2011, Standard & Poor's retained Vattenfall's long-term credit rating with A minus (A-). Rising carbon dioxide costs was included as a company weakness. Standard & Poor's noted that Vattenfall strengths were balanced by the company's high carbon intensity: 'the latter which results in rising costs from full exposure to payments for carbon dioxide emission allowances under Phase III of the EU ETS (Emission Trading System) from 2013' (Standard & Poor's 24 May 2011).

That such evaluations made an impact is indicated in Vattenfall's justification of its new 2010 reduced-carbon strategy: 'to reduce CO_2 exposure and thus also the financial exposure to the cost of CO_2 emission allowances' (Vattenfall, Annual Report 2010: 5). Standard & Poor's remarked similarly for RWE in 2010:

> Given its high carbon intensiveness, RWE is exposed to the EU's tightening carbon constraints: Under Phase II of the ETS (European Trading System, 2008–2012) its free allocation of CO_2 certificates has declined to about 60 per cent of its carbon emissions from about 90 per cent in Phase I. From 2013 under Phase III of the ETS, all CO_2 certificates will be auctioned, which should weigh on RWE's generation operations which are its key earnings drive. (Standard & Poor's 5 February 2010)

For RWE, Standard & Poor's had factored in the recent climate strategy changes on the positive side of its evaluation, including the acquisition of Dutch Essent, whose 'generation operations are also significantly less carbon-intensive than RWE's' (Standard & Poor's 5 February 2010).

As such, an entirely new mechanism by which the EU ETS would affect company climate strategies had now been established, involving third-party mediation of incentives.

Our data also suggest that the EU ETS affected company R&D expenditure decisions, shown in Figure 3.1 as a positive correlation between company exposure to EU ETS regulation (shortage in allowances) and change in R&D intensity.

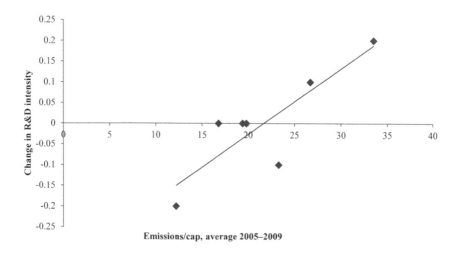

Figure 3.1   Shortage in allowances and change in R&D intensity:
Trendline for seven major allowance holders
in the electric power industry[54]

The trendline suggests that companies subject to the hardest regulations under the EU ETS were those with the greatest increase in R&D intensities. Companies facing less stringent regulation had reduced or kept their R&D intensity constant.[55] These findings should be interpreted with great caution because of the small set of

---

54   Only seven electric power companies had a level of R&D expenditure that made the EU Investment Scoreboard datasets for all years in the period 2004–2009. Figure 4.2 plots company scores on the average deficit of allowances for the years 2005–2009 (see Table 3.5) as a proxy of regulatory pressure and change in R&D intensity (change in the ratio R&D/net sales from 2004 to 2009).

55   Pearson's r = .85

companies for which panel data have been available. If we look more closely at two case companies RWE and Vattenfall, however, the data indicate clear changes in search behaviour, corresponding to the EU ETS gaining stringency on the companies – indicated by the shift in spending and composition of company R&D budgets in the second half of the past decade.

To sum up, data at various levels of aggregation indicate that the EU ETS was an important factor behind the climate strategy changes observed in the electric power industry during the first decade of the 2000s. Section 6 discusses in greater detail the mechanisms generating these effects.

4.2 The Broader EU Policy Explanation

As noted, the general trend observed in the industry, across companies and national contexts, needs explanation at the pan-European level. The EU ETS was not the only regulatory measure implemented at this level that provided signals relevant for company climate mitigation investments. To mention but a few, in 2001 the EU adopted the Renewable Electricity Directive (aimed at creating demand for renewable electricity); the 2003 Biofuels Directive (aimed at creating demand for renewable transport fuels); the 2004 CHP Directive (aimed at creating demand for more energy-efficient heat- and power-supply technology) and the 2006 Energy End-Use Efficiency and Energy Service Directive (aimed at strengthening demand for energy efficiency and promoting the supply side). Additionally, already in the 1990s the EU had started a process of gradual phase-out of state subsidies to coal as part of its internal energy market policy efforts. In 2002, efforts were geared up with Regulation 1407/2002, stating that the amount of aid to the coal industry 'shall follow a downward trend so as to result in a significant reduction' (EUR-Lex 2002).

However, none of these measures were legally binding on member states regarding the required volumes of renewables,[56] CHP,[57] energy efficiency[58] or reduced coal subsidies to be implemented within a specified time-frame. Their voluntary nature entailed strong asymmetries in implementation across the member states. As to the coal subsidy regulation, the European Commission in

---

56   The 2001 RES Directive set 22 per cent renewable electricity as target for 2010, up from 13.9 per cent in the base year 1997. It specified individual targets for each member state and required progress reports on these goals, but since the goals were only made indicative, results were highly asymmetrical among the member states (Eikeland and Sæverud 2007).

57   The 2004 CHP Directive set no quantitative targets but required member states to evaluate the potentials and to report on progress in implementing these.

58   The 2006 Energy Services Directive set the voluntary goal of 9 per cent energy savings for member states by 2016, and obliged them to develop national plans and ensure that national energy distributors, distribution system operators and/or retail energy sales companies provided proper information, started delivering energy services and refrained from activities that would impede the delivery of such services.

2010 evaluated it as exerting insufficient pressure on the member states.[59] Still, the total amount of subsidies was reportedly halved between 2003 and 2008, from €6.4 billion in 2003 to €3.2 billion in 2008 (European Commission 2010a).

The situation changed fundamentally from 2007 onwards, after the EU leaders adopted the '20–20–20' goals for the year 2020: reduction in CO_2 emissions, share of renewables, and improvement in energy efficiency. The 20 per cent goal for renewables was made *binding* for the member states as part of the 2009 Renewable Energy Directive. Also other relevant EU-level policies were adopted in the period; these included the 2007 Strategic Energy Technology Plan (SET plan) and the 2008 revision of state aid regulations, allowing more generous governmental support to the producers of renewable energy and CHP (Flåm 2009). The SET plan specified measures for extending and pooling resources to R&D in climate-related technology, spurring voluntary joint industry-governed R&D initiatives (European Industry Initiatives – EII) in various low-carbon technology fields. Adding supply-side support, the European Commission launched its 'NER 300' demonstration programme to fund low-carbon technologies through the sales of 300 million emission allowances saved for new entrants under the revised EU ETS.[60]

More recent additional EU Commission proposals reinforce the regulatory drive for renewable energy and energy efficiency. These include the 2011 proposals for a regulation on 'Guidelines for trans-European Energy Infrastructure', aimed at boosting the funding of and the speed-up of authorization for energy infrastructure (European Commission 2010b), and for a new Directive on energy efficiency (European Commission 2010c). The latter proposes requiring new electricity generation and district heating capacity to reflect the most efficient technology available, CHP wherever possible; and for requiring electricity distribution system operators to provide priority access for electricity from CHP. A third move by the European Commission that provided signals to the power industry that EU politicians talk business when demanding investments in climate-neutral solutions was the 2010 proposal to revise Regulation 1407/2002 to fully phase out coal subsidies by 2014 and to allow operating aid only in the context of a closure plan. The outcome of this process seems to have been postponement until 2018 for the full phase-out of coal subsidies in the EU.

---

59   €26 billion in state aid was approved for the hard coal sector between 2003 and 2008 (European Commission 2010a).

60   NER = New Entrants Reserve. The programme will provide seed money to encourage investment funding from the private sector and member state governments. By May 2011, 78 project proposals (13 CCS and 65 RES) had been submitted for the expected €4.5 billion fund, leveraging an equal amount of funding from the member states and industry.

5. Company External and Internal factors

Despite the general trend observed in climate strategies, individual company differences remained – in scale, scope and timing of response. This indicates that firm-specific factors have been at work. In this section we look into factors that could explain the differences observed between RWE AG and Vattenfall AB, significant at the start of the decade but becoming smaller as the decade progressed. Differences concerned attitudes towards the EU ETS regulation, but also the focus and timing of quantitative emissions reductions goals and investments in renewables and CCS. The greater convergence in strategies at the end of the decade indicates similar forces playing a strong role, whereas the considerable differences in the early decade would indicate that company-specific co-evolving factors were at work.

5.1 Company-internal Factors

As noted in the analytical framework of this book (Chapter 2), we would expect the strategic responses of companies to be conditioned by the inherent assets associated with specific carbon intensities and thus *climate-policy risks*. High-risk companies would be expected to take a more resistant stance on climate policy regulation than low-risk companies. On the other hand, once climate policies have been adopted, we would expect greater changes in climate strategies in high-risk companies.

Such patterns fit well with our observations. At the start of the decade RWE was more exposed to carbon policy risks with its carbon intensity double that of Vattenfall. Vattenfall actually had a surplus of allowances during the first EU ETS trading period 2005–2007, unlike RWE (see Table 3.5 above). RWE had thus more to lose than Vattenfall from climate policy regulations at the start of the decade, and took an early strategy of resistance to the EU ETS. Once the EU ETS had been adopted in 2003, RWE reacted more forcefully than Vattenfall with low-carbon investments to bring down its carbon intensity. Vattenfall appeared more complacent with its investment behaviour boosting carbon intensity. This led the climate policy risks of the two companies to converge over time, as shown in Table 3.5 by a tighter emissions/cap ratio for both companies in the second trading period. When the EU adopted its revised system in 2009 with a tighter cap and full auctioning of allowances for electric power companies from 2013, both companies saw their carbon risks rising to new levels, now also addressed by international credit rating bureaus in their evaluations of the companies' long-term credit worthiness. The converging climate-policy risk profiles of the two companies fits well with the decisions made by both companies at the end of the decade to pursue far more significant low-carbon climate strategies.

Yet, as noted in Chapter 2, companies are not entirely slaves of their inherited business strategies and assets. Managers may possess the capability to change strategies and re-configure assets and have leverage for making climate change a key premise for such change activities. Chapter 2 focused specifically on company dynamic capabilities: a firm's ability to integrate, build, and re-configure internal and external resources to address the new social demands for environmentally adapted business development (Teece et al. 1997), as a precondition for presenting innovative climate strategies. The data collected for this study have been insufficient for us to detect any strong variation at the start of the decade between the two case companies in terms of decisive dynamic capabilities. The fact that both companies had inherited major internal R&D resources and presented relatively high R&D-intensity figures in the 1990s indicates rather substantial similarities. This is further strengthened by the similar dynamics during the 2000s, where both companies first scaled down and later scaled up the amount of resources they reported using for search activities.

That said, our study has documented interesting cross-company variation and dynamics over time in another management factor that appears to co-vary with observed climate strategies: implementation of management routines for dealing with external social demands for environmental responsibility: environmental management systems (EMS).

At the start of the decade, Vattenfall AB had such routines far more firmly implemented than RWE AG. It had appointed a corporate environmental manager in the early 1990s, conducted life-cycle assessments of activities for environmental impacts, and reported externally on these from 1995. By the end of the 1990s, all Vattenfall plants were certified with EMS, including plans for further improvements (Eikeland 2013, forthcoming). By comparison, it was not until 1998 that RWE decided to implement a group-wide environmental management system, appointing an environmental manager and publishing its first separate environmental report (RWE AG, Environmental Report 2000).

Still, at the start of the decade these pre-existing environmental management routines appeared to have little impact on company investment behaviour, which focused on acquisitions of coal-based capacity as part of a larger internationalization strategy (Högselius 2009: 262). In fact, the Swedish General Accounting Office (*Riksrevisionen*) concluded in 2004 that Vattenfall had neglected its government-given environmental tasks early in the decade, resulting in failed investments in renewable energy (Riksrevisionen 2007: 12). Three years later, this body concluded that lack of environmental management was even more conspicuous for Vattenfall's subsidiaries abroad (Riksrevisionen 2007: 42).

Nevertheless, both companies established stronger environmental management routines later in the decade, corresponding in time with the convergences observed in climate strategies. This means that we cannot exclude a connection between internal environmental management routines and climate strategies. RWE AG undertook the gradual implementation of environmental management routines from early in the decade and onwards.[61] Vattenfall AB reinforced its

---

61   By the end of the decade, RWE reported 98 per cent of operations covered by the environmental management system, with third-party independent auditing and certification

environmental management routines, starting in 2007 when it took greater control over its German Vattenfall Europe AG subsidiary, replacing the CEO and subordinating the company to a new group-wide management system, with environmental governance routines integrated. As regards climate governance specifically, Vattenfall reported in 2008 to the Carbon Disclosure Project that the company had established one executive body with overall responsibility for climate change, including target-setting and follow-up. RWE AG had now caught up with Vattenfall: in fact, it joined the Carbon Disclosure Project in 2006, one year ahead of Vattenfall.

As further discussed below, however, our data suggest national government regulation as an important driver of company attention to environmental problems and solutions, preceding the implementation of environmental management routines. As such, environmental management routines should be seen as one possible intermediate variable between observed climate strategies and governmental regulation. The EU ETS here appears to have been less relevant, as both companies made decisions on implementing environmental management systems well before this system was adopted. However, we cannot exclude the possibility that the EU ETS at a later stage might have come to strengthen implementation efforts, given the concurrency in time of the two factors.

5.2 Company-external Factors

Despite the internationalization observed for our case companies over the past decade, they have continued to maintain headquarters in their traditional national home markets and kept up their resource interdependency relationships with national governments. This study finds support for national governmental influence on the company climate strategies directly through public ownership, and indirectly through national sector regulations more generally. The strongest influence is found towards the end of the decade for both companies, coinciding with the convergence seen in their climate strategies.

5.2.1 Vattenfall AB
Being fully state-owned, Vattenfall has since being established served as an important instrument for successive governments in the realization of industrial and environmental goals set by political Sweden, including the task of demonstrating Swedish-made technologies (Eikeland 2013, forthcoming). A good illustration was the 1997 national energy-political agreement in the Swedish Parliament that gave to Vattenfall an extensive strategic role in the transition of the Swedish energy system towards a lower-carbon future (Sveriges Regering 1997). The failure of Vattenfall to live up to these instructions was clearly documented by the Swedish General Accounting Office in 2004, which blamed the failure on blurred signals

---

made optional for the various units. Indeed, 38 per cent of the RWE Group had been certified to the ISO 14001 standard by the end of the decade.

given by the government, which had also instructed the company to operate strictly commercially and present high annual dividends.

In 2005, the Swedish Government therefore clarified its instructions to Vattenfall, specifying that 'the company shall within the restrictions set to operate on commercial principles, be the leading company in the transition to an ecologically and economically sustainable Swedish energy supply' (Riksrevisionen 2007). It added a short-term instruction for Vattenfall to add 5 TWh new renewable energy-generation capacity by 2005 as against the 2002 baseline year.

Another round of strong direct governmental influence targeted Vattenfall's operations abroad. As Högselius (2009) notes, the company's acquisitions of nuclear and coal-based power abroad met strong political scepticism in Sweden, which accelerated when successive reports came in on security breaches at nuclear plants and exploitation of wetlands for the mining of brown coal. In 2007, Swedish dailies were reporting that 'Vattenfall's ruthless exploitation of the environment' was incompatible with the governmental instructions given to the company (Aftonbladet 4 April 2007). Dissatisfaction grew stronger when the commercial results failed to live up to political expectations, as the company's return on equity experienced a declining trend. Added to this came media attention to the high leadership remuneration in Germany, despite unsatisfactory commercial results.[62] As noted above, Vattenfall AB responded by taking greater control over the German part of the company, but this proved insufficient to convince Swedish pressure groups and politicians of all parties that the top CEO would be the right man for the future. In 2010, the CEO of Vattenfall was replaced, and the new man at the top was to become instrumental in re-directing company climate strategy.

To a considerable extent, Swedish governmental instructions for Vattenfall reflected more overarching national energy and climate policies. Successive Swedish governments since the 1970s had adopted gradually more stringent environmental demands on the electricity industry, combining this with generous support for development and demonstration of new environmentally adapted solutions (Eikeland forthcoming 2013). In 1991, Sweden adopted comprehensive CO_2 taxes that made the country's major bioenergy resources competitive for national energy supply, followed in 2001 by a new green tax reform that switched taxation from labour to harmful emissions (Ghalwash 2007). Additional specific incentives for renewable energy investments came in 2003, when Sweden adopted a green certificate system aimed at adding 17 TWh of renewable electricity by 2016. Included in this energy and climate policy deal was cancellation of the adopted nuclear power phase-out programme. At the end of the decade (2009), Sweden responded to new EU policy signals by adopting the goal of 40 per cent cuts in GHG (compared to 1990) and the higher goal of 25 TWh for the renewable electricity certificate system by 2020.

---

62   When the CEO of Vattenfall Europe was fired in 2007, Swedish national media reported that the agreement included a SEK 32 million 'golden parachute', at that point equivalent to €3.5–4 million.

We can see a clear correlation between Vattenfall's climate strategy and Swedish policies. During the 1990s, Vattenfall had followed its government instructions and responded to the new carbon tax system with massive fuel conversion from fossil fuels to biomass. The company's acceptance of the EU ETS is hardly surprising, in view of the strong support to the system evinced by its owners, the Swedish state (Skjærseth and Wettestad 2008). In the early 2000s, the lower pace in renewable energy investments reflected mixed and blurred instructions from that same owner. Clarification of these instructions in 2005 was followed by major Vattenfall investments in wind power (Vattenfall, Corporate Social Responsibility Report 2005, published 2006). And governmental dissatisfaction with Vattenfall's environmental performance abroad was followed by stricter control over foreign subsidiaries and implementation of new group-wide environmental governance instructions. Despite this, towards the end of the decade the government still wanted a new CEO to head the re-direction of the company towards a lower-carbon business strategy, with the dual aim of improving the company's profitability/credit rating and its climate footprint.

5.2.2 RWE AG

RWE AG evolved as a semi-private company, with 30 per cent ownership held by municipalities in the state of North Rhine-Westphalia. These municipalities played a central direct role when RWE AG reshuffled its leadership and embarked on a new climate strategy in 2008. Prior to that, the municipal owners put forward a series of complaints against the CEO, Harry Roels, blaming him for failure to understand German political structures and a lack of networking with German central and state governments. The municipalities were particularly dissatisfied with RWE's poor internationalization strategy,[63] but also its environmental performance. At odds with the popular and political sentiments evolving in North Rhine-Westphalia and Germany at large, Roels had taken a clear pro-nuclear standpoint and showed unwillingness to transform the company from its high reliance on carbon-intensive lignite.[64] In 2006, the German Nature Conservation Society awarded Roels the 'Dinosaur of the Year' award for these reasons (SpiegelOnline 28 December 2006).

In 2005, the state government of North Rhine-Westphalia adopted a new energy and climate protection plan with ambitious quantitative targets set for

---

63   Consolidation and sell-out had brought increasing returns on capital employed, but the municipalities voiced fears that the good results would make the company vulnerable to a hostile takeover, after rumours that Russian Gazprom had shown an interest in the company.

64   The goals set were energy consumption 20 per cent below 2006 levels by 2020 and the share of renewables in electricity supply up from 2 per cent in 2006 to 12.5 per cent by 2010 and 20 per cent by 2020 (100TWh). This stance prompted the 'Dinosaur of the Year' award in 2006.

energy savings and increases in renewable electricity supply.[65] The main justification for the plan was compliance with the EU ETS: reportedly almost a quarter of German facilities covered by the system were located within the borders of North Rhine-Westphalia, accounting all in all for 57 per cent of the state's CO_2 emissions (Climate Group December 2007). The new plan included support for doubling the share of biomass as fuel in heat and electricity generation/ transportation, and an Energy Research Programme aimed at making North Rhine-Westphalia home to leading research in renewable energy and CCS for power plants (Climate Group December 2007). In 2010, North Rhine-Westphalia consolidated its position as regional frontrunner on climate policy, being the first German state to adopt a climate protection law that set *binding* goals for CO_2 reduction. The targets set were 25 per cent reduction by 2020 and 80–95 per cent reduction by 2050 compared to 1990.[66] The climate strategy shifts observed in RWE correspond in time with changes in local-level policies, including its stronger focus on renewable energy and CCS after 2005.

Direct influence from the municipal owners may also have been involved. After the intervention in 2007 to replace the CEO of RWE, his successor Jürgen Grossmann implemented a far more proactive and innovative climate strategy, reinforcing this greater internationalization through the acquisition of the less carbon-intensive Dutch company Essent. Still, even the new CEO strongly defended the company's nuclear power assets and contested the legality of the German government's shutdown decision. Out of touch with the political anti-nuclear movement gaining pace in Germany, not least in North Rhine-Westphalia, in August 2011 the RWE AG Board confirmed replacement also of Grossmann.

Yet, the RWE strategic changes are clearly linked also to the massive changes in German federal-level policies. In the 1990s, the German government stated clear and ambitious commitments to GHG reductions, but the demands on the industry were basically weak, based on voluntary action.[67] Moreover, various policies were sending out inconsistent signals to the electricity industry. The long-existing subsidization policy of domestic coal production continued, despite reductions in subsidies implemented after pressure from the environmental movement and

---

65　Expected results were a reduction of GHG emissions by about 31 $MtCO_2$ annually and an estimated 23,500 new jobs created by 2020.

66　This law was adopted by a new minority coalition government of the Green Party, die Linke and the Social Democratic Party. Local election winners in 2010 were the Greens and die Linke, with the worst ever results recorded for the ruling CDU party, which had pledged to extend the lifetime of existing nuclear power stations beyond 2022.

67　In the 1995 voluntary agreement with German industry associations, including the electricity supply associations, these industries declared their commitment to reducing specific energy consumption of their member companies by up to 20 per cent by 2005 compared to 1987, amended in 1996 with 1990 as base year and a target sharpened to 20 per cent (Jochem and Eichhammer 1996). In 1997, Germany committed to 21 per cent cuts in GHG emissions by 2008–2012 under the EU Effort-Sharing Decision to achieve its Kyoto Protocol goals.

EU internal energy market policies (Eikeland 2004). In one relevant policy field, however – support to renewable energy – the measures implemented sent out more forceful signals. The 1990 German Renewable Energy Law required electricity supply companies to connect independent producers of renewable electricity to the grid and to pay a fixed annual rate based on the utilities' average revenue per kWh sold (feed-in tariffs).

In the early 2000s, a series of policies strengthened the commitment of the German federal government to reducing GHG emissions.[68] A new Climate Change Policy Action Programme introduced financial support for CHP, and alteration of the feed-in tariff law for renewable electricity offered new opportunities to generators, with guaranteed tariffs based on different technologies, and extended the periods of eligibility for these tariffs. Yet another strong signal sent to the industry was the 2000 Government Agreement to phase out nuclear power by around 2022, altering the relative opportunities for renewables. In 2004, the federal government again reinforced these signals by setting the quantitative goal for renewables to fuel 20 per cent of total electricity generation by 2010, upscaling the 12.5 per cent commitment taken under the EU 2001 RES Directive (Lauber and Mez 2004). Still, continued subsidization of coal worked against these climate policy signals. Under the German first national allocation plan for the EU ETS, permits were allocated according to demand for allowances from existing technologies: this meant that coal-fired generators got the lion's share (Pahle, Fan and Schill 2011).

In the second part of the decade, federal climate policy signals to the industry grew far stronger, and eventually more consistent. Non-binding GHG emissions reductions targets were increased.[69] In 2007, a political agreement between the federal government and the state of North Rhine-Westphalia, home state of RWE AG and the bulk of Germany's remaining coal mines,[70] declared the phase-out of all coal subsidies by 2018. One year later, Germany accepted new and *binding* national goals for emissions reductions and renewable energy for 2020 under EU legislation.[71] However, Germany followed up with far more ambitious targets at the national level, aiming at 40 per cent reduction in CO_2/year and 35 per cent of total electricity consumption from renewables by 2020. Acting on the new drive in EU policy to set long-term goals also for 2050, the German federal government

---

68   Adopted by the SPD/Greens coalition government in power from 1998. The Climate Change Policy Action Programme set new non-binding emissions reduction targets at 45 million tonnes of CO_2/year.

69   In 2005, the target was set at 40 per cent reductions by 2020, adopted by the grand coalition government of the Christian Democrats (CDU) and Social Democrats (SPD).

70   In 2010, North Rhine-Westphalian coal mines produced 94 per cent of total German output of hard coal and 54 per cent of total lignite.

71   National targets under the Effort-Sharing Decision were set at 14 per cent (compared to 2005) and under the RES Directive, 18 per cent of total energy consumption by 2020.

in 2010 presented a plan for an 80 per cent reduction of CO_2 emissions, boosting the phase-in of renewable energy[72] and energy savings. Part of this plan involved reversing the nuclear power phase-out plan, granting plant owners and extension of 14 years beyond the previous 2022 deadline. However, the Fukushima nuclear power plant catastrophe in Japan in March 2011 was to make this decision short-lived. In June 2011, Germany adopted a new energy and climate package that upheld the ambitious transition plan to a basically renewable energy system without nuclear power.

Alongside political signals that increasingly pointed strongly towards renewable energy, another climate change solution for the electric power industry became hotly debated and contested in Germany at the end of the decade. This was carbon capture and storage (CCS), under development by RWE AG, Vattenfall and other German companies. Massive local opposition to onshore CO_2 storage facilities had led several German states to decide not to permit any exploration for storage, with subsequent postponement by the German federal government in getting implemented the Carbon Storage Law demanded by the EU under the 2009 CCS Directive. A draft law passed by the German Parliament in July 2011 would allow the federal states the right to designate/exclude areas for CCS pilot projects but was then rejected by the Bundesrat, the legal body that represents the German federal states, in September 2011. The draft law postponed the final decision on whether the CSS technology should be used as a full-scale climate mitigation option in Germany to 2017.

We observe strong matches between signals provided in German federal-level policies and RWE AG climate strategy responses. First of all, the company's initial resistance to the EU ETS was firmly rooted in the alternative German political tradition of voluntary agreements, making also the German government initially sceptical towards the EU ETS. Later acceptance by the German government corresponded with greater acceptance also by RWE. Next, the company's gradual strengthening of its engagement in renewable energy and its later suspension of coal power expansion plans matches well the increasingly stronger and more consistent climate policy signals from German federal climate and energy policies – including the decision to phase out coal subsidies in the medium term, and the generous support regime evolving for market introduction of renewable energy. Yet, we can also see how specific elements of RWE's climate strategy interacted with national policies such as its emerging CCS investments. The CEO of RWE responded promptly to the German draft CCS Law, giving veto right to state governments concerning storage sites, by declaring: 'CCS won't be able to help here in Germany – There's no more effective way to stop CCS in Germany'

---

72 The phase-in plan for renewable energy set the following targets and deadlines: 30 per cent of total energy consumption by 2030, 45 per cent by 2040 and 60 per cent by 2050. For total gross electricity consumption, the following targets were set: 50 per cent by 2030, 65 per cent by 2040 and 80 per cent by 2050 (Federal Ministry for the Environment, Nature Conservation and Nuclear Safety 2010).

(Bloomberg 20 April 2011). With Vattenfall's internationalization strategy also bringing this company heavily into coal- and nuclear-based power production in Germany, German politics surely affected the similar climate strategy changes observed for this company towards the end of the decade. For example, in December 2011 Vattenfall announced its decision to cancel its CCS demonstration project in Jänschwalde and its plans for exploring storage facilities in eastern Germany, because of what it claimed was 'lack of political will' to provide the legislation needed for CCS in Germany.[73] The project had been awarded funding under the European Energy Programme for Recovery, and had applied for funding from the EU NER300 mechanism.

5.3 Summary of Company-external Drivers

Summing up, direct political steering from the Swedish state and German local governments and regulatory signals provided by national policies co-evolved with the EU ETS to affect the changes observed in company climate strategies. The convergence in Vattenfall's and RWE's climate strategies towards the end of the decade matches the fact that both Swedish and German energy and climate policies had evolved with increasingly ambitious targets for emissions reductions – but also the fact that Germany was emerging as the most important market for both companies, with German energy and climate policies particularly important premises for company climate strategy-making. We can note interesting connections between the EU ETS and national climate policy development, as when North Rhine-Westphalia, home to municipalities that held ownership shares in RWE and most of the remaining German coal industry, formulated local climate policy goals in 2005, citing the EU ETS as justification since the state hosted more than half of the German installations covered by the system. When North Rhine-Westphalia in 2010 formulated ambitious binding short- and long-term GHG reduction goals, climate strategy changes in RWE became an important part of the solution.

Thus, towards the end of the decade, local, federal and EU policies combined in putting pressure on the companies to adopt new proactive and innovative low-carbon strategies for the future.

6. Discussion

The similar direction in strategy shifts observed for many electric power companies across national contexts strengthens our conclusion that EU-level factors, notably the EU ETS, played a role. Chapter 2 outlined several models and expectations

---

73   Vattenfall press release 5 December 2011 'Uncertainties with CCS Law stop Vattenfall investment in demo plant', http://www.vattenfall.com/en/ccs/news-archive.htm ?newsid=7217C9BD4BC04D8CBC7F55B5DE07505E

concerning how a regulatory instrument like the EU ETS might affect company climate strategies. Model I involved expectations of reluctant adaptation in the form of resistance to the EU ETS, and the implementation of only short-term low-cost measures once the system was adopted. The underlying assumption was cost minimization as the dominant behavioural mechanism. This model has found support in our company data. Initial resistance to the EU ETS on the part of RWE, but not Vattenfall, matches far higher initial risk exposure for the former company, in turn reflecting differences in carbon intensities. Once the ETS was adopted, the cap was not really restrictive for either company (costs of allowances more than offset by revenues from higher electricity prices), and adaptation could well be categorized as not too proactive in the first trading period, again supporting Model I. A clear example is seen in the continued investments in coal-based power capacity in the period.

There were new rounds of open opposition from RWE when the EU ETS gained stringency for the power industry in the second and third trading periods, again supporting the model. However, particularly after the EU revision in 2009, we note far-reaching shifts in climate strategies. These shifts appear very similar in both case companies, as they basically start to announce the end of investments in coal-based power and a re-focus on renewable energy. Also this convergence could well be explained by Model I. Earlier lack of cost pressure during Phase I and Phase II of the EU ETS had led Vattenfall to invest heavily in coal-based capacity. Thus, its carbon intensity and thereby carbon risk for Phase III saw a converging trend towards that of RWE, which made the two companies experience more similar cost pressure from the system due to start in 2013. This pressure was now tied not only to the cost of production but also capital costs after credit-rating companies started to include carbon costs in their evaluations of company long-term credit worthiness. With this, an entirely new mechanism by which the EU ETS would affect company climate strategies became established through third-party mediation.

The second model outlined in Chapter 2 involved the expectation that the EU ETS would trigger attention to and stimulate innovation and pursuit of profit opportunities formerly not apprehended by the companies. It followed from the Porter Hypothesis, based on the assumption of company behaviour as only boundedly rational, in the sense that companies are seen as myopically pursuing only well-known profit opportunities. Yet a qualifying assumption made by Porter was that regulation must be stringent and appropriate, if it were to stimulate such kind of behaviour. Also this model finds support in our observations, notably the upscaling of low-carbon technology R&D in the electric power industry and stabilization (for some companies even an increase) in a long-term trend of falling R&D intensity. In line with the Porter Hypothesis, data suggest the strongest increase in R&D intensity for companies with the highest deficit of allowances (regulatory pressure) and most significant changes towards the end of the decade, after the EU ETS revisions made the power industry face stronger regulatory pressure from 2013 onwards. Most significantly, we see that this model is supported by the substantial number of CCS projects reported by power companies, a technology field entirely new to this industry.

Model III generated the proposition that regulation in the form of cap-and-trade is likely to 'crowd out' social norms of responsibility. This proposition sees companies as having an even broader repertoire of behavioural motivations, including acting rightly according to internalized social norms. Our study has not found any convincing support for this proposition. To be sure, the data suggest that Vattenfall appeared to pay less attention to environmental performance in its operations and investments abroad in the early 2000s than what had been the case in its home market Sweden in the 1990s. However, we find no indication that these changes were connected to the EU ETS.

Yet, our study observes another shift in industry behaviour that points towards rejection of the crowding-out proposition – stronger engagement in long-term *collective action* for promoting and developing climate solutions (common vision set for carbon neutrality in 2050 and new joint long-term R&D initiatives). This indicates a break with the earlier trend of unwillingness to share business information and collaborate with competitors towards a future common good. Here, we can note clear indications of the EU ETS as a trigger of these new patterns of collective behaviour – as when Eurelectric headed allowance trade simulations early in the decade and started joint scenario projects that showed new opportunities for industrial expansion and profits from committing to major emissions reductions. Once the joint industrial long-term goal of carbon-neutral electricity supply had been formulated, pooling of capital and human resources represented a new rational solution for the companies to share the enormous capital costs. If stabilized, such new behavioural patterns could also develop as new internalized social norms of information-sharing and collaboration: a crowding-in of new norms. Companies would still be competitors in the electricity market, but their shared long-term vision of solving the climate problem *through market extension* could, if implemented, reduce the competitive pressure ('a bigger pie for all') and make continued collaboration a viable strategy.

Despite the evidence of the EU ETS as a driver of company climate strategies, its causal power emerges as less clear, relative to that of other EU regulations, when we look at specific strategy elements, such as investments in renewable energy generation capacity or CCS. The recent stronger focus on deployment of renewables in company strategies could also be attributed to the more ambitious and *binding* EU 2020 targets on renewables, adopted in 2009 and replacing the less ambitious and non-binding targets in place since 2001. It could even be argued that, in the medium term, the Renewable Energy Directive will take over the role of the EU ETS as main driver of company climate strategies. Higher shares of renewables in the market will reduce the demand for allowances and press allowance prices down, dampening the incentive effect of the EU ETS for investments in low-carbon solutions more generally.

Concerning the role of CCS in company strategies, the EU ETS regulation seems to have played a dominant role in providing market demand signals (still insufficient to cover the costs of CCS demonstration) and after the EU ETS revision, also support for deployment of CCS technology through the NER300 mechanism. However, the stronger involvement in joint industrial CCS initiatives should also be seen as triggered by the more general developments in EU technology supply policy, through the Zero Emissions Platform established under the EU Strategic Energy Technology Plan. This plan aims at pooling national and industrial resources for the development and commercialization of various zero-carbon solutions – an institutional framework directly supporting the fostering of new norms of collaboration seen in the electric power industry.

Identifying clear causal links between the EU ETS and climate strategies is difficult because the former did not develop in a vacuum. Some member states preceded the EU in developing climate policy regulations, while a central role of the EU has been to induce laggard states to catch up. We have seen that energy companies adapt to different national regulatory traditions, providing variation in steering signals and room for manoeuvre. Both our case companies were under public ownership influence over the past decade, whether through direct instructions or through board decisions to employ new leadership expected to pay more attention to the climate change problem. However, public ownership might work in several directions, depending on the strength of national commitments to solving the problem of climate change. Both the Swedish national owners (with Vattenfall) and local German owners (with RWE) held strong commitments. The Swedish government had made ambitious commitments even before the EU ETS was adopted, making it a clear independent force for Vattenfall's strategic work. The government of North Rhine-Westphalia justified its stronger climate policy by adaptation pressure from the EU ETS, which would suggest public ownership as an intermediate factor between the EU ETS and the strategic changes observed in RWE.

7. Concluding Reflections

We have noted a clear shift in studied power companies towards more proactive and innovative climate strategies since the turn of the millennium. We find empirical support for the proposition that EU ETS has affected company strategies through various mechanisms. Still, it has been impossible to establish the explanatory power of the EU ETS relative to that of other EU- and national-level policy drivers. At the national level, state/municipal ownership played a clear role, as did national climate and energy policy regulations more indirectly. Our study has focused on two companies where Germany and Sweden are main national contexts for their operations, countries keen to take on frontrunner roles in the development of climate policies. Therefore, the synergies between national and EU-level policies were probably stronger than with other companies and other national contexts. If the focus is shifted to specific parts of company climate strategies, like engagement in renewable energy and CCS, synergies with other EU regulations will have to be brought in on the explanatory side.

Lack of policy synergies – the incoherence of European climate and energy policies – has been placed firmly on the agenda by the electric power industry in Europe in its current strategic thinking on how to achieve the EU low-carbon goals set for 2050. The main message sent out by Eurelectric and many other analysts and lobbyists in Brussels is that the adoption in 2009 of the Renewable Energy Directive and energy-efficiency policies to be implemented will make the EU ETS less effective as least-cost guidance for company climate strategies. These other policies instruct specific technological solutions that, once implemented, will reduce the demand for EU ETS allowances and thereby the price signals sent through the system for continued development of other long-term solutions, such as CCS technologies applied on fossil fuel-based electric power generation. The industry has argued strongly for the EU ETS as the technology-neutral driver of long-term development and for politicians to keep their hands off, spreading the message that such political interference entails the risk of short-term solutions being implemented that could lock in unabated high carbon or uneconomic technologies for a long time. Instead, the industry seeks to buy time for developing solutions to be more extensively implemented only from 2025 onwards.

The industry may have good reasons for fearing the over-hasty development of intermittent forms of renewable energy in the sense that supply stability becomes less predictable, but also because of commercial threats to their already operating base-load plants. Renewable energy plants have high capital costs but lower operating costs than plants where the fuel is not cost-free. With a growing share of renewable energy in the system, the industry may well fear poorer commercial conditions for its dominant base-load capacity.

Still, investment in renewable energy remained high in Europe in 2009 and 2010, apparently one sector less affected than others by the ongoing economic and financial crisis. In fact, this may indicate that renewable energy is increasingly viewed by European politicians as part of the solution to the economic crisis, as one sector that should be politically stimulated to grow and create new employment. In that case, the established electric power industry is unlikely to sit still and let others do the job: instead, we may expect further consolidation of climate strategies, with a growing role for investments in renewable energy. This will exert further downward pressure on EU ETS allowance prices in the short term unless the EU decides to cut the supply of allowances through set-aside, proving the industry right in its fears that the system will reduce its role as driver of climate emissions reductions in Europe in the near future.

References

Aftonbladet. 4 April 2007. *Vattenfall bedriver rovdrift på miljön.* [Online]. Available at: http://mobil.aftonbladet.se/debatt/article10918903.ab?partner=www [Accessed: 19 June 2012].

Bloomberg. 20 April 2011. *RWE Says German Federal States' CCS Veto Rights to Hinder Technology Use.* [Online]. Available at: http://www.bloomberg.com/news/2011-04-20/rwe-says-german-federal-states-ccs-veto-rights-to-hinder-technology-use.html [Accessed: 19 June 2012].

Cames, M. 2010. *Emissions Trading and Innovation in the German Electricity Industry.* Dissertation, Berlin: Technische Universität Berlin.

Carbon Disclosure Project, Vattenfall Report 2008.

Climate Group, The. December 2007. *Low Carbon Leader: States and Regions.* [Online]. Available at: http://www.theclimategroup.org/_assets/files/Low_Carbon_Leader.States_and_Regions.pdf [Accessed: 19 June 2012].

Deutsche Welle. 14 July 2011. Energy giant RWE puts brakes on new coal and gas plants. [Online]. Available at: http://www.dw-world.de/dw/article/0,15231271,00.html [Accessed: 19 June 2012].

Eikeland, P.O. 2004. *The Long and Winding Road to the Internal Energy Market – Consistencies and Inconsistencies in EU Policy.* FNI report 8/2004. Lysaker, Norway: FNI.

Eikeland, P.O. Forthcoming 2013. *Environmental Innovation in Vattenfall and Statkraft – A Multilevel and Longitudinal Approach to Explaining Company Differences.* PhD dissertation to be submitted. Oslo: The Norwegian School of Management.

Eikeland, P.O. and Sæverud, I.A. 2007. Market diffusion of new renewable energy in Europe: Explaining front-runner and laggard positions. *Energy & Environment*, 18(1), 13–37.

EEI (Edison Electric Institute), FEPC (the Federation of Electric Power Companies of Japan), and UNIPEDE. 1998. *Common EEI/FEPC/UNIPEDE Position Paper for COP-4 (Buenos Aires). Flexibility Mechanisms under the Kyoto Protocol*, Brussels: Eurelectric.

Eurelectric. December 1994. *Joint Eurelectric and UNIPEDE Position Paper on Joint Implementation Within the Framework of the Convention on Climate Change.* Brussels: Eurelectric.

Eurelectric. March 1996. *Eurelectric Statement on the EC's Amended Proposal for a Council Directive Introducing a Tax on CO_2 Emissions and Energy.* Brussels: Eurelectric.

Eurelectric. 15 May 2002. *Eurelectric Comments to Parliament on the Commission's proposal for a Community greenhouse gas emissions trading scheme.* Brussels: Eurelectric.

Eurelectric. October 2002. *Energy Wisdom Programme (EWP), Global Report 1990–2000.* Brussels: Eurelectric.

Eurelectric. January 2003. *R&D in EURELECTRIC Countries Overview of R&D organization, 2002 Update*. Ref: 2003-030-0214, Brussels: Eurelectric.

Eurelectric. June 2004. *The Impact of Emissions Trading on Electricity Prices, Task Force Emissions Trading & Electricity Pricing*. Brussels: Eurelectric.

Eurelectric. September 2005. *Proposals to Amend the Emissions Trading Directive (2003/87/EC) and the Linking Directive (2004/1010/EC)*. Brussels: Eurelectric.

Eurelectric. March 2007. *The Role of Electricity – A New Path to Secure, Competitive Energy in a Carbon-Constrained World*. Brussels: Eurelectric.

Eurelectric. July 2007. *Position Paper on the Review of the EU Emissions Trading directive (2003/87/EC) and the Linking Directive (2004/10/EC)*. Brussels: Eurelectric.

Eurelectric. April 2008. *Position Paper, Review of the EU Emissions Trading Directive (2003787/EC)*. Brussels: Eurelectric.

Eurelectric. 2009. *Power Choices – Pathways to Carbon-neutral Electricity in Europe by 2050, Full Report*. Brussels: Eurelectric.

Eurelectric. 2010a. *Energy Wisdom Programme 2010–2011 Edition – Improving Energy Efficiency and Reducing Carbon Emissions*. Brussels: Eurelectric.

Eurelectric. 2010b. *Analysis of the EURELECTRIC Survey Questionnaire on the Role of R&D in Power Companies*. Brussels: Eurelectric.

European Commission Directorate for Environment, McKinsey & Company and Ecofys. 2006. *EU ETS Review, Report on International Competitiveness*. Brussels: European Commission.

European Commission. 2010a. *Frequently Asked Questions – Coal Regulation*. MEMO/10/348, 20 July 2010. Brussels: European Commission.

European Commission. 2010b. *Energy Infrastructure Priorities for 2020 and Beyond – A Blueprint for an Integrated European Energy Network*. COM(2010) 677 final, Brussels, 17 November 2010.

European Commission. 2010c. *Energy Efficiency Plan 2011*. COM(2011) 109 final. Brussels, 8 March 2011.

European Commission Joint Research Centre – Institute for Prospective Technological Studies. *EU 2004–2010 EU Industrial R&D Investment Scoreboards*. Brussels: European Commission.

European Environment Agency. 2012. Why did greenhouse gas emissions increase in the EU in 2010? In *Annual European Union Greenhouse Gas Inventory 1990–2010 and Inventory Report 2012*. Technical Report 3/2012. Brussels: EEA.

EUR-Lex. 2002. Council Regulation (EC) No 1407/2002 of 23 July 2002 on State aid to the coal industry. Brussels: European Union Publications Office.

Flåm, K. 2009. EU environmental state aid policy: Wide implications, narrow participation? *Environmental Policy and Governance*, 19(5), 336–49.

Ghalwash, T. 2007. Energy taxes as a signaling device: An empirical analysis of consumer preferences. *Energy Policy*, 35(1), 29–38.

Hermann, H., Graichen, V., Gammelin, C., et al. 2010. *Free Allocation of Emission Allowances and CDM/JI Credits Within the EU ETS – Analysis of Selected Industries and Companies in Germany*. Berlin: Öko-Institut.

Högselius, P. 2009. The internationalization of the European electricity industry: The case of Vattenfall. *Utilities Policy*, 17(3–4), 258–66.

ICIS. 15 February 2011. Dutch CCS in disarray as 'on land' storage ruled out. [Online]. Available at: http://www.icis.com/heren/articles/2011/02/15/9435644/dutch-ccs-in-disarray-as-on-land-storage-ruled-out.html [Accessed: 19 June 2012].

Industrial Info Resources. 14 September 2010. RWE Cancels Polish Coal-fired Power Plant. [Online]. Available at: http://www.advfn.com/news_RWE-Cancels-Polish-Coal-Fired-Power-Plant-an-Industrial-Info-News-Alert_44370052.html [Accessed: 19 June 2012].

Jochem, E. and Eichhammer, W. 1996. Voluntary agreements as an instrument to substitute regulation and economic instruments? Lessons from the German voluntary agreement on CO_2 reduction. Paper submitted to the Conference Economics and Law of Voluntary Approaches in Environmental Policy Venice November 18/19, 1996. [Online]. Available at: http://www.feem.it/userfiles/attach/Publication/NDL1997/NDL1997-019.pdf [Accessed: 19 June 2012].

Johansen, K. and Damm, N. 2004. *Research and Development in Transition – The Effects of Market Liberalisation*. Thesis. Stockholm: Kungliga Tekniska Högskolan.

KPMG. 2011. *Power Sector Development in Europe – Lenders' Perspectives 2011*. [Online]. Available at: http://www.kpmg.com/Global/en/IssuesAndInsights/ArticlesPublications/Documents/power-sector-development-europe.pdf [Accessed: 19 June 2012].

Lauber, V. and Mez, L. 2004. Three decades of renewable electricity policies in Germany. *Energy & Environment*, 15(4), 599–623.

NyTeknik. 25 April 2007. *En halv promille til miljön*. 25 April 2007. [Online]. Available at: http://www.nyteknik.se/nyheter/energi_miljo/energi/article252982.ece [Accessed: 19 June 2012].

Pahle, M., Fan, L. and Schill, W.-P. 2011. How emission certificate allocations distort fossil investments: The German example. *Energy Policy*, doi:10.1016/j.enpol.2011.01.027.

Platts. 3 April 2012. *Verified EU ETS CO_2 Emissions Fall 2.4% in 2011: Partial Data* [Online]. Available at: http://www.moneyscience.com/pg/newsfeeds/MondoVisione/item/307132/platts-eu-emissions-trading-system-co2-emissions-fall-24-in-2011-partial-data-analysis [Accessed: 10 September 2012].

Power Engineering International. 21 January 2009. *RWE: No New Coal Plants in Western Europe, Looks East*. [Online]. Available at: http://www.powergenworldwide.com/index/display/articledisplay/350919/articles/power-engineering/industry-news-2/2009/01/rwe-no-new-coal-plants-in-western-europe-looks-east.html [Accessed: 15 September 2011].

Power Engineering International. 14 August 2002. *RWE Chief Seeks Changes to EU Emissions Regime*. [Online]. Available at: http://www.powergenworldwide.com/

index/display/articledisplay/152040/articles/power-engineering-international/
  industry-news-2/2002/08/rwe-chief-seeks-changes-to-eu-emissions-regime.
  html [Accessed: 19 June 2012].

Reuters. 22 June 2007. *German Parliament Seals CO₂ Deal, Utilities Protest.*
  [Online]. Available at: http://www.reuters.com/article/2007/06/22/environment-
  germany-co2-law-dc-idUSL2285341220070622 [Accessed: 19 June 2012].

Riksrevisionen (the Swedish General Accounting Office). 2007. *Vattenfall – med
  vind i ryggen?* RiR 2007: 299, Stockholm: Riksrevisionen.

Rogge, K.S., Schneider, M. and Hoffmann, V.H. 2010. *The Innovation Impact
  of EU Emission Trading – Findings of Company Case Studies in the German
  Power Sector.* Working Paper Sustainability and Innovation No. S 2/2010.
  Köln, Heidelberg, Mannheim, Karlsruhe: Fraunhofer ISI.

Rudsar, K., Eikeland, P.O., Krumbacher, F. and Martinsen, G. 1999. *Rettslige
  rammebetingelser i Danmark og Tyskland og deres innvirkning på
  strukturutviklingen i elindustrien* (The legal framework in Denmark and
  Germany: Implications for structural development in electricity industry). FNI
  Report 11/99. Lysaker, Norway: The Fridtjof Nansen Institute.

RWE AG. *Annual Reports* (2001, 2004, 2008–2010). Essen, Germany: RWE AG.

RWE AG. Corporate Responsibility Report 2005. [Online]. Available at:
  http://www.rwe.com/web/cms/mediablob/en/105276/data/316928/4/rwe/
  responsibility/cr-reports/archive-cr-reports/unsere-verantwortung-engl-2005.
  pdf [Accessed 19 June 2012].

RWE AG. Environmental Report 2000. [Online]. Available at: http://www.rwe.
  com/web/cms/mediablob/en/614620/data/316928/1/rwe/responsibility/cr-
  reports/archive-cr-reports/rwe-environmental-report-2000.pdf [Accessed 19
  June 2012].

RWE AG. Environmental Report 2001. [Online]. Available at: http://www.rwe.
  com/web/cms/mediablob/en/614618/data/316928/1/rwe/responsibility/cr-
  reports/archive-cr-reports/rwe-environmental-report-2001.pdf [Accessed: 19
  June 2012].

RWE AG. Our Responsibility, Status 2008. [Online]. Available at: http://www.
  econsense.de/sites/all/files/RWE_Our_Responsibility_2008_0.pdf [Accessed:
  19 June 2012].

RWE AG. Our Responsibility. Report 2010. [Online]. Available at: http://www.
  rwe.com/web/cms/mediablob/en/625102/data/10122/3/rwe/rwe-group/RWE-
  Group-CR-Report-2010.pdf Corporate Social Responsibility Report 2010
  [Accessed: 19 June 2012].

RWE Innogy Web Presentation Wind – A Power Coming of Age. [Online] Available
  at: http://www.rwe.com/web/cms/en/87264/rwe-innogy/renewable-energies/
  wind/ [Accessed 19 June 2012].

RWE Innogy Web Presentation Biomass and Cogeneration. [Online]. Available
  at: http://www.rwe.com/web/cms/en/87098/rwe-innogy/renewable-energies/
  biomass/ [Accessed: 19 June 2012].

Sandoff, A., Schaad, G. and Williamsson, J. 2011. *Carbon Mitigation Strategies in the Heat and Power Sector and the Efficiency of the European Union Emission Trading Scheme* (EU ETS), 13th annual SNEE conference European Integration in Swedish Economic Research, Mölle, Sweden, 17–20 May 2011. [Online]. Available at: www.snee.org/filer/papers/660.pdf [Accessed: 19 June 2012].

Sijm, S.J., Bakker, A., Chen, Y., et al. 2005. *CO$_2$ Price Dynamics: The Implications of EU Emissions Trading for the Price of Electricity*. ECN report ECN-C--05--081, Petten, the Netherlands: ECN.

Skjærseth, J.B. and Wettestad, J. 2008. *EU Emissions Trading: Initiation, Decision-making and Implementation*. Aldershot: Ashgate.

SpiegelOnline. 28 December 2006. *Negativpreis, RWE-Chef Roels ist der 'Dinosaurier 2006'*. [Online]. Available at: http://www.spiegel.de/wirtschaft/ negativpreis-rwe-chef-roels-ist-der-dinosaurier-2006-a-456831.html [Accessed: 19 June 2012].

Standard & Poor's. 5 February 2002. *Vattenfall AB Ratings Removed from Credit Watch and Affirmed; Outlook Negative*. Press release.

Standard & Poor's. 5 February 2010. *Global Credit Portal – Ratings Direct*, RWE AG.

Standard & Poor's. 24 May 2011. *Global Credit Portal – Ratings Direct*, Vattenfall AB.

Sterlacchini, A. 2010. *Energy R&D in Private and State-owned Utilities: An Analysis of the Major World Electric Companies*. Departmental Working Papers 2010–29, Department of Economics, Business and Statistics, Milano, Italy: Università degli Studi di Milano.

Sveriges Regering (Government of Sweden). 1997. Proposition 1996/97: 84 *En uthållig energiförsörjning*, Stockholm: Närings och Handelsdepartementet (Ministry of Industry).

Teece, D.J., Pisano, G. and Shuen, A. 1997. Dynamic capabilities and strategic management. *Strategic Management Journal*, 18(7), 509–533.

UNIPEDE. December 1998. *R&D in UNIPEDE Countries, Overview of R&D organization and projects, 1998 Update*. Ref: 1998–150–0005, Brussels: Eurelectric.

Vattenfall. 2010a. *Bridging to the Future*. Newsletter on Carbon Capture & Storage at Vattenfall No. 17, December 2010.

Vattenfall. 2010b. *R&D Magazine* Vattenfall Research and Development Magazine No. 2, June 2010. Vattenfall Annual Reports 1999–2010. Stockholm: Vattenfall AB.

Vattenfall. 'Biomass'. [Online]. Available at: http://www.vattenfall.co.uk/en/ biomass.htm [Accessed: 19 June 2012].

Vattenfall. Corporate Social Responsibility Report 2003. [Online]. Available at: http://www.vattenfall.com/en/file/Corporate-Social-Responsibili_8458309. pdf [Accessed: 19 June 2012].

Vattenfall. Corporate Social Responsibility Report 2004. [Online]. Available at: http://www.vattenfall.com/en/file/Corporate-Social-Responsibili_65934_8458695.pdf [Accessed: 19 June 2012].

Vattenfall. Corporate Social Responsibility Report 2005. [Online]. Available at: http://www.vattenfall.com/en/file/Corporate-Social-Responsibili_75063_8458424.pdf [Accessed: 19 June 2012].

Vattenfall. Corporate Social Responsibility Report 2006. [Online]. http://www.vattenfall.com/en/file/Corporate-Social-Responsibili_8459766.pdf [Accessed: 19 June 2012].

Vattenfall. Corporate Social Responsibility Report 2008. [Online]. Available at: http://www.vattenfall.com/en/file/Corporate-Social-Responsibili_100589_8459767.pdf [Accessed: 19 June 2012].

Vattenfall. Corporate Social Responsibility Report 2010. [Online]. Available at: http://www.vattenfall.com/en/file/Corporate_Social_Responsibility_report_CSR_2010.pdf_17546076.pdf [Accessed: 19 June 2012].

Vattenfall. Sustainability Report 2001. Stockholm: Vattenfall AB

Vattenfall. Sustainability Report 2002. [Online]. Available at: http://www.vattenfall.com/en/file/Sustainability-report-2002_8458425.pdf [Accessed: 19 June 2012].

Appendix Table 3.1   Electric power industry data. CO_2 emissions of the 20 major electric power companies under the EU ETS, 2005–2009

Company	HQ	Emissions 2005	Emissions 2008	Emissions 2009	Change 2005–2008 %	Change 2005–2009 %
RWE	Germany	149,277,603	145,750,761	131,916,995	-2.4	-11.6
E.ON	Germany	114,419,189	107,507,810	93,625,295	-6.0	-18.2
Enel	Italy	105,233,244	93,245,880	70,963,842	-11.4	-32.6
Vattenfall	Sweden	88,238,237	85,316,894	84,907,164	-3.3	-3.8
EDF	France	80,162,305	76,276,523	66,862,560	-4.8	-16.6
PGE	Poland	57,947,239	58,589,544	55,659,881	1.1	-3.9
PPC	Greece	52,836,347	52,372,047	50,006,935	-0.9	-5.4
Endesa	Spain	49,831,098	44,516,103	34,433,430	-10.7	-30.9
GDF SUEZ	France	45,682,714	43,090,126	40,545,068	-5.7	-11.2
CEZ	Czech Rep.	37,091,654	36,430,315	34,656,278	-1.8	-6.6
Electrabel	Belgium	33,405,063	31,313,522	30,012,692	-6.3	-10.2
Iberdrola	Spain	31,087,646	26,701,889	25,450,331	-14.1	-18.1
EDP	Portugal	29,126,801	20,538,586	20,679,732	-29.5	-29.0
Edison	Italy	25,440,773	27,700,796	21,360,898	8.9	-16.0
SSE	UK	24,495,786	15,725,839	17,476,809	-35.8	-28.7
RWE npower	UK	23,079,402	24,628,106	16,514,113	6.7	-28.4
Drax Power	UK	20,771,624	22,299,778	19,851,702	7.4	-4.4
PKE	Poland	19,238,536	18,416,601	17,296,279	-4.3	-10.1
EnBW	Germany	19,040,613	16,826,021	14,456,430	-11.6	-24.1
ScottishPower	UK	15,052,499	15,363,348	15,137,420	2.1	0.6

Source: CarbonMarketData.

Appendix Table 3.2   Electric power industry data. R&D intensity (R&D/net sales) among European power companies consistently listed in the EU Industrial R&D Investment Scoreboards in the period 2004–2009

Group	Country	Rank among the 1000 top	R&D. 2008 (€ mill.)	R&D/net sales					
				2004	2005	2006	2007	2008	2009
EdF	France	57	421	0.9	0.8	0.7	0.5	0.7	0.7
RWE	Germany	88	210	0.3	0.3	0.3	0.3	0.4	0.5
Vattenfall	Sweden	112	155.76	0.7	0.8	0.6	0.8	1.0	0.7
GdF SUEZ	France	139	127	GDF: 0.5 SUEZ: 0.2	GDF: 0.3 SUEZ: 0.2	GDF: 0.3 SUEZ: 0.2	0.2	0.2	0.3
E.ON	Germany	165	102	0.1	0.0	0.0	0.1	0.1	0.1
Iberdrola	Spain	206	73.10				0.4	0.3	0.4
Enel	Italy	219	68	0.1	0.1	0.1	0.1	0.1	0.1
Energie Baden	Germany	358	29.30	0.1	0.2	0.2	0.3	0.2	0.2
Fortum	Finland	378	27	0.2	0.4	0.4	0.5	0.5	0.6
EDP	Portugal	411	24					0.2	
Pohjolan Voima	Finland	430	22.2	3.0	3.6	1.9	2.3	2.4	2.4
Teollisuuden Voima	Finland	452	20.6			6.5	7.4	8.0	6.9
Verbund	Austria	526	16.10		0.2	0.2	0.3	0.4	
British Energy (part of EdF)	UK	542	15.52	0.9	0.6	0.4	0.4	0.5	
International Power (part of GdF)	UK	641	11.38	0.4	0.2	0.1	0.2	0.3	0.2
MW Energie	Germany	726	9	0.2	0.2		0.4	0.3	0.2
Scottish and Southern Energy – SSE	UK	977	4.55			0.1	0.0	0.0	0.0

Appendix Table 3.3   Electric power industry data. Selected data on climate change strategies for major European electric power companies

	Climate strategic goals	ETS position	Long-term strategy	Energy efficiency	Renewable energy	CCS
RWE[1] Germany	2008: halving specific emissions in 2020. Long-term: carbon-neutral by 2050 (signed 2009 joint industry declaration).	Opposition to EU ETS in early decade but later supportive.	Recent strong growth in R&D expenditure.	Strong focus on energy efficiency, at own and customers' facilities. Quantitative targets set early in the decade. Established a new energy-efficiency company in 2009.	New renewable unit, RWE Innogy from 2008. Strong growth in wind power investments. Part of the Desertec solar PV consortium.	Engaged in various R&D&D projects in Germany and abroad.
E.ON[2] Germany	2007: 50 per cent cut in emissions by 2030. Long-term goal: carbon-neutral by 2050 (signed 2009 joint industry declaration).	Supportive of a well-functioning market for CO_2 allowances. In favour of auctioning if harmonized at the EU level.	Modest resources committed to R&D: €17 million spent on university R&D in 2009. 2006: set-up of E.on Aachen Research Centre funded with €40 million over ten years.	E.on determined to build more efficient fossil fuel power plants and provide energy-efficient solutions for end-use customers.	By 2009: 13 per cent renewables with short- and long-term goals: 18 per cent by 2015, 33 per cent by 2030. 2008–2009: 1 GW wind power added. 2009: set up of solar energy as second renewable energy pillar. Part of the Desertec consortium.	2009: operational testing at a unit in Germany, various demonstration projects initiated in 2009.

	Climate strategic goals	ETS position	Long-term strategy	Energy efficiency	Renewable energy	CCS
Enel[3] Italy	Retrospective reporting. No quantitative short-term goals. Long-term: carbon-neutral by 2050 (signed 2009 joint industry declaration.	Costs of CO_2 internalized among the production costs. The company lists regulatory risks (and opportunities) created by the EU ETS. Enel promotes itself as a leader in the global market for CDM.	2009: Technological innovation plan for 2009–2013 (€650). € 86 spent in 2009. 46 per cent spent on fossil fuels (CCS, hydrogen, improving plant efficiency) and 50 per cent on renewables.	No quantitative targets set but continuous commitment: 'Enel is constantly increasing the efficiency of its production plans and networks.'	2009: 7.4 GW installed capacity. Goal to double the capacity in a few years. In 2008, Enel Green power established to run its 500 plants based on hydro-power and other renewable energy.	Claims Enel a pioneer in Italy. Signed agreement in 2008 with ENI on a pilot project the Brindisi power plant.
Vattenfall Sweden	Long-term goal: signed 2009 joint industry agreement on carbon neutrality by 2050. Fifty per cent reduction in specific emissions by 2030 compared to 1990-levels. Short-term goals: 2 per cent cut in emissions for the years 2009–2011.	Strong supporter of the EU ETS from the start.	High R&D intensity, down in the early and up again in the late decade. 2009: R&D refocused on renewables, CCS, efficiency, nuclear power, and new energy conversion technology.	Very high focus on energy-efficient services since the 1990s in Sweden – one of the first companies starting up delivering energy efficiency services Investments in new efficient fossil fuel-technologies.	Strong inherited hydropower and biomass-based CHP assets the end of 2009, Vattenfall. Major investments in wind power from 2006 and co-firing coal and biomass.	In 2008, Vattenfall inaugurated the first pilot power plant in the world testing out CCS in Germany, with work started on the project in 2001.

	Climate strategic goals	ETS position	Long-term strategy	Energy efficiency	Renewable energy	CCS
EDF[74] France UK	Remain the European energy group with lowest emissions of CO_2: 2008: 30 per cent reduction of CO_2 emissions in France from 1990 to 2020. Cut specific emissions by 50 per cent in the same period. Achieve 8 per cent reduction from non-industrial sector in the period 2007–1012. Long-term goal: signatory to the European industrial agreement for no carbon emissions in 2050.	The group is fighting for an ETS system harmonized across Europe and based on auctioning (Sustainability Report, 2008). The groups wants zero-emissions projects, in particular large nuclear and hydropower schemes to be accepted under the flexibility mechanisms.	Total R&D expenses in 2009: €438 (20 per cent to environmental protection including energy efficiency and renewable energy.		Strong inherited hydropower assets. Deployment of a diversified portfolio of renewable energy generation capacity towards the end of the decade. EDF Energies Renouvelles established 2004. EDF Energy Renewables (UK) established 2008 with major investments in wind power.	No specific mention of CCS in sustainability report 2008 and annual report 2009. In 2010, EDF, Veolia and Alstom announced the start of a project to build and operate CCS demonstration plant at EDF's coal-fired power plant at Le Havre.

	Climate strategic goals	ETS Position	Long-term Strategy	Energy efficiency	Renewable energy	CCS
PGE Poland	No quantitative targets and not a signatory to the European industrial agreement of no carbon emissions in 2050.	By 2012: Claims a short-term strategy for balancing emission allowances and factual emissions in the years 2008–2012.	By 2012: Claims the most important R&D project demonstration of CCS.	By 2012: Claims large investments in the construction and modernization of conventional generation capacity, including CHP.	By 2012: Claims over 20 per cent of total investment expenditures on the development of renewable sources of energy.	The Belchatow CCS initiative supported by EEPR and submitted by the Polish Government to the European Investment Bank for funding through NER300.
PPC Greece	Only retrospective reporting on reduction in specific emissions. Long-term goal: Signatory to the European industrial agreement of no carbon emissions in 2050.	Cover 50 per cent of allowances in the 2005–2012 period through 5 carbon funds engaged in CDM projects.		By 2012: claims upgrading and modernization of the existing facilities and operation on the basis of the Best Available Techniques.		Reports no major involvement in demonstration but aims to use CCS when the technology is well-tested.

	Climate strategic goals	ETS position	Long-term strategy	Energy efficiency	Renewable energy	CCS
GDF SUEZ France Belgium (through ownership of Electrabel)	Combating climate change as one of four major strategic goals No quantitative goals set up on reduction of CO_2.		Signed the Grenelle Plan on the Environment with the French government in 2007. 2008: €203 devoted to R&I, 40 per cent of R&D expenses to sustainable development projects.		Goal to double share of renewables in generation portfoilio from 7.5 per cent in 2008 to 15 per cent in 2015.	Pilot project starting in 2004 on re-injection of CO_2 in a depleted gas filed in the North Sea. Together with Vattenfall, the company has another pilot project on injection in Germany (Altmark field).
CEZ Czech Republic	Plan of Action to reduce CO_2 emissions by 2020, which has been posted by the firm management includes the objective of a 15 per cent reduction of GHG emissions. Goal to reduce specific emissions from 0.55t CO_2/MWh (2005) to 0.47t CO_2/MWh (2020).	The ČEZ Group considers the use of an economic instrument (EU ETS) to reduce emissions a suitable solution, particularly when compared to the potential alternatives (administrative restrictions, instruments without flexibility).		In the short-term and middle-term perspective, the action plan stipulates improving the efficiency of coal power plants (power plant renewal programme) as important to reach the GHG emissions reductions.	CEZ Group is planning to triple the production from renewable energy resources by 2020 (5.1 TWh as compared to 1.7 TWh in 2005).	In the long term: investments in developing advanced combustion technologies with low CO_2 production, CO_2 separation and its storage in geological structures.

Note: [1] RWE data based mainly on the report 'Our Responsibility – Report 2009' published in 2010; [2] E.on data based mainly on web-based CR reporting: http://www.eon.com/en/responsibility/35071.jsp; [3] Enel data based mainly on 2008 Sustainability Report; [4] EDF data based mainly on the 2008 Sustainability Report.

Chapter 4

Oil Industry

Jon Birger Skjærseth

1. Introduction

The EU ETS was the first international climate policy instrument that directly targeted the CO_2 emissions from major oil companies in Europe. The oil industry, which makes a living from some of the key sources of anthropogenic climate change, stands out as an important sector for achieving a low-carbon economy in Europe and other parts of the world. To what extent, how and under what conditions has the ETS affected the climate strategies of major multinational oil companies? Oil companies can be affected in the short term by holding emissions allowances, particularly in the refining sector, and more strategically in the longer term by carbon pricing and the possibility of diffusion beyond Europe.[1]

Studies of the oil sector have indicated that the impact of the ETS has been limited to date (e.g. Lacombe 2008, Ellerman, Convery and de Perthuis 2010). These studies have focused mainly on short-term results. This chapter, however, shows that the main importance of the ETS in this sector concerns the long-term strategic consequences of carbon pricing. First of all, oil companies base their long-term strategies on forecasts or scenarios. The ETS has affected these forecasts by changing company beliefs about a future in which concerted governmental action, a price on carbon emissions and the need for more low-carbon energy will occupy a more prominent place. Second, the impact of the ETS has been most significant for those companies that have lagged furthest behind the 'leaders'. This has had the aggregate effect of making the position and role of the European industry association for the oil industry more proactive – temporarily. The recession, structural changes and the EU's long-term plan of decarbonizing Europe have made the oil industry increasingly reluctant to any new, binding policies beyond 2020. Finally, we will see that the EU is developing a technology supply side to its climate policy in the form of technology platforms to stimulate innovation on low-carbon solutions. These initiatives are linked to the ETS in several ways, including the financing mechanism based on sale of ETS allowances. The success of these initiatives will be important for future GHG emissions from the oil industry, and in particular efforts to make carbon capture and storage (CCS) commercially viable. In addition to the long-term consequences, this chapter analyses how the ETS

---

1   This chapter covers both effects, but does not deal specifically with specialized refining companies.

works together with other company-internal and company-external factors to co-produce various outcomes. The EU ETS represents an important driver for climate strategies, but still only one of many, as the oil industry is composed of high-profile multinational companies exposed to a wide range of risks and opportunities around the world.

The chapter begins by presenting the climate strategies of the major companies covered by the ETS and the changing role and position of the European industry association, Europia. In section three, ExxonMobil and Shell are studied in depth. These major multinational companies represent traditional 'leaders' and 'laggards' in climate strategies. Initially, they responded to the ETS in completely different ways, but their long-term strategies show signs of convergence. The fourth and fifth sections analyse the consequences of the ETS, the mechanisms through which the system works, and why Exxon and Shell responded so differently at first.

2. Oil Company Climate Strategies

The oil and gas industry covered by the EU ETS is dominated by major multinationals like Total, ENI, Shell, ExxonMobil, Repsol/YPS and BP. ETS covers the activities of these companies, including upstream and downstream emissions from exploration, production and refining. Of these, refining is the major source of emissions covered by the ETS.

2.1 Oil Refining

The refining sector is essential to the European economy. Petrol, diesel and aviation fuels make Europe work. The sector generated 33 billion euros in 2004, 8.6 per cent of the total EU tax revenues in 2005 and employed 140,000 persons (Lacombe 2008). The European part of the industry represents nearly 20 per cent of the world's production of petroleum products and contributed approximately 3 per cent of the EU's emissions of CO_2 in 2005 and about 7 per cent of the sectors covered by the ETS (Europia 2007). Energy consumption and CO_2 emissions in EU refineries have risen since 2000 and are expected to increase towards 2020, unless countermeasures are applied (CONCAWE 2008).

There are essentially four ways in which oil refining companies can reduce their GHG emissions: (1) increasing energy conservation and efficiency; (2) switching to fuels with lower carbon content; (3) investing in renewable sources of energy; (4) applying carbon capture and storage (CCS). According to its own reports, the sector boosted its energy efficiency by about 13 per cent between 1990 and 2005. First to be chosen were the 'no-regret' abatement options, like more energy-efficient pumps and compressors. A significant part of the improvement has been achieved by using combined heat and power (CHP) to replace simple steam boilers and imported electricity. To a significant extent, these improvements will be able to compensate for an increase in energy requirements until 2020 (CONCAWE

2008: 32). This balancing means that CO_2 emissions will not necessarily go down in the short term.

Another mitigation option is fuel switching. EU refineries have already replaced heavy fuel oil by natural gas, amounting to 5–10 per cent of refinery energy use. Further replacement of liquid fuel by natural gas is considered as an option for reducing direct CO_2 emissions from a refinery site. Refineries can also use lighter crude oil, which requires less processing energy. Finally, there is the CCS option for EU refineries, but here opinion varies. Some analysts see the sector as a prime candidate for CCS (Ellerman, Convery and de Perthuis 2010: 216). For various reasons, the EU refining sector does not consider the sector ideal for capture, transport and storage. For one thing, only a small portion of total CO_2 emissions come in concentrated form. A recent study of the potential for CCS in EU oil refineries (CONCAWE 2011) concludes that individual refineries will require alliances with other large CO_2 producers, particularly the power industry, in order for CCS to be commercially justified. The study also underscores the need for new breakthrough CCS technologies.

2.2 Oil Companies and Europia

Prior to the 1997 Kyoto Protocol, the oil majors were almost united in their opposition to binding climate targets and measures. In 1996, British Petroleum adopted a more proactive strategy, followed by Shell in 1997. These strategy changes were 'voluntary' in the sense that they occurred in the absence of any coordinated mandatory climate policy. Also other companies softened their opposition, except for ExxonMobil, which continued to work actively against mandatory climate policy internationally as well as within the EU (Skjærseth and Skodvin 2003).

Divergent strategies were reflected in the position of the European Petroleum Industry Association, Europia, which represents the European refining and marketing industry. Europia's 18 member companies account for around 80 per cent of EU petroleum refining capacity and some 75 per cent of EU motor fuel retail sales.[2] In the early 1990s, Europia was one of the most aggressive opponents of any kind of mandatory EU climate policy, particularly the proposed EU carbon/ energy tax (Skjærseth 1994). By the end of that decade, Europia's opposition had mellowed somewhat and it came to accept the emergent EU ETS as a learning exercise and part of the process of establishing an international system under the Kyoto Protocol (Europia, 1999). Europia welcomed the idea of emissions trading, but preferred voluntary participation, with no penalties for non-compliance, and it made its support conditional on the entry into force of the Kyoto Protocol.

---

2   BP, Chevron, Cepsa, ConocoPhillips, Eni, ExxonMobil, Galp Energia, Hellenic Petroleum, MOL, Neste Oil, Omv. PKN Orlen, Petroplus, Repsol YPF, Saras, Shell, Statoil, Total.

Europia's cautious acceptance of the ETS can be seen as a compromise between the reactive and proactive companies in its membership.

A main concern for the sector has been to minimize harm to the competitiveness of the European petroleum industry. New refineries in China and South East Asia are expected start exporting high-quality fuels to Europe and the USA in the near future. Refined products are currently traded within and outside the EU. Europe has become increasingly dependent on trade due to growing imbalances between supply and demand. About 29 per cent of the petrol produced by European refineries is exported, mainly to the USA, in the face of declining demand in Europe. Diesel is mainly imported from Russia, amounting to 22 per cent of EU demand (Europia 2010a). Apart from trade for structural reasons, transport costs and logistics have kept refining markets fairly local (European Commission 2006).

When the European Commission proposed to include refining as part of the energy sector in the revised 2009 ET Directive, this was strongly opposed by Europia. What the sector did achieve through its lobbying was to change its status from 'energy sector' to 'energy-intensive sector' in the final Directive. This modification guaranteed a longer period of free allowances from 2013. Accordingly, the manufacture of refined petroleum products and extraction of crude petroleum and natural gas were included on the EU list of sectors deemed to be exposed to significant risk of 'carbon leakage' (*OJEU* 2010). This refers to the risk that companies in sectors facing significant international competition may relocate production from the EU to third countries where climate policies are more lenient.[3] Installations in sectors exposed to carbon leakage are to receive allowances based on benchmarks that reflect the average performance of the most efficient 10 per cent of installations in terms of GHG emissions in the EU sectors during 2007–2008. Only the most efficient installations will receive all their allowances for free: thus, the most energy-efficient companies will enjoy a competitive advantage (see below).

Since the initiation of the ETS, the oil majors covered by the system have converged around more proactive strategies (Skjærseth 2011). The most significant changes have been announced by the companies that have traditionally lagged behind the 'leaders', like ExxonMobil.[4] The oil majors have generally adopted relative targets for emissions reduction, and short-term measures like energy efficiency and flaring. For the longer term, they have also adopted strategies for low-carbon solutions, varying significantly in scale and stage of development. Advanced biofuels and CCS are common in most of the companies. However, these are still petroleum companies, and they foresee expansion in fossil fuels rather

---

3   The list will apply until 2014 and may be revised if the risk of carbon leakage is lessened by a new international climate agreement. Sectors deemed at risk account for about 25 per cent of the total emissions covered by the ETS and almost 80 per cent of the total emissions from manufacturing industry.

4   However, the sector also includes specialized European refining companies, such as the Polish PKN Orlen, which appear reluctant to adopt any climate policy.

than a low-carbon future. Quite illustrative is the latest 2030 forecast produced by one of the most proactive companies, BP (BP 2011). It foresees increased economic growth, global energy demand, energy efficiency and a gradual shift to a less carbon-intensive fuel mix. Still, it expects fossil fuels to continue to dominate the energy mix, and greenhouse gases (GHGs) to rise significantly. This forecast informs current investment decisions. In 2010, BP spent USD 32 billion on capital investment and exploration of fossil fuels, a level of spending that the company aims to maintain in the foreseeable future (Forbes 2011).

Changes in the climate strategies of oil companies and a more harmonized ETS have affected the position of Europia. In its 2008 position paper on the ETS revision proposal, the association for the first time explicitly accepted the ETS in light of concerns about climate change (Europia 2008). Although various concerns are noted, this is a generally more positive response than in the earlier position papers: 'There are many aspects of the EU ETS Directive proposal ... that are welcomed by EUROPIA ... a level playing field ... a long term trajectory' (EUROPIA 2008:1). Working with CONCAWE, the oil and gas industry association for scientific and technical research, Europia has explored various abatement opportunities for the refining sector. One such opportunity is to capture and store CO_2 at the refineries. In 2008, the European refining industry united for the first time in a public initiative to reduce GHG emissions from transport fuel. Still, most of the climate-related analyses and activities in this sector are driven by the individual members of these organizations, the major multinational oil companies.

Most companies, as well as CONCAWE and Europia, refer to the EU ETS as an important driver shaping their own climate strategies. According to CONCAWE, 'one of the most important features of the future EU industry including refining will be the increasing cost attached to CO_2 emissions, reflected in a CO_2 market price under the EU Emissions Trading Scheme (EU ETS)' (CONCAWE 2008: 26). In addition, companies mention the EU targets and the 2008 EU climate and energy package as regulatory drivers for CCS and advanced biofuels specifically. Moreover, most companies link their climate strategies to the Intergovernmental Panel on Climate Change (IPCC) and corporate responsibility for problem-solving.

In light of the 2008 recession and the EU's plans for decarbonizing Europe by 2050, Europia has recently painted a dark picture of the future of the European refining industry. A 'dramatic' and challenging future is foreseen in various analyses and forecasts published by Europia (Europia 2010a 2010b, 2011a, 2011b), mirrored also in the European Commission's recent report on the status of the European refining industry (European Commission 2010b). These analyses cite the decreasing demand for petroleum products in recent years and reduced refining margins, with the trend expected to continue over the next decades. On the basis of data from the International Energy Agency (IEA), Europia foresees a decline in oil demand by between 30 per cent and 55 per cent by 2050, depending on the specific climate-policy assumptions. In addition to the EU ETS, tighter fuel specifications, restrictions on car emissions and support for the development of non-fossil fuels are mentioned as factors that will stimulate this decline. This means that a scaling-down of the European refining industry appears inevitable. Divestments and shutdowns have already been carried out in EU refineries since the start of the recession in 2008. Still, as oil is expected to remain important, particularly in the transport sector, Europia argues that it is in the EU's interest to maintain refining capacity in Europe for reasons of security of supply, technological leadership in setting global fuel specifications, as well as jobs and tax revenues.

A second challenge noted is the mismatch between diesel and petrol, resulting in expected high trade deficits. Reducing this mismatch will require significant investment in additional refining conversion capacity and reduction of petrol production capacity. Since additional capacity for diesel and other middle distillates (jet fuel, kerosene) is more energy-intensive, CO_2 emissions will increase and thus also the costs under the ETS. Yet, more stringent product specifications (particularly on marine gasoil, due to new IMO requirements) may lead to higher CO_2 emissions despite falling demand. CO_2 emissions may increase by 12 per cent between 2005 and 2020, and 6 per cent between 2005 and 2030 (European Commission 2010b).

As noted, the oil refining industry successfully moved refining from the prospect of full auctioning to the continuation of free allowances, to be allocated on the basis of benchmarks. The European Commission accepted the CWT (Complex Weighted Tonne) approach developed by the oil refining industry. Alternative proposals, such as allocation based on crude input, were successfully defeated by the industry. However, the oil refining industry did not manage to gain acceptance of its proposal for progressive implementation of the benchmark in the period 2013 to 2020.

According to the industry, benchmarking will have significant consequences (Europia 2010b). Companies will on average have to pay for 25 per cent of the allowances in addition to an expected pass-through of CO_2 costs in electricity purchased, representing an additional 5 per cent costs. These average costs, claimed to increase cash costs by 13 per cent, will affect refineries differently, because of the significant differences in energy efficiency. Increased costs from benchmarking and the structural challenges facing the sector will lead to improvements in energy efficiency but also to shutdowns. Big multinational oil companies tend to have efficient as well as less-efficient refineries as well as upstream activities. Thus they are likely to be less affected than specialized European refining companies, particularly from the new EU member states that still lag behind in energy efficiency.

All in all, as a result of these changes and expectations, the oil refining industry has become more cautious in its support of the EU ETS and EU climate policy. Europia has expressed 'strong concerns' about initiatives for raising the unilateral reduction target from 20 per cent to 30 per cent (Europia 2010b: 25). In its long-term outlook to 2050, the association argues that the EU should focus on achieving the goals of the climate and energy package for 2020, but not yet set binding policy targets or instruments beyond 2020 because there is too much uncertainty;

furthermore, the EU should apply flexible, predictable and realistic policies that will not prematurely close down future pathways (Europia 2011a: 45).

Thus we may conclude that the major oil companies covered by the ETS have signalled more proactive long-term climate strategies based on innovative elements. This change has in turn affected the industry association Europia, which for a while became increasingly supportive of the ETS. This took place alongside the strengthening of the system and the formulation of a more ambitious EU climate policy up to 2020. Europia also acknowledged its responsibility for taking action on climate change. The oil companies refer particularly to the ETS as a main driver behind this development, but they also mention wider EU climate policies. However, future expectations in light of the recession, falling demand, trade imbalances and the long-term plan of decarbonizing Europe have now made the sector reluctant to any new binding policies beyond 2020.

3. Shell and ExxonMobil

Shell and ExxonMobil, the two biggest publicly traded oil companies, are high-profile multinational companies with operations all over the world. The climate strategies of these titans are of great importance to the whole industry. They also play a significant role in Europia, occupying most positions on the Board of Directors and internal working groups (Europia 2010b). Exxon and Shell are the third and fifth largest oil companies, respectively, under the ETS. In 2005, when the EU ETS was launched, Shell had 56 installations covered by the system (22,143 allowances) and Exxon had 48 (18,825 allowances).[5]

3.1 Shell's Climate Strategy

Shell is the result of an alliance made in 1907 between the Royal Dutch Petroleum Company and the 'Shell' Transport and Trading Company, where the two companies agreed to merge their interests on a 60:40 basis while keeping separate identities. Shell was also one of the 'seven sisters', the influential oil company cartel during the first half of the twentieth century. The company currently employs 101,000 people and conducts business in more than 90 countries in oil, gas, chemicals, refining and renewable energy sources. Shell figures as a global group of energy and petrochemical companies conducting business in upstream, downstream and projects/technology. Its vision or strategy is to reinforce the position as a leader in the oil and gas industry in order to provide a competitive shareholder return while helping to meet global energy demand in a responsible way.[6]

---

5   Community Transaction Log/Carbon Market Data. Available at: http://www.carbonmarketdata.com/

6   www.shell.com

Shell acknowledged the problem of human-induced climate change and responsibility for helping to solve the problem in 1998 (Shell, 1998). The company first aimed for a 10 per cent reduction of GHG emissions from 1990 levels by 2002. According to the company, that target was met with time to spare (Shell 2002: 24), and in 2002, a new 5 per cent reduction by 2010 from (new and higher) 1990 levels was adopted.[7] The absolute target was adjusted to growth in production due to new acquisitions. Shell has decided not to adopt a new company-specific GHG target extending beyond 2010. However, it apparently supports the plan to increase the EU target from 20 per cent to 25 per cent by 2020, including a more stringent cap to keep carbon prices stable.

In January 2000, Shell followed BP and launched an internal GHG emissions trading system, the Shell Tradable Emission Permit System (STEPS), aimed at using tradable emissions permits to help meet the self-imposed emissions targets. The same year, the company incorporated future costs of CO_2 emissions into its financial planning of, and decisions on, major projects. In 2001, Shell created an Environmental Products Trading Business (EPTB) within Shell Trading. The EPTB team within Shell trading was the first to execute a trade in the EU ETS, in February 2003; it was also the first to execute a trade in the second EU ETS period (2008–2012), and the first to receive Certified Emission Reductions into an account in the UNFCCC Secretariat's Clean Development Registry.

Shell has monitored and reported its GHG emissions data from 1998. From 2002, Shell has worked on *energy efficiency* through its Energise™ programme. Nevertheless, energy efficiency at Shell refineries has declined since 2004, especially rapidly in the 2008/09 (Shell Report 2009). Shell has described these results as 'disappointing', and plans to adopt more advanced monitoring and maintenance procedures and information systems to help operators to improve plant efficiency (Shell Report 2008:29). With regard to installations under the EU ETS, verified emissions have grown from about 22 million tons in 2005 to nearly 23 million tons in 2009.[8]

3.1.1 Long-term, low-carbon Solutions

In 1997, Shell established Shell International renewables as a fifth core business activity. Shell aimed to become an energy company by capturing a 10 per cent share of the renewables market by 2005. Shell divested itself of its forestry business, and in 2006 sold its entire silicon-based solar panel business, officially because next-generation thin-film technologies had a better chance of cutting costs faster (Shell 2005: 13).[9] Shell is involved in 11 wind projects in Europe and North America

---

7   For the 1990 baseline year, 114 million tonnes CO_2 eq. was amended to 123 million tonnes CO_2 eq. (Shell Report 2002: 24).

8   Community Transaction Log/Carbon Market Data. Available at: http://www.carbonmarketdata.com/

9   The most active oil companies in the EU solar technology platform are ENEL and BP.

with a total generating capacity of 1,100 MW (550 MV Shell share). In Europe, Shell is a 50:50 partner in three projects in Spain, Germany and the Netherlands (Shell Fact 2009: 48). In 2009, the company announced its intention to terminate all new investments in wind, solar and hydrogen energy and focus on oil, gas and biofuels (*The Times* 2009).

Over the past five years, Shell has spent USD 2 billion on CCS and alternative energies, including biofuels (Shell, CDP 2010). According to the company, this represents a significant increase in R&D on CCS, but not necessarily on low-carbon non-petroleum activities. The main change in low-carbon R&D investments has been an adjustment from renewables to CCS. Shell plans to step up its activities in the area of advanced biofuels. Carbon capture and storage is an increasingly important aspect of the company's climate strategy. Shell is the second-largest patent holder of carbon capture technologies, with 414 patents (Lee, Iliev and Preston 2009: 39). Carbon capture is a necessary first step for storage, but has mainly been used in industrial contexts such as enhanced oil recovery.

Shell is involved in a range of demonstration projects for developing CCS technologies, and is also involved in plans for full-scale CCS projects. For example, Shell and the governments of Alberta and Canada have announced a joint funding agreement to capture over 1 million tonnes of CO_2 a year; the final investment decision awaits regulatory approval. Shell's involvement in oil sands in Canada, representing 2.5 per cent of the company's total oil and gas production, is more CO_2-intensive than conventional oil production and has drawn considerable criticism. Shell also has CCS plans for its Pernis refinery in Rotterdam, The Netherlands. The aim has been to capture 400,000 tonnes of CO_2 a year, which at a CO_2 price of 20 euros per tonne would deliver annual revenues of 8 million euros (Shell CDP 2010: 12).[10] Pernis is Europe's biggest refinery, and has a track record of innovative solutions. In 2005, the refinery started supplying pure CO_2 to local greenhouses (Shell 2005: 10).

In Shell's opinion, the future of CCS depends on its becoming financially viable and widespread, which in turn hinges on a sufficiently high CO_2 price. Shell has collaborated increasingly with other companies, academic institutes and governments to make CCS commercially viable. One example is the European Technology Platform for Zero Emissions Fuel Power Plants (ZEP, see below), where Shell's Graeme Sweeney is chairman of the ZEP Advisory Council. In addition to collaboration on innovative technologies and policies, Shell is involved in a range of voluntary CSR initiatives in the fields of emissions trading, climate change and environmental protection. For example, the company participates in the International Emissions Trading Association (IETA), which promotes emissions trading around the world. Shell's climate change advisor is vice-president of the Association.

---

10   In November 2010, The Netherlands ended Shell's CCS project in Barendrecht, mainly because of local opposition and lack of public acceptance of the project.

Shell has produced long-term scenarios since the 1970s. Its long-term scenario 'Energy 2050', envisions two possible futures (Shell, Energy 2008). In the Scramble scenario, GHG emissions are not dealt with seriously until there are major climate shocks. In the Blueprint scenario, such emissions are addressed by putting a price on a critical mass of emissions, which stimulates the development of clean technologies. Shell has indicated a preference for the interaction or collaboration between state and society actors that underlies the Blueprint scenario. The company is convinced that this scenario is achievable with the right mix of policy and technology, although this will admittedly not be easy.

3.2 Exxon's Climate Strategy

Exxon Corporation started out as Standard Oil in 1882, mainly as a refinery company. In 1888 it began to internationalize its downstream assets, and in the 1920s it invested heavily to become a fully integrated oil company. Exxon was, like Shell, one of the 'seven sisters', the oil cartel that controlled the world oil trade in the first half of the twentieth century. In November 1999, Exxon and Mobil merged to form ExxonMobil Corporation. Exxon's takeover of Mobil was the largest in history. The company conducts business in upstream, downstream and chemicals throughout most of the world, with headquarters in Irvine, Texas. Exxon is committed to being the world's premier petroleum and petrochemical company. To that end, the company aims to continuously achieve superior financial and operating results while simultaneously adhering to high ethical standards'.[11]

Until recently, Exxon remained extremely sceptical on the question of human-induced climate change (Skjærseth and Skodvin 2003: 47–8). No GHG emissions reduction targets for its own operations were therefore put in place. ExxonMobil was also active in lobbying against international GHG regulations and against US ratification of the Kyoto Protocol. From the establishment of the Washington-based Global Climate Coalition until its deactivation in 2002, ExxonMobil was a key member, spending large resources on PR campaigns to influence public opinion and policy-making on the issue.

From around 2001, Exxon began to soften its position on climate change somewhat. This culminated in its 2007 acknowledgement of responsibility in helping to alleviate the problem (Exxon Citizen Report 2007: 15). The company declared a stop to its funding of several public policy research groups whose 'position on climate change could divert attention from the important discussion on how the world will secure the energy required for economic growth in an environmentally responsible manner' (Exxon Citizen Report 2007: 39).[12] It also accepted carbon pricing by international taxation, but not cap-and-trade.

---

11   www.exxonmobil.com

12   The company claims that it has stopped funding such organizations; Greenpeace has challenged this claim (Webb 2007).

ExxonMobil has not adopted GHG emission targets, basically because the company is actively seeking to increase its production of oil and gas. In 2002, the company adopted a 10 per cent energy efficiency target between 2002 and 2012 across worldwide refining and chemical operations and a 2008 target to reduce flaring by 20 per cent over 'the next few years' (Exxon, CDP 2010: 15). Since 2003, ExxonMobil has reported all direct GHG emissions from own operations (Exxon CDP 2008: 5).[13] In 2000, the company adopted a Global Energy Management System (GEMS), which has identified opportunities for improving energy efficiency by 15 to 20 per cent at the company's refineries and chemical plants. Since 2004, Exxon has invested more than USD 1 billion in co-generation projects, and it plans to increase capacity to 5,000 MV in the near future (Exxon CDP 2009: 27). In contrast to Shell, ExxonMobil has been working successfully to improve energy efficiency at its refineries and chemical plants since 2006 (Exxon CDP 2009: 27, Exxon Citizen Report 2009:32).[14] Verified emissions under the ETS have remained almost stable at about 17 million tons from 2005 to 2009.[15]

3.2.1 Long-term, low-carbon Solutions

ExxonMobil is not active in renewable energy such as wind and solar. Since 2009, however, it has collaborated with a biotech firm in California on algae-based biofuel, and plans to invest over USD 600 million in this research over the next 10 years (Exxon Citizen Report 2009: 35). The project has now entered into a second phase with the opening of a new greenhouse research and testing facility (Lamp 2010). In 2009, Exxon bid USD 41 billion for XTO, the largest natural gas producer in the USA and a leader in the development of unconventional oil and gas supply. The merger was completed in 2010.

Like Shell, ExxonMobil invests in basic R&D to reduce GHG emissions in the long term. This includes gasification, biofuels, CCS and hydrogen. ExxonMobil is now the world's tenth-largest patent holder of clean coal technology (Lee, Iliev and Preston 2009: 36). It is a founding sponsor of the Global Climate and Energy Project (GCEP) at Stanford University, which focuses on breakthrough energy technologies. Since 2005, the company has invested USD 1.3 billion in activities to improve energy efficiency and reduce GHG emissions (Exxon, CDP 2010: 32) – a significant increase, according to the company.

Exxon was the first oil company to explore CCS technology (Tjernshaugen 2010). It is also the largest carbon capture patent-holder in the world, with 978 patents (Lee, Iliev and Preston 2009: 39). ExxonMobil participates in the CO_2 Remove project under the Sixth EU Framework Programme for Research, which is aimed at developing and demonstrating methods for monitoring CO_2 storage in

---

13   The company started reporting its GHG emissions in 1998.

14   Data on energy efficiency cover world-wide operations and are not directly comparable with those from Shell.

15   Community Transaction Log/Carbon Market Data. Available at: http://www.carbonmarketdata.com/

geological reservoirs. In contrast to Shell and despite its own lengthy experience with CCS, Exxon has been more hesitant than its competitors to explore CCS in collaboration with public authorities and less eager to promote commercialization (Tjernshaugen 2010). The company does not participate in the EU-led Technology Platform for Zero Emissions Fuel Power Plants (ZEP) or the International Emissions Trading Association (IETA).

Exxon produces forecasts in the form of Outlook for Energy, used for assessing the business environment and future investments. The major change to be noted in these forecasts is that climate policy has gained a prominent place. Exxon anticipates a shift from coal to natural gas in particular, but also nuclear and renewable fuels through to 2030, driven by environmental policies, not least carbon prices. Exxon foresees a carbon price of USD 30 per tonne of CO_2 by 2020 and USD 60 by the year 2030 (ExxonMobil Outlook for Energy 2009).

3.3 Comparing Exxon's and Shell's Climate Strategies

The developments in the climate strategies of Shell and Exxon are somewhat mixed, but can generally be characterized by *convergence* over time, including elements of innovation. Exxon has shifted from a reactive stance into a more proactive strategy – particularly as regards monitoring and reporting, energy efficiency, specific goals, acceptance of international carbon pricing and responsibility for contributing to problem-solving. For the long term, Exxon foresees higher carbon prices and has stepped up its investment in low-carbon R&D, advanced biofuels, CCS and natural gas. With Shell, the picture is less clear-cut. On the one hand, Shell has tempered its once-innovative ambition to become an oil industry leader in renewables and has dropped new GHG mitigation goals. On the other hand, the company has stepped up its low-carbon R&D and collaboration on making CCS commercially viable and is still engaged in wind power.

4. Linking the ETS to Exxon's and Shell's Climate Strategies

The EU ETS is, as noted, the first mandatory international regulation that directly targets a significant share of the CO_2 emissions of oil companies. Both Exxon and Shell have responded actively to the introduction of the system, but in completely different ways.

4.1 ETS and Short-term Strategy

When the EU ETS was initiated in the late 1990s, Shell responded enthusiastically by shifting attention to the emerging new carbon market. The Shell Tradable Emission Permit System (STEPS) and incorporation of carbon prices in 2000 were implemented explicitly *in anticipation* of future regulated carbon markets (Shell

CDP 2008: 6). Shell participated actively in the initiation of the ETS at the time, making the upcoming system important for the company's anticipatory actions.

ExxonMobil, by contrast, opposed the introduction of the EU ETS. Private US interests lobbied intensively against ratification of the Kyoto Protocol and the emerging European emissions trading system. Shortly after the ET Directive was proposed in 2001, the American Council for Capital Formation (ACCF) published studies predicting the adverse consequences for the European economy and employment that would accrue if Kyoto Protocol targets were to be met by emissions trading in Europe. As recently as 2009, ExxonMobil contributed USD 25,000 to the ACCF (Exxon Contributions 2009), and the company has continued to oppose the EU ETS (Exxon, CDP 2009: 36). Cap-and-trade in general and the EU ETS in particular have been characterized as unnecessarily costly, complex and ineffective, with the allowance market seen as shifting the emphasis away from the goal of reducing carbon emissions (Tillerson 2010).

For the first (2005–2007) and second (2008–2012) trading periods under the ETS, both Shell and ExxonMobil received all allowances for free. The two companies have also tended to hold positions close to actual emissions or long positions from 2005 to 2010 with a long position indicating a lenient cap. Nevertheless, Shell had anticipated fewer allowances than expected emissions (Shell Report 2005: 8), which forced some facilities to invest in emissions reductions and encourage trading in surplus allowances. When the EU ETS was officially adopted in 2003, Shell stated that the system would affect nearly one third of the company's global GHG emissions. Shell was confident that its own energy-efficiency programmes and experience of Shell Trading would position the company well in the new market (Shell Report 2004: 8). Since 2003, Shell had been preparing for the EU ETS by 'developing the business processes required, *identifying potential emission reduction projects* and building capacity in Shell Trading' (Shell Report 2005: 9, emphasis added).

In 2005, the ongoing energy efficiency programme (Energise™) was integrated into the new Business Improvement Review (BIR). A three-year capital investment programme was launched specifically to boost energy efficiency at refineries (Shell Report 2005: 9). This would indicate that the EU ETS had actually strengthened ongoing abatement programmes. Still, energy efficiency at Shell's refineries declined after 2004, in part because extra energy is required to produce more environmentally-friendly lower-sulphur fuels (Shell Report 2009). This point is related to a conflict between EU requirements on fuel quality and CO_2 emissions. This conflict may be aggravated by the 2009 EU Fuel Quality Directive, which specifies the quality of petrol, diesel and gas/oil. As noted, more energy is needed to produce more environmentally-friendly fuels that are lower in sulphur.[16]

---

16 Other reasons for low efficiency include unplanned shutdowns requiring extra energy to re-start, and drops in demand linked to the global financial crisis, causing installations to run below full production capacity and hence less efficiently (Shell Report 2009).

Exxon opposed the ETS, but had to adapt since the system was mandatory. The initial impacts of the mandatory ETS cap on Exxon were related to emission monitoring, reporting and verification. In 2003, the company began systematic reporting of its GHG emissions. The 'appropriate measurement of overall emissions', stated the company, is part of 'preparatory work' regarding regulations (Exxon Citizen Report 2004). Exxon's Rotterdam refinery sought to play a leadership role in establishing monitoring protocols for the refining industry in Europe. Esso Nederland BV worked closely with Dutch authorities to develop monitoring protocols for CO_2 and NO_x in preparation for the EU ETS (Exxon Citizenship Report 2004: 40). The German petroleum industry adopted the protocol's general structure (Exxon Citizenship Report 2005). Since that time, the EU's approach has become more harmonized for the second and third trading periods.

Like Shell, ExxonMobil is heavily exposed to the EU ETS: as of 2005, the system covered nearly 90 installations operated by Exxon or its joint-venture partners (Exxon Citizen Report 2005: 38). Exxon was able to meet its obligations without purchasing extra allowances. Gains from the sale of allowances in 2005 (before the carbon price collapsed) were offset by the cost of administrative procedures and rising costs of electric power (Exxon, CDP 2008). In 2007, Exxon consumed energy at a cost of USD 10 billion, representing 15 per cent of the company's total operating expenses.

ExxonMobil has improved energy efficiency at its refineries since launching the Global Energy Management System (GEMS) in 2000. According to Exxon, energy efficiency has grown two to three times faster than the industry average (Exxon, CDP 2010: 11) not least because Exxon's refineries are over 60 per cent larger than the industry average (ExxonMobil Financial & Operating Review 2009: 79). Larger capacity means more flexibility to optimize operations and produce high-quality products with lower feedstock and operating costs. Unlike Shell, Exxon does not espouse the argument according to which extra energy is needed to produce more environmentally-friendly lower-sulphur fuels at refineries. The ETS has not been the primary driver of energy efficiency programmes in Exxon, but has provided additional incentives for programmes already in place prior to the ETS.

Although this is difficult to verify, Exxon's cautious change in strategy in 2007 appears to have been related to the EU ETS, European climate policy and the significant attention to climate change that year. On 21 June 2007, the Texas-based company announced its strategy shift at the Royal Institute for International Affairs in London. The location was not accidental, as the European audience was clearly the prime target. According to the company, Exxon's acceptance of the problem of human-induced climate change as a reality, and responsibility for helping to find a solution, was linked to new and deeper knowledge about the causes and consequences of climate change, which became apparent between the 2001 Third IPCC Assessment report and the Fourth Assessment report on climate change in 2007 (ExxonMobil 2007).

As noted, the oil industry, including Exxon and Shell, has been generally satisfied with the new benchmarks for the period 2013 to 2020. However, benchmarking may act to phase out the least efficient installations over time. Most of Exxon's and Shell's installations will probably qualify for the most efficient 10 per cent of installations in terms of GHG emissions in the EU sectors 2007/2008. As to GHG targets, the EU ETS and EU climate policy goals have made Shell unwilling to adopt new company-specific targets extending beyond 2010.

4.2 ETS and Long-term Strategy

The main importance of the EU ETS for the oil industry lies in the long-term strategic consequences of political agreement on carbon pricing. The EU ETS has shown that governments have the willingness and ability to develop international climate policies that send out a price signal. Oil companies must now take into account that EU policies can be widely copied elsewhere over the longer term, despite short-term setbacks in, e.g., the USA.

Shell introduced carbon pricing in projects in 2000 in anticipation of regulated carbon markets. The system prices CO_2 emissions consistently into project economics while recognizing that each investment decision is unique (Shell, CDP 2009). In 2010 and 2011, the planning premise for CO_2 was USD 40 per tonne. While this figure is not a forecast, it is used by Shell in economic modelling before taking investment decisions. It helps to lower CO_2 emissions by building measures into the design of new projects to avoid expensive retrofits. The planning premise shows that 'over-allocation' of allowances in the second phase of the ETS has not led Shell to downward adjustment of future carbon prices.

The main consequence of incorporating carbon costs linked to the EU ETS is that it makes low-CO_2 projects in Europe more attractive. This approach has, according to Shell, helped to drive the design of new facilities or expansion of existing facilities towards optimal performance (Shell, CDP 2008: 18). With a price on carbon, new installations will be designed to be more energy-efficient than otherwise. For a highly concrete example of innovative consequences, and problems, we can take the CCS plans for Shell's Pernis refinery in Rotterdam. This project is technologically feasible and economically profitable because of the price on carbon, but has not proven politically acceptable due to local opposition.

The EU ETS has also stimulated Shell's increasing activity in CCS. First, the EU ETS is the principal policy instrument for facilitating CCS projects within the EU (IEA/OECD 2008). The EU ETS recognizes captured emissions as CO_2 that is not emitted. In addition, the revised system mandates the use of 300 million allowances from the New Entrants Reserve (NER-300) to support up to 12 CCS demonstration projects and projects demonstrating renewable energy technologies. These resources are in turn linked to EU technology supply in connection with CCS technology (ZEP 2008).

In June 2010, the EU launched the first-ever European Industrial Initiatives (EII) based on the Zero Emissions Platform and including CCS. Wind, solar (photovoltaics and concentrated solar power) and electricity grids are also part of EII. The network is open for interested projects and companies, and is based on a knowledge-sharing protocol to facilitate such exchange not only in Europe, but also the global CCS community. The EU ETS will, it follows, underpin Shell's CCS strategy by learning from demonstration projects as part of a broader package of measures to facilitate CCS. Shell has been cited as one of the architects behind the NER-300 and, as noted, currently holds the chairmanship in ZEP. Shell cites its motivation for cooperation with competitors on CCS innovation as relating mainly to the economic opportunities. This is also the case concerning the general increase in cooperation with governments at various levels and NGOs, which may contribute to shaping acceptance and legitimacy for new commercial (low-carbon) solutions. Shell's failed CCS project at the Pernis refinery exemplifies the importance of building social acceptance.

Turning to ExxonMobil, we see that the impacts of the EU ETS long-term strategy in carbon pricing have been more indirect than in the case of Shell. Exxon has not reported factoring in a specific CO_2 price. However, the company does conduct 'sensitivity' analyses in the evaluation of capital projects when GHG emissions and regulations are also included in the analysis (Exxon, CDP 2009: 32).

The EU ETS and expectations of diffusion to the USA and other parts of the world are an important reason why Exxon now includes climate policy and carbon prices in its forecasts. According to Exxon, nothing will happen without a carbon price. Climate change and an expected increase in carbon prices underpin the company's R&D strategy with regard to CCS, energy efficiency and natural gas (Exxon, Outlook for Energy 2009). At USD 30 per tonne of CO_2, natural gas would become the most economical alternative for new power plants. As the price increases, Exxon anticipates fuel switching from coal to natural gas (by running natural gas plants at higher load factors), as well as the building of new natural gas plants and retiring old coal plants. The recent acquisition of the largest natural gas producer in the USA, XTO, was in line with Exxon's Energy Outlook highlighting the increasingly important role of natural gas in supplying the world's energy needs that also can help reduce CO_2 from power generation (Lamp 2010). This fits well with the considerable commercial opportunities in exploiting the huge shale gas reserves in the USA.

The EU ETS is directly linked to the purpose of the CO_2 Remove Project. Exxon does not participate in ZEP, but is involved in the CO_2 Remove Project funded by the European Commission under the Sixth Framework Programme for research. An important goal is to formalize a European methodology for qualifying CCS in the EU ETS.[17] Exxon's participation includes financial support and the provision of expert technical guidance. The project also seeks to disseminate information about the results to scientists, industry, decision-makers and other CCS stakeholders.

---

17 Available at: http://www.co2remove.eu/

In conclusion, then, we may say that the EU ETS helped turn Shell's attention to a new area of opportunities. These opportunities were mainly in the new carbon market, where Shell aimed to act as an 'early mover'. The ETS also strengthened short-term abatement programmes, but other EU policies and factors not directly related to climate policy have conflicted with CO_2 reduction. The ETS has facilitated long-term investment, particularly in CCS, by setting a price on carbon and facilitating learning through EU technology programmes. As for Exxon, we have seen how it initially opposed the ETS and came round only reluctantly. Still, the ETS has prompted the company to adopt a more proactive strategy. First, it did so by mandating Exxon to monitor, report and verify its GHG emissions. Second, the ETS has contributed to the company's stated acceptance of international carbon prices and facilitated learning and diffusion of knowledge on monitoring and CCS. Finally, and perhaps most importantly, the ETS has contributed in changing Exxon's views as to the future, with a price on carbon emerging as an important premise of long-term forecasting.

5. Company-internal and Company-external Factors

The development of climate strategies in Shell and ExxonMobil has, as noted, has been characterized by convergence over time, and the EU ETS has been one important factor here. But we have also seen some striking differences. Why did Exxon and Shell respond so differently to the EU ETS? Why did Exxon initially oppose the system and then adopt a more proactive stance? Why did Shell lose some of its enthusiasm, particularly in the area of renewables? And why is Shell more engaged than Exxon in joint EU innovation aimed at commercializing CCS?

As both Shell and ExxonMobil have received relatively generous allocations of allowances for free, differences in regulatory pressure are not a likely explanation of the varying responses. Let us start by looking at external factors other than the ETS. Both Exxon and Shell are high-profile multinational oil companies exposed to wide range of risks and opportunities around the world. They both have to deal with regulatory diversity in all the countries they operate in, and have experienced significant public criticism related to social and environmental incidents and accidents (Skjærseth and Skodvin 2003, Skjærseth et al. 2004). High exposure to societal and public activity may be important for understanding how these companies behave, but seem unlikely to explain differences in climate strategies and ETS response. However, the national context of ExxonMobil and Shell's home-base countries has previously been found to be an important explanatory factor for Shell's adoption of a proactive climate strategy and for Exxon's reactive strategy before the ETS was introduced (Skjærseth and Skodvin 2003). In Europe, relatively high public demand and governmental supply of climate policies based on public–private participation informed Shell's proactive strategy. Shell is also a Dutch–British company – based in two countries which have supported the idea of emissions trading in Europe since the ETS was planned. In the USA, the national context facilitated Exxon's strategy of resistance. Different strategies adopted prior to the ETS obviously fed into these companies' initial responses to the new system.

Also, later US climate policy discussions can shed light on Exxon's opposition to cap-and-trade. During the 2008 election campaign, presidential candidate Obama spoke of making the USA a leader in global efforts to combat climate change; his ambition was to establish an economy-wide cap-and-trade system in the USA (Skodvin and Andresen 2009). There had already been several state-led initiatives of the same type; Exxon had opposed these initiatives and programmes. Responding to Exxon's tax proposal, Greenpeace called it a tactical manoeuvre in the debate on cap-and-trade, as no US politician would propose something with the word 'tax' in it (Climate Scam, 2009). There may be significant merit to this argument in light of the failure of the British Thermal Unit (BTU) tax proposed by the Clinton/Gore administration in the early 1990s. The BTU tax was based on the heat content of the fuel and aimed to stimulate energy efficiency and cut federal deficit. The tax was, however, rejected in the Senate, despite the Democratic majority in Congress. The energy tax provoked the US oil industry, which played a crucial role in killing it (Skjærseth and Skodvin 2003: 118). Exxon's position has also split the US oil majors, as Chevron and ConocoPhilips support emissions trading through their membership in the International Emissions Trading Association.

Company-internal factors are equally important for understanding climate strategies. ExxonMobil is often described as a 'super-tanker' that keeps to its core business and changes course only gradually. It is also a very centralized company, with all major investment or strategy decisions taken at company HQ in Irvine, Texas. Between 1993 and 2006, ExxonMobil was led by Lee Raymond, known for his top-down management style. He has been described as a conservative corporate leader notoriously sceptical of government intervention. He has reportedly described European suggestions that Americans should use smaller cars as neo-colonialism: 'Most Americans like to make that decision themselves – that's why they left [Europe]' (Skjærseth and Skodvin 2003: 97). In 2006, Raymond was replaced by Rex Tillerson, who now serves as both chairman and CEO. The green movement saw some hope in Tillerson: he couldn't be worse, they believed, than Raymond. That same year, Tillerson said the company could have done a better job of putting the climate case forward (Sox First 2006). Another difference is that Tillerson to a larger extent communicates the reasons for his decisions, placing more emphasis on knowledge about climate change. Exxon's more proactive strategy can thus partly be understood as an exercise in adjusting company rhetoric to actions and knowledge, for instance linking energy efficiency and CCS to climate-change mitigation.

A 'culture' of *innovation* is likely to stimulate an innovative response to new regulation. Shell and Exxon are highly innovative companies within the petroleum industry. Both increased their R&D investments steadily from 2003 to 2008. In 2009, Shell was investing most in R&D of the EU oil and gas majors, ranking number 33 among the top 1000 EU companies, independent of sector. The corresponding figures for ExxonMobil are number 2 of the non-EU oil and gas companies and 74 in the top 1000 non-EU companies independent of sector (EU Industrial R&D Investment Scoreboard 2010).[18] High R&D expenditures are in line with observations of both companies' involvement in innovation, particularly CCS and advanced biofuels. However, this factor can hardly explain the differences in their climate strategies and ETS responses.

A related explanation of why Shell has adopted a more cooperative and commercial attitude to CCS innovation than ExxonMobil may be because of different attitudes to government intervention. Exxon is generally sceptical of government intervention, but accepts public support to pilot projects. Still, the company strongly believes that 'market prices should drive the selection of solutions' (Exxon, CDP 2009: 35). This is also reflected in the words of the CEO, Rex Tillerson. '[P]ublic policy must not ... add market uncertainties by picking winners and losers. Good policy sets aspirational goals ... and then provides the broad framework for entrepreneurs and innovative thinkers to achieve these goals' (Exxon, CDP 2010: 17). This argument is also common among the European oil majors, but Shell does not stress technology neutrality to the same extent. Shell is also more of an 'outward-looking' company than Exxon. For example, Shell responded more intensively to the widening Corporate Social Responsibility (CSR) agenda before the introduction of the ETS by emphasizing the fight against corruption, support of the universal Declaration of Human Rights, and cooperation with NGOs and international organizations (Skjærseth et al. 2004).

Finally, carbon intensiveness and exposure to global competition are not likely to explain the differences in response to the ETS. Exxon and Shell have quite similar portfolios inasmuch as their key business areas are oil and gas exploration, production and refining. With regard to exposure to global competition, the main markets of both companies are the USA and Europe. Their exposure to the EU ETS in Europe is, as noted, quite similar. However, the global financial crisis widened the gap between the two companies' uneven economic performance.

Shell results show a drop in income from USD 26 billion in 2008 to USD 13 billion in 2009. Returns on average capital employed decreased from 18.3 per cent in 2008 to 8.0 per cent in 2009 (Shell Annual Review 2009:6). In the words of the CEO, 'Our 2009 earnings were sharply reduced by the recession, despite Shell's self-help programs and USD 2 billion of costs savings. Although oil companies have been cushioned from the recession by OPEC's action on quotas and oil prices, Shell has been disadvantaged recently, due to our higher exposure in refining and natural gas, where margins are hard-wired to the economy' (Shell Media Releases 2010). Shell responded to the downturn by restructuring the company and making it more competitive (Shell Annual Review 2009). Its divestment of renewables started before the financial crisis, but the decision to drop new investments in renewables was accelerated by the recession. According to Shell's head of gas

---

18   Available at: http://iri.jrc.es/research/scoreboard_2010.htm

and power activities, 'wind and solar struggle to compete with other investment opportunities we have in our portfolio' (*The Times* 2009).

ExxonMobil's superior financial strength has made it one of the few public companies with the highest credit rating in decades. In 2009, it reported strong earnings of USD 19.3 billion and industry-leading returns on average capital employed of 16 per cent. The corresponding figures for 2008 were record earnings of USD 45.2 billion and returns on average capital employment of 34 per cent. Even though the 2008–2009 decline was significant, Exxon sees the 2009 results as representing a unique competitive advantage in 'today's challenging economic environment' (ExxonMobil Financial & Operating Review 2009: 5). Exxon's acquisition of XTO exemplifies this advantage.

In conclusion, we have seen how climate strategies and responses to the ETS are conditioned by other external and company-internal factors. Differences in home-base countries, attitudes to governmental regulation, economic strength, the financial crisis as well as changes in leadership can all shed light on change and differences in climate strategies and responses to the ETS.

6. Discussion

Our first observation is that major oil companies have responded to the EU ETS by changing or adjusting their climate strategies. At the aggregate sector level, we see a pattern of convergence in the companies' declared long-term, low-carbon solutions, particularly CCS. The industry association Europia and the major companies place greater emphasis on long-term problem-solving, linking these initiatives to changes in the regulatory environment, particularly the ETS but also the EU's climate targets and the 2008 EU climate and energy package. Our study of Shell and Exxon Mobil has shown the companies responding by variously adjusting their short-term and long-term strategies. They have strengthened monitoring and reporting, incorporated carbon costs in investment decisions, scaled up low-carbon R&D and CCS projects and included carbon costs in forecasts used to inform their investment decisions. The gradual change in Exxon's climate strategy towards a more proactive approach has also been conducive to Europia's more proactive position, at least temporarily. For the foreseeable future, however, the main bulk of investments will still be made in fossil fuel exploration and production, to meet rising global demands for energy.

It is extremely difficult to say precisely to what extent these changes can be attributed to the ETS. As high-profile global companies significantly exposed to environmental risks, many oil companies had climate strategies in place before the ETS was adopted. The ETS has mainly worked together with other drivers and strengthened these existing strategies, particularly in the companies that lagged furthest behind when the ETS was introduced. Second, it is also extremely difficult to separate the effect of the ETS from other elements of EU climate and energy policy, particularly with regard to long-term strategies. Still, the main impact of the EU ETS on oil company climate strategies appears to be the long-term consequences of the political willingness to put a price on carbon.

Various expectations have been presented, describing how environmental regulations like the ETS would affect company strategies (see Chapter 2). The first expectation was that the ETS would lead to resistance. A reactive strategy, based on the least-costly option, would be likely to express itself as reluctant adaptation to the ETS over the shorter term. This expectation gains some support in our material. Exxon's initial opposition to the ETS and the company's implementation of monitoring and reporting procedures are an expression of reluctant adaptation to mandatory requirements. However, Exxon's active engagement in developing monitoring procedures for the sector, its participation in EU research on CCS and the incorporation of carbon prices in its future planning all represent something more (long-term) than this proposition would indicate.

The second expectation, based on the Porter Hypothesis, states that the ETS will stimulate companies to search for new opportunities and promote more proactive and innovative climate strategies. Bearing in mind the relatively short time that has passed since the introduction of the ETS, this expectation gains support, in that the response of the industry association and major companies has been increasingly proactive and innovative as more activities in CCS and advanced biofuels were signalled. Most of these activities are not entirely new to the companies, but they include new elements – as with EU CCS pilot projects, Shell's CCS plans for the Pernis refinery, supply of CO_2 to local greenhouses and allowance trading. Moreover, Shell and Exxon have increased or changed their spending on low-carbon R&D. Still, we cannot conclude that the ETS has led to any comprehensive large-scale innovation efforts as yet.

The Porter Hypothesis is based on the assumption that regulation is 'stringent' and 'appropriate'. Cap-and-trade like the EU ETS is used by Porter as an example of appropriate regulation. The EU ETS has also been significantly strengthened over time. Still, thus far the system has been too lenient, too recent and too narrow to generate any large-scale, concerted and radical low-carbon change in the sector. The caps have not been sufficiently stringent; allowances have been distributed for free, and the rules on import of credits have been lenient. In addition, the carbon price has fluctuated significantly (mainly between 10 and 30 euro/tonne) and the global financial crisis has led to a surplus of allowances and downward adjustment in carbon price expectations until 2020. In addition, we must bear in mind that the ETS is a regional system that covers only a part of oil company activities worldwide.

The mechanisms through which the EU ETS works are clearly linked to economic incentives, costs and benefits. Incentives work, but how they work and to what effect can be harder to predict. Our analysis has shown that the interaction between the ETS and other policy instruments is important. The most concrete example of a negative effect is that more stringent regulations for improving fuel quality at refineries will require more energy, in turn leading to higher CO_2 emissions. The system also works though learning and diffusion. The ETS has, for example, instigated learning through the EU's technology platforms on CCS. Abatement knowledge and measures have been diffused by the companies spreading monitoring techniques, incorporating carbon prices and cap-and-trade in other parts of the world.

The final expectation posited that regulation in the form of cap-and-trade is likely to affect norms of responsibility. As noted in Chapter 2, the relationship between regulation and social norms of responsibility is extremely difficult to assess empirically. ExxonMobil has accepted the EU ETS as a reality, but still opposes emissions trading in general and the EU ETS in particular. Interestingly, the company substantiates this position by arguing that the allowance market shifts the emphasis away from the goal of reducing carbon emissions. This line of reasoning resembles the logic underlying the 'crowding out' effect (see Chapter 2), but we have found no indications that this has actually happened.

The case of Shell illustrates the complexity of studying the link between regulation and norms of responsibility. In the late 1990s, Shell aimed to become an energy company by capturing a 10 per cent share of the renewables market by 2005. This ambition formed part of a proactive climate strategy that included the ambition to be a trading leader within the oil industry. Since then, Shell has divested itself of renewables, and in 2009 the company stated that it intended to terminate all new investments in renewable energy. This is clearly not linked to the company's perception of the ETS, as Shell has strongly supported the ETS. The most reasonable interpretation is that Shell's engagement in renewables represented a strategic decision in line with the Porter Hypothesis. The divestment of renewables has also been a result of business considerations. The ETS has apparently not significantly affected norms of responsibility either way, and the system has been too weak and narrow to create sufficient incentives to counter the renewables divestment. What we can say with reasonable certainty is that the ETS has not 'crowded-out' voluntary initiatives that had been adopted before the system was introduced.

Why did ExxonMobil and Shell adopt different climate strategies and respond differently to the ETS? Four tentative observations that are compatible with different perspectives of company behaviour can be noted. First, the US climate policy context is important for understanding ExxonMobil's strategy of resistance prior to the EU ETS and its opposition to the new system. Similarly, the European climate policy context is important for understanding Shell's proactive strategy prior to the EU ETS and its opportunity-based response to the system, once implemented. Second, the change in Exxon leadership appears to have had some effect on climate strategies. Third, different principles on governmental intervention seem to have affected the companies' positions on CCS cooperation aimed at commercialization. Finally, the global economic crisis represented an 'external shock' that affected Exxon and Shell differently, due to the companies' differing economic performance and portfolios. As a result, Shell decided to halt new investments in renewable energy despite a price on carbon and high EU ambitions as to renewables.

7. Concluding Reflections

The EU aims to put Europe on track toward a low-carbon economy by 2050. This will require long-term low-carbon strategies in the private sector and large-scale joint innovation based on a combination of individual commercial interests and social responsibility for collective problem-solving. The key climate policy instrument in this striking challenge is the EU Emissions Trading System (ETS), the main purpose of which is to reduce GHG emissions in a cost-effective way. This chapter has explored the consequences of this system to date on the climate strategies of oil companies, with a specific focus on the oil majors ExxonMobil and Shell. Stimulating such major oil companies to adopt innovative solutions will be important for realizing a low-carbon economy in Europe and abroad.

The following main conclusions can be drawn. First, oil companies base their investments on long-term strategies in the form of forecasts or scenarios. These long-term outlooks show that global energy demand and GHG emissions will increase significantly outside Europe and the OECD. Accordingly, oil company investments for the foreseeable future will be made mainly in fossil fuels, to meet this demand. Nevertheless, carbon prices have emerged as important for the type of energy mix foreseen in the crystal bowl. The ETS has affected these companies' beliefs and outlooks about the future, and has shown that 27 EU governments have had the willingness and capacity for carbon pricing. The companies now expect that carbon pricing through trading or taxes will diffuse around the world over the long term. As these companies are global, international carbon prices will be needed to move this industry significantly towards low-carbon solutions. Second, the impact of the ETS has been most significant for the companies that have lagged furthest behind the leaders. This has contributed to make the position and role of the industry association, Europia, more proactive – and this aggregate effect is important, as Europia represents the oil industry in the EU. On the other hand, this may well be only a temporary effect. The recession, structural changes and the EU's long-term plan for decarbonizing Europe have made the oil industry increasingly reluctant to adopt any new binding policies beyond 2020.

Third, the EU is developing a technology-supply side of its climate policy in the form of technology platforms to stimulate innovation in low-carbon solutions. These initiatives are linked to the EU ETS through the financing mechanism based on sales of ETS allowances. Such initiatives can be used as a 'sweetener' to get companies to engage in joint innovation. Finally, most companies in this sector have now acknowledged the problem of climate change, they admit their responsibility for helping to deal with it, and have initiated various voluntary initiatives. Still, this study has not been able to establish any significant relationship between the EU ETS and social norms of responsibility. Understanding how cap-and-trade and other types of policy instruments can 'crowd-in' or 'crowd-out' social norms of responsibility is important for achieving EU long-term goals. This topic represents a promising avenue for future research.

Interviews

Chris Beddoes, European Petroleum Industry Association, Europia (15 September 2011, personal).

David Hone, Shell Climate Change Advisor (29 June 2011, telephone).

Hans van der Loo, Head European Union Liaison, Shell International (13 April 2011, personal).

Ingvild Skare, Environmental Advisor, ExxonMobil Exploration and Production Norway AS (02 March 2011, telephone).

International Europe (22 February 2011, telephone).

Norbert Herlakian, ExxonMobil, R&S Climate Change Advisor, EMEA Biofuels Venture Mgr. Brussels (12 April 2011, personal).

Tomas Wyns, EU ETS policy officer, Climate Action Network Europe (19 April 2011, telephone).

Trym Edvardson, Environmental Discipline Specialist, Shell Upstream Yvon Slingenberg, European Commission DG Climate Action (12 September 2011, personal)

References

BP (British Petroleum). 2011. BP Energy Outlook 2030. Available at: http://www. bp.com/sectiongenericarticle.do?categoryId=9035979&contentId=7066648

Climate Scam. 2009. Exxon vill ha koldioxidskatt. Available at: http:// theclimatescam.se/2009/01/10/exxon-vill-ha-koldioxidskatt/

CO$_2$ REMOVE. Available at: http://www.co2remove.eu/

Community Transaction Log. Available at: http://ec.europa.eu/environment/ets/ oha.do

CONCAWE (The oil companies' European association for environment, health and safety in refining and distribution). 2008. *Impact of Product Quality and Demand Evolution on EU Refineries at the 2020 Horizon – CO$_2$ Emissions Trend and Mitigation Options*. Brussels: CONCAWE, December 2008.

CONCAWE. 2011. *The Potential for Application of CO$_2$ Capture and Storage in EU Oil Refineries*. Brussels: CONCAWE, October 2011.

Ellerman, A.D., Convery, F.J. and de Perthuis, C. 2010. *Pricing carbon. The European Union Emissions Trading Scheme*. Cambridge: Cambridge University Press.

EU Industrial R&D Investment Scoreboard. 2010. Institute for Prospective Technological Studies (IPTS). Available at: http://iri.jrc.ec.europa.eu/research/ scoreboard_2010.htm

European Commission. 2006. *EU ETS Review: Report on International Competitiveness*. Brussels: DG Environment, 2006.

European Commission. 2010a. Draft decision of criteria for the financing of commercial CCS demonstration projects. D007721/02. Brussels.

European Commission. 2010b. Commission staff working paper on refining and the supply of petroleum products in the EU. SEC(2010) 1398/2. Brussels.

Europia. 1999. *Activity Report.* Brussels: European Petroleum Industry Association.

Europia. 2007. *EUROPIA position paper on the review process of the EU ETS Directive.* Brussels: European Petroleum Industry Association.

Europia. 2008. *Position paper on the proposed directive amending Directive 2003/87/EC.* Brussels: European Petroleum Industry Association.

Europia. 2010a. *White paper on EU Refining: A contribution of the refining industry to the EU energy debate.* Brussels: Europia.

Europia. 2010b. *Annual Report 2010.* Brussels: Europia.

Europia. 2011a. *Europia Contribution to EU energy pathways to 2050.* Brussels, Europia.

Europia. 2011b. *Europia White Paper on Fuelling EU Transport.* Brussels: Europia.

Exxon Citizen Report. 2005. Available at: http://exxonmobil.com/corporate/

Exxon Citizen Report. 2007. Available at: http://exxonmobil.com/corporate/

Exxon Citizen Report. 2009. Available at: http://exxonmobil.com/corporate/

Exxon Contributions. 2009. *Exxon Mobil Corporation 2009 Worldwide Contributions and Community Investments: Public Information and Policy Research.* Available at: http://exxonmobil.com/corporate/

ExxonMobil. 2007. *ExxonMobil's response to publication of the IPCC Fourth Assessment Report Climate Change 2007: Climate Change Impacts, Adaptation and Vulnerability.* 6 April 2007. *Exxon Citizen Report.* 2004. Available at: http://exxonmobil.com/corporate/

ExxonMobil CDP. 2008. *Carbon Disclosure Project: Company Response.* Available at: https://www.cdproject.net/en-US/Pages/HomePage.aspx

ExxonMobil CDP. 2009. *Carbon Disclosure Project. Company Response.* Available at: https://www.cdproject.net/en-US/Pages/HomePage.aspx

ExxonMobil. 2009. *Financial & Operating Review 2009.* Available at: http://exxonmobil.com/corporate/

ExxonMobil. 2009. *The Outlook for Energy: A View to 2030.* Available at: http://www.exxonmobil.com/Corporate/files/news_pub_eo.pdf

ExxonMobil CDP. 2010. *Carbon Disclosure Project. Company Response.* Available at: https://www.cdproject.net/en-US/Pages/HomePage.aspx

Forbes, A. 2011. Low-carbon economy: Not in sight yet. *European Energy Review*, 3 February.

Gunningham, N., Kagan, R.A. and Thornton, D. 2003. *Shades of Green: Business, Regulation and Environment.* Stanford, CA: Stanford University Press.

IEA/OECD. 2008. CO_2 *Capture and Storage: A Key Carbon Abatement Option.* Paris: IEA/OECD.

IPIECA (The global oil and gas industry association for environmental and social issues). 2003. Available at: http://www.ipieca.org/

Lacombe, R.H. 2008. *Economic Impact of the European Union Emission Trading Scheme: Evidence from the Refining Sector*. Master's thesis, MIT, Cambridge MA.

Lamp, The. 2010. Biofuels program advances with opening of new facility. 2010 Number 2, Irving, Texas: ExxonMobil.

Lee, B., Iliev, I. and Preston, F. 2009. *Who Owns Our Low Carbon Future? Intellectual Property and Energy Technologies*. London: Chatham House.

OJEU (Official Journal of the European Union). 2010. Commission Decision of 24 December 2009, determining pursuant to directive 2003/87/EC of the European Parliament and of the Council, a list of sectors and subsectors which are deemed to be exposed to a significant risk of carbon leakage, 5 January 2010.

Porter, M.E and van der Linde, C. 1995. Green and competitive: ending the stalemate. *Harvard Business Review*, September/October, 120–34.

Schumpeter, J.A. 1934. *The Theory of Economic Development*. Cambridge, MA: Harvard University Press.

Shell. 1998. *The 1998 Shell Report*. Available at: http://www.shell.com/

Shell. 1999. *The 1999 Shell Report*. Available at: http://www.shell.com/

Shell. 2002. *The 2002 Shell Report*. Available at: http://www.shell.com/

Shell. 2004. *The 2004 Shell Report*. Available at: http://www.shell.com/

Shell. 2005. *The 2005 Shell Report*. Available at: http://www.shell.com/

Shell. 2008. *The 2008 Shell Report*. Available at: http://www.shell.com/

Shell. 2009. *The 2009 Shell Report*. Available at: http://www.shell.com/

Shell Annual Review. 2009. Available at: http://www.shell.com/

Shell CDP. 2008. *Carbon Disclosure Project. Company Response*. Available at: https://www.cdproject.net/en-US/Pages/HomePage.aspx

Shell CDP. 2009. *Carbon Disclosure Project. Company Response*. Available at: https://www.cdproject.net/en-US/Pages/HomePage.aspx

Shell CDP. 2010. *Carbon Disclosure Project. Company Response*. Available at: https://www.cdproject.net/en-US/Pages/HomePage.aspx

Shell Energy. 2008. *Shell Energy Scenarios to 2050*. Available at: http://www.shell.com/

Shell Fact. 2009. *Five-year Fact Book*. Available at: http://www.shell.com/

Shell Media Releases. 2010. *Royal Dutch Shell plc Updates on Strategy to Improve Performance and Grow*. 16 March 2010. Available at: http://www.shell.com/

Shell NZ. 2009. *Shell New Zealand's Submission to the Emissions Trading Scheme Review Committee*. 13 February 2009. Available at: http://www.shell.co.nz/

Skjærseth, J.B. 1994. The climate policy of the EU: Too hot to handle? *Journal of Common Market Studies*, 32(1), 25–45.

Skjærseth, J.B. 2011. Towards a low carbon economy? EU emissions trading, corporate climate strategies and the oil industry. Paper prepared for presentation at the International Studies Association 52th Annual Convention, Montreal, Canada, 16–19 March.

Skjærseth, J.B. and Skodvin, T. 2003. *Climate Change and the Oil Industry: Common Problem, Varying Strategies.* Manchester: Manchester University Press.

Skjærseth, J.B., Tangen, K., Swanson, P. et al. 2004. *Limits to Corporate Social Responsibility: A Comparative Study of Four Major Oil Companies.* FNI Report 7/2004. Lysaker: Fridtjof Nansen Institute.

Skodvin, T. and Andresen, S. 2009. An agenda for change in US climate policies? Presidential ambitions & congressional powers. *International Environmental Agreements: Politics, Law and Economics*, 9(3), 263–80.

Sox First, Management and Compliance. 2006. Exxon Mobil, climate change and the reputation wars, 20 May. Available at: http://www.soxfirst.com/50226711/exxon_mobil_climate_change_and_the_reputation_wars

Steger, U. 1993. The greening of the board room: How German companies are dealing with environmental issues, in K. Fischer and J. Schot (eds), *Environmental Strategiesfor Industry: International Perspectives on Research Needs and Policy Implications.* Washington, DC: Island Press.

Tillerson, R. 2010. Remarks by Rex W. Tillerson, chairman and CEO. Available at: http://www.exxonmobil.com/Corporate/energy_climate_views.aspx

Times, The (London). 2009. Anger as Shell reduces renewables investment. 18 March.

Tjernshaugen, A. 2010. Exxon planla storstilt CO_2-lagring. *Klima* 2–2010. Oslo: CICERO.

Webb, T. 2007. *Strategy: Titans clash on a shifting battleground – The War Goes on for Exxon Mobil.* Greepeace, ClimateChangeCorp. Available at: http://www.climatechangecorp.com/content.asp?ContentID=4858

Winter, S.G. 2000. The satisficing principle in capability learning. *Strategic Management Journal*, 21, 981–96.

Worrel, E. and Galitsky, C. 2005. *Energy Efficiency Improvements and Cost Saving Opportunities for Petroleum Refineries.* Energy Analysis Department, Environmental Energy Technologies Division, Ernest Orlando Lawrence Berkeley National Laboratory, University of California at Berkeley.

ZEP. 2008. *EU Demonstration Programme for CO_2 Capture and Storage (CCS).* Brussels: European Technology platform for Zero Emission Fossil Fuel Power Plants (ZEP).

Chapter 5

Pulp and Paper Industry

Lars H. Gulbrandsen and Christian Stenqvist

1. Introduction

The EU ETS was the first international policy instrument to introduce regulation of CO_2 emissions of pulp and paper companies with installations in Europe. Out of 11,500 installations introduced to the system about 900 were pulp and paper mills. In terms of allocated emission allowances (EUAs) the pulp and paper industry (hereafter PPI) represents two per cent of the EU ETS (Hyvärinen 2005: 40). The ability of the ETS to induce companies to adopt more proactive and innovative abatement strategies represents a critical test of the EU's ability to realize a low-carbon economy. This chapter examines to what extent and how the ETS has influenced the climate strategies of major pulp and paper companies. Thus far, there have hardly been any in-depth studies of the impact of the ETS on the PPI. One notable exception is a study by Rogge et al. (2011). Based upon an examination of survey data of paper producers and technology providers in Germany, this study found that innovation activities in the PPI are mainly governed by market factors and are hardly affected by the ETS and other climate policies. However, given that the ETS is the first regulation directly targeting the CO_2 emissions from pulp and paper companies in Europe, we were puzzled by the finding that the ETS had apparently so little effect on innovation activities. We suspect that an examination of survey data may miss critical aspects of corporate responses to the ETS.

In our study, therefore, we conduct a closer examination of the effect of the EU ETS in the pulp and paper sector by analysing changes in aggregate climate strategies and changes within two comparable, yet different, pulp and paper companies. These aggregate changes are analysed by investigating the EU ETS position of the Confederation of European Paper Industries (CEPI), the industry association of the European PPI, and the climate strategies of the ten largest pulp and paper companies in the EU ETS. The purpose of this sector study is to examine the sector response to the ETS as well as trends in the development of climate strategies within the sector.

The sector study is followed by a study of two pulp and paper companies – Svenska Cellulosa Aktiebolaget (SCA) and Norske Skogindustrier ASA (Norske Skog) with their respective headquarters in Sweden and Norway – to shed more light on the short- and long-term effects of the EU ETS. Both companies seem to have quite progressive climate strategies, having been named the best Swedish and the best Norwegian company in terms of CO_2 emission accounting in a recent appraisal by the Carbon Disclosure Project (CDP 2010). Studying these two environmental frontrunners enables us to examine both direct and more indirect, subtle and complex causal effects of the ETS on corporate climate strategies. Another reason for selecting these two companies is that they also seem to display a certain variation in climate strategies and development over time, with SCA apparently experimenting more with innovative abatement projects than Norske Skog does. This variation enables us to explore under what conditions different climate strategies emerge.

We find that increasing electricity prices are perceived to be the strongest influence on the scheme. For both SCA and Norske Skog, company-wide CO_2 emission objectives existed prior to the introduction of the EU ETS, as did systems for site-specific emissions monitoring. However, due to its influence on electricity prices, the EU ETS has reinforced commitments to improve energy efficiency and reduce emissions. Reporting of emissions has also been extended to include more staff hours at the mills. The value of CO_2 emissions is recognized and accounted for by SCA and Norske Skog, but the carbon price tag still represents only a minor factor out of a larger set of factors that underpin industrial investment decisions.

In section two, we provide some sector characteristics, focusing on the pulp and paper production processes, CO_2 emissions and abatement options. Section three examines changes in the EU ETS position of CEPI and the climate strategies of the ten largest pulp and paper companies in the EU ETS. In section four, the climate strategies of SCA and Norske Skog are presented and discussed in greater detail. Section five examines the link between the EU ETS and changes in corporate climate strategies, while section six discusses factors that have conditioned corporate responses to the system. In section seven, we discuss our findings in light of the analytical framework of this book. The closing section offers some conclusions and reflections on the outlook of EU emissions trading and the European PPI.

2. Production Processes and Abatement Options

Globally, the PPI is the fourth largest industrial energy user. However, its greenhouse gas (GHG) emission intensity is relatively low since half of the fuels come from biomass sources (IEA 2007). In 2009, the European PPI had a total fuel consumption of 1.18 EJ, composed as follows: biomass 53 per cent, natural gas 37 per cent, fuel oil 4 per cent, coal 4 per cent, and other fuels 2 per cent. Electricity use was 102 TWh of which 49 TWh was produced on-site. Energy-related fossil CO_2 emissions were 45 $MtCO_2$, of which 34 $MtCO_2$ were direct emissions at the mills and 10.5 $MtCO_2$ were indirect emissions from purchased energy. Since 1990, despite a 40 per cent increase in production, fossil CO_2 emissions have been cut by 15 per cent, in absolute figures (CEPI 2011a).

The PPI includes both integrated pulp and paper mills able to carry out the whole value chain from wood to final product, and stand-alone pulp mills that produce market pulp for separate paper mills. Pulping separates the wood fibres from each other. The two main processes are chemical and mechanical, accounting for 68 and 31 per cent, respectively, of European pulp production (CEPI 2011a). In chemical pulping, the wood chips are cooked in chemicals to dissolve the lignin, which gives a yield of 50 per cent fibre from wood input. The lignin goes into a stream of spent pulping liquor for reuse of the cooking chemicals and combustion in a recovery boiler, which generates both electricity and steam. Additional steam demand can be supplied from internal resources like bark and tar oil, or from external fuels like biomass, oil and natural gas. The main use of chemical pulp is for producing high-quality fibres for packaging paper and sanitary products that requires high absorbency capacity and brightness.

In the electricity-intensive mechanical pulping, the wood structure is decomposed in large grinders, which gives a high yield of up to 95 per cent fibre from wood input. Most of the electricity used for grinding the wood chips is transformed to heat which can be recovered and used for drying the pulp or paper at a later stage (IEA 2007). The main use of mechanical pulp grades is for production of newsprint and magazine paper. In addition to virgin pulp, half of the fibre input to European paper production comes from recovered paper, which involves less energy-intensive processing (CEPI 2011a).

More detailed categorization of paper and board products is possible, but here the point is that all paper machines involve the same essential steps: wet end, press section and drying section (IEA 2007). Large amounts of steam are required to bring the moisture content down in the diluted pulp. After the pulp mixture has been spread out on a screen and dewatered by gravitation, the remaining water is removed stepwise in the press and drying sections. In integrated mills the heat recovered from the pulping process is utilized for this purpose. Movement through the long machine is facilitated by a large number of electric motors (IEA 2007). At the end of the machine the paper can be coated in various ways before being prepared for shipment.

The use of carbon-neutral biomass fuels (black liquor, bark etc.) is highest in countries like Sweden and Finland where there is ready access to biomass feedstock. Also, due to the large share of chemical pulp production, which facilitates internal use of biomass for energy generation, fossil fuels account for only 10–20 per cent of the PPI's fuel consumption in these countries. In Italy, the UK, Spain and Germany, the PPI is largely dependent on natural gas. Access to biomass fuels is limited and the main production capacity consists of stand-alone paper mills that use recovered fibre and market pulp as feedstock (Ecofys 2009). In essence, the energy demand and CO_2 emissions of the PPI are determined by the paper grade, requirements as to fibre quality, and the access and choice of fuels. Pulp and paper can be produced without fossil CO_2 emissions if internal biomass sources are converted to electricity with high efficiency and biomass fuels are used for the remaining energy needs (IEA 2007). As yet there are only a few examples of chemical pulp mills that are self-sufficient in heat and electricity while being net exporters of energy carriers (e.g. electricity, district heating, refined bio-fuels)

(Bengtsson 2010). Even these mills may use some fossil fuels to cover specific or periodical demand and therefore keep fuel oil boilers remaining flexible. Hence, there is a technical potential for further phasing out of oil in favour of bio-fuels, less carbon-intensive fuels (like waste or natural gas) or electric boilers for steam generation. In 2003, parts of the European PPI committed to increase the biomass share of total primary energy use from 49 per cent to 56 per cent by 2010. Due to the closure of some chemical pulp mills, the industry missed the target by a few per cent (CEPI 2003, 2011b). Being both a large user and supplier of renewable energy, the industry association CEPI portrays the PPI as a key enabler for the EU to meet its renewable energy target of 20 per cent by 2020 (CEPI 2007a). Indirect CO_2 emissions from purchased electricity belong to the power sector of the EU ETS, but need also to be considered in the quest for a carbon-neutral PPI. Investments in renewable power assets (like on-site backpressure turbines and wind power) can be an option to ensure a reliable electricity supply. Net production of electricity can be sold to the grid and may also receive support from policy schemes based on feed-in tariffs.

Reducing specific energy use is important for energy-intensive industries like the PPI. Since production requires both electricity and steam there is a considerable potential for highly efficient on-site co-generation of heat and power (CHP). Estimates of PPI in several countries indicate that CHP can provide 20–50 per cent of the on-site electricity demand (IEA 2007). Further deployment could have positive impact in terms of reduced CO_2 emissions, especially if low-temperature heat generated from fossil fuels is replaced (IEA 2007). Black liquor is the main internal biomass fuel for CHP production as it is utilized by the recovery boilers in chemical pulping. In the Swedish PPI it accounted for 63 per cent of total fuel consumption in 2007 (Wiberg 2007). There are prospects for the alternative technology of black liquor gasification to increase the efficiency of energy recovery and also enable production of synfuels (e.g. Bio-DME, methanol), which can substitute for fossil automotive fuels. The International Energy Agency (IEA) estimates the global mitigation potential from applying black liquor gasification technology in the PPI to be 30–75 $MtCO_2$/year (IEA 2008). There are also prospects for gasification technology to facilitate introducing carbon capture and storage (CCS) in the PPI (Möllersten 2002). By applying CCS to biogenic CO_2, the PPI could contribute to negative emissions. Black liquor gasification technology has been tested in pilot plants for many years but with no commercial breakthrough as yet.

3. Analysing Changes in Aggregate Climate Strategies

In this section, we begin by examining the position of the Confederation of European Paper Industries (CEPI) on the 2003 ETS Directive, the 2008 revised ETS proposal, the 2009 revised ETS Directive, and the benchmarking process for the third trading period (2013–2020). We then go on to look into the strategies of the major paper companies in the ETS, including their stance on acknowledging the problem of climate change, their short-term abatement strategies, and long-term low-carbon solutions.

3.1 CEPI's Position on the EU ETS: From Opposition to Acceptance

All the major pulp and paper companies in Europe are organized under CEPI, which serves as their main collective interest representation in EU policymaking and implementation. CEPI represents some 800 companies producing pulp, paper and board and 1000 mills across Europe, together accounting for 40 Mtonne (20 per cent) of global pulp production and 96 Mtonne (24 per cent) of global paper and board production. Major players in CEPI are companies based in Sweden and Finland, which account for 30 per cent and 27 per cent respectively of pulp production, whereas companies in Germany, Portugal, Spain, France, Norway and Austria each account for 4–7 per cent of pulp production. In terms of paper and board production, Germany is the largest, with 24 per cent of the total CEPI production. Sweden and Finland each account for 12 per cent while a few other countries each account for some 4–9 per cent (CEPI 2011a).

According to CEPI, the initial design of the EU ETS was flawed because it allowed electricity producers to pass on almost the full cost of emission allowances to their customers. Since prices for products from the pulp and paper industry are determined on the international market, not locally as in the electricity markets, the pulp and paper industry cannot pass on these extra costs to their customers. The environmental director of CEPI thus claimed that the main effect of the EU ETS was a large transfer of revenues from the energy-intensive industries to electricity producers. She characterized the scheme as 'a very expensive trial that will not meet the environmental expectations placed on it but will put a disproportionate burden on EU manufacturing industries' (Hyvärinen 2005: 41).

During the first years of operation of the EU ETS, CEPI's position shifted from strong opposition to reluctant acceptance of the scheme. Part of the explanation may be the fact that the consequences of the scheme for the competitiveness of the European pulp and paper industry proved less severe than the industry had anticipated. As a result of over-allocation of allowances and lower demand for pulp and paper products in the international market, European pulp and paper mills could sell allowances that they did not need. According to CEPI, however, these sales of excess allowances could not compensate for the sharp increase in electricity prices (Mensink pers. comm.). This meant that the industry could gain from improvements in energy efficiency, switching to low-emissions energy sources, and investing in energy generation. The main concern for pulp and paper companies was therefore the effect of the EU ETS on power prices.[1]

---

1   When emissions trading started in 2005, the potential impact on power prices was substantial, as EUA prices increased from 7 euros to 20–30 euros within a year. However, when member states reported their verified emissions for 2005, the situation of over-

Comparison of the Commission's 2008 ETS proposal and the revised 2009 EU ETS Directive with CEPI's positions reveals some degree of convergence. According to the Commission's 2008 ETS proposal, every installation would begin with zero allowances and would have to buy the required allowances on the allowance market (auctioning). By contrast, CEPI advocated that the EU ought to allocate free allowances to installations on the basis of their past emissions ('grandfathering'), as the EU member states did in the first trading period. The revised 2009 EU ETS operates with free allocation based on benchmarking for the best-performing installations in every sector or sub-sector in the Community, and is thus more in line with CEPI's position than the Commission's 2008 ETS proposal had been. The energy-intensive industries were quite influential in the EU policymaking process, and managed to prevent full or major auctioning of allowances (Slingenberg pers. comm.).

The revised EU ETS Directive required the Commission to develop rules for free allocation of CO_2 allowances to the most GHG-efficient and energy-efficient installations in a sector or sub-sector in the third trading period (2013–2020), based on *ex-ante* benchmarks. The ETS benchmarking processes caused concern in the pulp and paper industry that the outcome could incur significant costs and threaten their competitiveness. Because most CO_2 emissions from pulp and paper mills are related to fuel combustion, the fuel mixes used at different mills greatly influence emission levels, as explained above. Emissions vary considerably among mills and regions, depending on the source of electricity and the type of fuel used to produce process heat. For this reason, the pulp and paper industry wanted the European Commission to take the fuel mix into account when setting the benchmarks (Ecofys 2009). By contrast, the Commission insisted that in order to encourage greater emission cuts, benchmarks would have to be fuel- and technology-neutral. Such technical and political disagreements made it challenging for the Commission to draft benchmarks that were acceptable to the industry. A real concern, not only in the industry but also in the Commission, was that strict benchmarks could result in 'carbon leakage in the industry'– relocation to regions outside Europe with less stringent carbon constraints. Another challenge in devising benchmarks was that some EU member states treat energy-producing installations related to the pulp and paper industry as separate energy activities, whereas in other member states they are considered part of the pulp and paper industry.

In December 2010, the EU member states adopted 52 product benchmarks and one additional benchmark (for heat and fuel). Industrial sectors deemed at

---

allocation became evident, and that caused a significant price drop. This was strengthened by the general prohibition against transferring surplus allowances from the first to the second trading period (Convery and Redmond 2007); in 2007 the EUA price reached rock bottom at around one euro. After an initial high of 30 euros, it has stayed at more firm levels of 10–15 euros over the second trading period (Kossoy and Ambrosi 2010). Studies of the impact from EU ETS on power prices indicate that a major part of the carbon cost is passed over to the wholesale power market (Sijm et al. 2006).

significant risk of carbon leakage, including the pulp and paper industry, will receive up to 100 per cent of their benchmarked allowances for free in 2013, compared with a maximum of 80 per cent for industries not deemed at risk of carbon leakage. In total, about 900 installations in the European PPI are covered by the EU ETS. Although the PPI represents only some 2 per cent of emissions covered by the scheme, 11 product benchmarks out of 53 were developed for this industry. The Commission initially proposed only two benchmarks – one for pulp and one for paper – but a main concern for the industry was the need to differentiate among a wide range of products. According to CEPI, the industry should ideally have 64 product benchmarks, one for each type of pulp and paper product; but 'politically that was out of the question', and in the end CEPI agreed that 11 product benchmarks would be acceptable (Mensink pers. comm.).

In 2010, the European Commission called for stakeholder engagement in developing its roadmap for a low-carbon economy on how to achieve the EU target of 80–95 per cent GHG emissions reduction by 2050 compared to 1990 (ENDS 29 October 2010). CEPI was the first industry association to announce it would engage and give input on possible contributions from the industry it represents (Wyns pers. comm.). In November 2011, in order to influence the Commission's roadmap, the implementation of EU's climate and energy package, and member states' policies, CEPI published its own 2050 roadmap for a low-carbon bio-economy (CEPI 2011c). This roadmap envisions a transformation of the European forest fibre industry (the pulp, paper and wood manufacturing industry) to become the central hub in a biomass-based economy, which is depicted as a broader framework than the low-carbon society alone. The main measures in CEPI's projection for an 80 per cent reduction of the sector's CO_2 emissions include:

- a partial shift in current fuel/heat use towards more and decarbonized electricity use;
- continuous application of the best available technology whenever equipment is replaced;
- fuel-mix changes in favour of biomass lime kilns, gasification of biomass and waste and lignin extraction;
- application of new and as yet unknown breakthrough technologies to lower heat demand due to reduced water use and improved methods for drying.

CEPI's roadmap relies on the conditions of the Commission's 'global action scenario with available technology' and further underlines the importance of pushing the development of breakthrough technology to become available by 2030. In the call to policymakers it is suggested to use EU ETS auctioning revenues from 2013 and beyond to finance the transformation of relevant industries. Thereby, CEPI claims, the EU ETS can 'become an innovation instrument that yields double dividends rather than costs alone' (CEPI 2011c).

To conclude, CEPI has gradually become more accepting of the EU ETS. This growing acceptance seems related to three factors: experience gained during the first trading period, demonstrating to the pulp and paper industry that the economic consequences of the scheme were less severe than anticipated; accommodation of some of the industry's main concerns in the revised 2009 EU ETS; and increasing attention to climate change internationally and in the EU and its member states. That said, there is little to indicate that the ETS has triggered industry-wide innovation activities. There are no joint innovation programmes for developing climate-friendly technologies within the industry. So far, CEPI has been more concerned with the risks of the EU ETS in terms of carbon leakage and reduced competitiveness in the European PPI. Little consideration has been given to the potentials of the EU ETS to stimulate the development of emerging and new technologies. To some extent, the recent CEPI roadmap may indicate a change of perspective, as it encompasses a vision based on the Commission's scenario, where the European PPI can actually benefit in a low-carbon and bio-based society.

3.2 Strategies of the 10 Largest Pulp and Paper Companies in the EU ETS

Since the early 2000s, and especially after 2006/2007, the ten largest pulp and paper companies covered by the EU ETS have gradually adopted more proactive and converging climate strategies.[2] The companies recognize that they have a role to play in reducing GHG emissions and seem to be taking responsibility for problem-solving. Wood, their raw material, is both recyclable and renewable, and an important aspect of corporate climate strategies is to promote wood and paper products as climate-friendly alternatives to competing materials. They also highlight that well-managed, sustainable plantations and forests serve as carbon sinks.

Most of the companies examined have adopted short- and medium-term targets for reducing CO_2 emissions. These targets are often expressed in relation to production level and encompass reductions of 15 or 20 per cent by 2020 against the base year (e.g. 2005). Almost all companies share the two main measures for CO_2 reduction: using *less* energy through increased energy efficiency, and *changing the type* of energy used – for example by switching from fossil fuels to biofuels. It can be difficult to distinguish between short- and long-term measures in this sector, but long-term research is often focused on technologies that minimize, or change the type of, energy consumption. In many companies the share of biomass energy is already high (around 60 per cent), and their climate strategies aim to increase this share further. In general, the PPI appears to focus mainly on continuous improvements of operations and reductions in energy use, rather than long-term innovative solutions.

Many of the companies are on track to achieve their short-term targets. Still, none of them have announced explicit long-term targets beyond 2020. Most companies

---

    2   Thanks to Anne Raaum Christensen for helpful assistance with compiling data on the climate strategies of these companies.

identify the EU ETS as an important driver behind their climate strategies, often describing it as a regulation they simply have to accept and comply with as part of the new business reality. Many companies report that they factor in carbon costs in investment decisions. The ETS is described as a regulatory risk rather than an opportunity for creating competitive advantages through innovation; most companies fear higher costs in the third phase. The expectations of increased costs (directly through the need to buy allowances, and indirectly through increased electricity prices) can, according to many companies, explain the stepped-up focus on improving energy efficiency. In other words, companies change their practices partly because they want to avoid the increased costs associated with the EU ETS.

All companies reviewed have received a surplus of emission allowances (EUAs) over the first and second period. For a few companies this has amounted to a factor as high as 1.4 (CITL 2011). Primarily this has been a consequence of overly generous national allocation procedures. Likewise, internal emissions targets seem to have been rather slack, as some companies met their targeted reduction levels several years in advance of the target year. In general, company-internal targets do not seem to be more ambitious than an expected development in CO_2 emissions in a business-as-usual scenario. CO_2 intensity in the PPI should normally fall because of fuel switching and decreased energy intensity. This has been demonstrated in the European PPI, which has decreased its specific CO_2 emissions by 40 per cent since 1990 (CEPI 2011b).

Some companies appear more progressive than others. In particular, the review of corporate climate strategies indicates that the Nordic forest companies – Stora Enso, SCA and UPM – have adopted more ambitious climate policies and programmes than companies in other countries have. This finding corroborates earlier studies showing that the Nordic forest companies have traditionally been technological and environmental frontrunners in the industry (Smith 1997). That said, the similarities in corporate climate strategies within the PPI are more striking than the differences. Since most companies have become good at 'talking the talk', it is hard to identify specific characteristics or significant differences on the basis of a general examination of corporate climate strategies. That is why we need to supplement the sector study with an in-depth study of corporate climate strategies.

In conclusion, the major pulp and paper companies covered by the ETS seem to be taking responsibility for problem-solving. Judging from their company reports and reports to the CDP, most companies – especially those with headquarters in Northern Europe – seem to be knowledgeable on issues of climate change, and they are working to reduce CO_2 emissions. Increased energy costs in recent years have strengthened the case for energy efficiency improvement and self-reliance in energy supply. However, it is hard to discern the long-term picture and fully grasp how the individual companies will in fact respond to a stricter carbon regime. Many companies mention the uncertain regulatory environment and refer to the EU ETS as a risk, which explains their hesitation to move beyond business-as-usual practices. The climate strategies of the ten largest pulp and paper companies covered by the ETS are summed up in the Appendix Table 5.1.

4. Corporate Climate Strategies: SCA and Norske Skog

4.1 The Climate Strategy of SCA

Svenska Cellulosa Aktiebolaget (SCA) was founded in 1929 through a merger of several Swedish forest companies. The internationalization of the company started in the 1960s; today it is one of the world's leading forest industry companies. It develops, produces and markets a broad portfolio of products within the main segments: personal care (e.g. baby diapers and incontinence care); tissue (e.g. toilet paper and napkins); packaging material; publication papers; and solid-wood products (SCA 2011). SCA operates some 250 production facilities, of which 45 are large pulp and/or paper mills, in 60 countries; the products are sold in more than 100 countries.[3] Europe represents a strong base, with 75 per cent of the company's total net sales (which were 11.45 billion euros for 2010), 75 per cent of its total number of 45,000 employees, and 75 per cent of group-wide energy use (fuel, heat and electricity) (SCA 2011, Isaksson pers. comm.)

SCA had been aware of the problem of human-induced climate change many years before the introduction of the EU ETS in 2005. Fuel shifting (from oil to biofuels and electricity) was undertaken already in the 1970s, but then chiefly aimed at reducing dependency on oil. One interviewee recalled how LUFT '90, a 1990s action programme of the Swedish Environmental Protection Agency against air emissions and acidification, came to serve as a starting point for continued interaction with authorities on reducing oil use at production sites (Fält pers. comm.). Since then, anthropogenic climate change has become one of the driving forces for reducing the use of fuel oil under the guidance of the environmental management system (Fält pers. comm.). In 1998, SCA published its first environmental report. Here, as in later annual reports, the challenge of climate change is recognized. Asked what would be the most important environmental issue for the coming decade, the president and CEO answered: 'Without any doubt it has to be climate change. I think this will completely dominate the debate. It's becoming obvious that human life on earth cannot go on this way. Conferences like Kyoto make it clear that we must all accept the facts and accept humanity's share of the burden ... SCA is contributing towards the achievement of the Kyoto goals in its operations' (SCA 1999: 5).

From 1998 onwards, SCA has monitored and reported its CO_2 emissions. This has been facilitated by the internal Resource Management System (RMS), developed for better control over the company's resource use and emissions to air and water (Isaksson pers. comm.). The RMS has been a prerequisite for SCA's environmental target-setting, monitoring and reporting (Strandqvist pers. comm.). Since 2009, sustainability reporting has been prepared according to level A or A+ (intended for advanced reporting organizations) of the Global Reporting

---

3   Recently, SCA announced it will divest itself of its main operations in the packaging segment, which will alter the production portfolio compared to the 2010 reporting.

Initiative (GRI) guidelines. SCA has a vision of being a leader within the area of sustainability (SCA 2011) and has become increasingly confident after being awarded several environmental sustainability honours. With regard to measuring and reporting GHG emissions and corporate climate strategies, the CDP has rated SCA as the number one Swedish company (SCA 2011). According to interviewees this is because SCA is good at systematically monitoring its activities (Isaksson pers. comm.).

In 2001, SCA manifested its position on climate change by making a group-wide commitment to reduce CO_2 emissions from fossil fuels in relation to production levels (SCA 2002). In 2008, a quantified group-wide target was formulated: 'SCA will reduce its carbon emissions from fossil fuels and from the purchase of electricity and heat, relative to the production level, by 20 per cent by the year 2020, using 2005 as a reference year' (SCA 2009: 4). In 2010, SCA reported that the group's fossil fuel derived CO_2 emissions were 4.48 million tonnes, covering most although not all production sites. Direct CO_2 emissions from on-site fossil fuel consumption were 2.59 million tonnes, while indirect emissions from the purchase of electricity were 1.79 million tonnes. SCA makes downwards adjustments to exclude CO_2 emissions that arise from electricity and heat that is generated on-site to be sold to external customers. The environmental reporting does not clarify how the carbon intensity of this (generally biofuel-derived) co-generation is evaluated. Assumptions about the carbon intensity of different energy carriers can have significant impact on the emissions accounting, especially since SCA supplied almost 0.45 TWh electricity and 0.25 TWh thermal energy for district heating in 2010 (SCA 2011). SCA's target monitoring showed that, by the end of 2010, specific CO_2 emissions had decreased by 4.2 per cent compared to 2005 (SCA 2011).

As an energy-intensive company, SCA sees opportunities for reducing CO_2 emissions by maintaining and improving its installations with the most suitable technology in terms of fuel usage and energy performance (Strandqvist pers. comm.). In its external communication, SCA highlights several projects related to the energy supply and demand of its business activities. The group-wide energy savings programme ESAVE provides a structured approach to identifying and implementing smaller-scale measures to improve energy efficiency. This effort is also guided by the quantified target of reducing specific energy consumption by 7.5 per cent in 2012, as compared to 2005 (SCA 2011). Since 2003, this has resulted in the implementation of 1300 energy-saving measures, bringing an estimated annual reduction in CO_2 emissions of 72,000 tonnes (SCA 2009, 2011).

Among the larger projects are investments in new or retrofitted energy installations that can offer immense potential for cutting CO_2 emissions. In 2006 the Östrand chemical pulp mill under SCA Forest Products in Sweden made a 160 million euro investment in a new recovery boiler and a 75 MW back-pressure turbine which doubled the capacity for auto-produced electricity and made the mill self-sufficient in electricity and heat (SCA 2009). More recently the same mill made a 50 million euro investment in a new lime kiln. The plan is for the installation to be fueled entirely by crushed sawdust pellets, reducing oil consumption at the mill by 17,000 m^3/year and CO_2 emissions by 50,000 tonne/year, or 80 per cent (Fält pers. comm.). The project is innovative with regard to the large volumes of fuel oil replacement and the specific requirements of the combustion process. A survey of a few pulp mills with lime kiln installations that use either solid or gasified biomass shows that fuel oil still covers 35–65 per cent of the fuel demand at these kilns (Wadsborn et al. 2007). In general, the process is viewed as fossil fuel dependent (Ecofys 2009), but in the CEPI roadmap the biomass lime kiln is mentioned as a long-term solution up to 2050. If the Östrand installation proves successful it can point the way to further installations.

Another rather recent project is the Witzenhausen containerboard mill under SCA Packaging in Germany. An external partner has invested 127 million euros in a CHP plant that incinerates industrial byproducts and refuse-derived fuel (RDF) from household waste in this densely populated region (Isaksson pers. comm.). Here SCA Packaging and the Witzenhausen mill have outsourced the electricity and heat production, and contracted the owner and operator of the CHP plant to ensure a long-term energy supply. The Witzenhausen mill has been able to cut its direct CO_2 emissions by almost 100,000 tons, or 80–90 per cent (CITL 2011).

A major current project is the joint venture Statkraft SCA Vind AB, which has been formed to realise planned wind power installations with a total of 1140 MW capacity (Vindkraft Norr 2011). The Norwegian power company Statkraft is to answer for the 1.6 billion euro investment while SCA provides the necessary land area. Through a long-term agreement SCA is ensured an affordable electricity supply for its electricity-intensive Ortviken pulp and paper mill (SCA 2011). It is new for SCA and the PPI in general to engage in large-scale wind power projects like this one. It is also relatively new to erect wind farms in forest areas (SEA 2008), which could give rise to fewer conflicts of interest and thus speed up the permit process. The consortium estimates that the total wind power installation will reduce annual CO_2 emissions by about 2 million tons (Vindkraft Norr 2011).[4]

To summarize, SCA recognized the climate change challenge early on, as seen in the company's first environmental report from 1998. Since then, the company has monitored its CO_2 emissions along with other emissions and resource use. In 2001, SCA undertook a commitment to reduce group-wide CO_2 emissions in relative terms; in 2008 this objective was concretized by a quantified target. With its high energy demand, SCA sees efficient use and generation of energy as well as increased use of renewable energy and less carbon-intensive fuels as its main abatement alternatives. These are areas where SCA can take responsibility for problem-solving. In recent years and for the short-term future, SCA communicates several examples of energy-related projects and investments, from small- to large-

---

4   This estimate is probably based on the assumption that wind power will replace marginal production from coal condensing power. Assuming a replacement of Nordic electricity mix instead would generate a much lower result due to the large share of hydropower.

scale, together implying significant reductions in CO_2 emissions. Some of these measures can also be seen as long-term and innovative solutions.

4.2 The Climate Strategy of Norske Skog

Established by Norwegian forest-owners in 1962, Norske Skog is the only major pulp and paper company in Norway. The company was established to exploit timber resources in the central part of the country, and a newsprint mill with two paper machines and mechanical pulp technology was built in Skogn, starting production in 1966. Until around 1990, the company expanded in Norway, acquiring businesses in paper production, paper pulp and wood-based construction material. Many of these businesses had been in operation much longer than Norske Skog itself. During the 1990s, the company grew internationally, beginning on the European continent and expanding further through the acquisition of other newsprint and magazine paper companies around the world (Sæther 2004). In recent years, however, Norske Skog has sold assets to reduce its debts and has shut down several mills to cut costs in a difficult market with surplus production capacity (Norske Skog 2011).

Although Norske Skog is a much smaller company than SCA, it is one of the world's largest producers within its segment of publication paper. It has 13 wholly-owned mills located in 10 countries, annual sales of some 2,600 million euros, and 5,300 employees worldwide. In 2010, the company produced four million tons of publication paper. With three mills in Norway (Follum, Saugbrugs and Skogn) and one in France (Golbey), in Austria (Bruck), in Germany (Walsum) and in the Netherlands (Parenco), the European part of Norske Skog's business – accounting for 70 per cent of its total production capacity – is covered by the ETS.[5] In addition to its seven European mills, the company has mills in Australia, New Zealand, Thailand, Brazil and Chile as well as one partly-owned mill in Malaysia (Norske Skog 2011).

Norske Skog has a long history of dealing with local air and water pollution at its mills, but systematic efforts to reduce CO_2 emissions are more recent. Still, an examination of annual reports shows that the company was a frontrunner in acknowledging the problem of climate change and taking responsibility for contributing to reduce GHG emissions. Norske Skog published its first environmental report for the calendar year 1990, when the company had completed its expansion in Norway. In this report, the company made clear that it would seek to replace the use of fossil fuels in the production process by biofuels and other environmentally-friendly energy sources (Norske Skog 1991: 12). Since 1996, Norske Skog has reported on CO_2 emissions for the entire company (Norske Skog 1999). The company has also supported the work for a binding international

---

5   In March 2012 Norske Skog entered into an agreement to sell Follum to the Norwegian forest-owner association Viken Skog; for the time period under study here, Follum was fully owned by Norske Skog.

climate treaty. In its 2001 Environmental Report, for example, Norske Skog stated that it 'supports international climate efforts', highlighting that '[t]he Kyoto protocol must be seen as a first step towards reducing the climate threat'. In the same report, the company observed: 'Human activity has led to an accelerating increase in greenhouse gases in the atmosphere during the past few decades. The most important cause is the emission of carbon dioxide (CO_2) from fossil fuel' (Norske Skog 2002: 27).

In 2007, Norske Skog adopted a GHG emissions reduction target for the entire company, covering direct emissions from pulp and paper production and indirect emissions from purchased energy. The target is 25 per cent reduction of GHG emissions (CO_2-equivalents) by 2020, with 2006 as the baseline year. This is an *absolute* rather than *relative* target, meaning that GHG emissions are to be reduced by 25 per cent regardless of production volumes (Carlberg pers. comm.). In contrast, the emissions reduction targets of most companies in the PPI, including SCA, relate to the volume of produced tons of paper (see Appendix). In addition to CO_2, Norske Skog's GHG emissions inventory includes methane (CH_4) and nitrous oxide (N_2O) that together account for less than 0.5 per cent of the CO_2-equivalent emissions (Norske Skog 2010). The GHG emission accounting is guided by the WRI/WBCSD Greenhouse Gas Protocol Pulp and Paper Workbook, developed by the World Resources Institute and the World Business Council for Sustainable Development.[6]

Calculated on the basis of all 13 wholly-owned Norske Skog mills for the 2009 calendar year, Norske Skog operations emitted a total of 2.38 million tons of fossil fuel-derived CO_2 equivalents in 2009 (Norske Skog 2010). This represents an absolute reduction of 18 per cent compared to the 2006 base year. Not least between 2008 and 2009 emissions fell by 14 per cent due to declining paper production. Production increased again in 2010, resulting in a 6 per cent increase in total emissions compared to the previous year. Norske Skog operations thus emitted a total of 2.52 million tons of fossil fuel-derived CO_2 equivalents in 2010, of which 774,000 tons were direct emissions.[7] As of 2010, the total emissions from the company's mills were 13 per cent below the 2006 base year (Carlberg pers. comm.).

Although Norske Skog's target is formulated as an absolute target, the company also manages its specific emissions levels by calculating the carbon footprint of its products. In the absence of an internationally approved method for calculating the carbon footprint of paper products, Norske Skog estimates its carbon footprint by

---

6   In this protocol, 'scope 1 emissions' are direct emissions from the combustion of fossil fuels in boilers, CHP plants, infrared drying equipment, mobile machinery, and other mill site-based equipment. 'Scope 2 emissions' are indirect emissions from the purchase of electricity or heat from external sources.

7   Hence, more than 70 per cent of the company's total GHG emissions are indirect emissions from externally purchased energy and thereby belong to the sector of energy installations under EU ETS.

calculations based on the *Framework for the Development of Carbon Footprints for Paper and Board Products* developed by CEPI in 2007 (CEPI 2007b).

Figures for emissions per ton of paper production vary considerably between mills and regions, ranging from 557 kg CO_2 equivalents/tonne of paper in Europe to 1,616 kg CO_2 equivalents/tonne of paper in Australasia (Norske Skog 2011: 28). The differences in the carbon footprint among mills and regions are due to variation in the source of electricity and the type of fuel used to produce process heat. Norske Skog's main strategies for reducing GHG emissions involve reducing the consumption of energy, changing the source of energy, and optimizing the use of process chemicals and transport (Norske Skog 2010: 101). Energy costs represent some 20 per cent of the company's operational costs, and are thus a key consideration for the operational and strategic management of its business (CDP 2011). Norske Skog reports that costs and savings related to climate change are incorporated into the company's operational cost information and projections. This information is taken into account in energy purchasing and generation strategies. Climate change impacts and opportunities are also taken into consideration in assessing potential investment options at the mill, local and company levels.

Since 2008, Norske Skog has participated in the Carbon Disclosure Project (CDP). It was named the best Norwegian company by the CDP 2010 Nordic report, which ranks companies on the basis of their climate change strategy and emissions reporting. Norske Skog reports that climate change issues are integrated into its business strategy in various ways, including the management and projection of operational costs; the identification of investment options, as noted above; its engagement with employees, customers and other stakeholders; and its engagement with governments and regulators (CDP 2011). The company considers it necessary to develop both short- and long-term strategies in order to achieve its emissions reduction target. As examples of strategic business decisions that are influenced by climate change, the company highlights several issues including:

- Participation in a consortium investigating the possibilities to develop and produce second-generation biofuel based on the unused growth of biomass in Norway.[8]
- Several mills are conducting feasibility studies into the increased use of biofuel either as investments in new assets or as upgrades of existing assets.
- Reduced energy use and GHG emissions by increasing the capacity of the company's Skogn mill in Norway to incorporate clay fillers into its paper products.

---

8 This project has been stopped due to financial considerations, but Norske Skog reports that the knowledge base that has been build up will be valuable for similar projects in the future.

To summarize, Norske Skog was among the first companies in the pulp and paper sector to recognize the problem of climate change and express its responsibility for problem-solving. In 2007, Norske Skog adopted an absolute GHG emissions reduction target for the entire company, and the company reports that climate change issues are well integrated in the overall business strategy. Among abatement measures, energy efficiency improvements are accorded main priority.

4.3 Comparing SCA and Norske Skog

Summing up the climate strategies of SCA and Norske Skog, we first observe that both companies accepted the climate change problem at an early stage, as well as their responsibility for contributing to problem-solving. Because both companies have a long history of handling local air and water pollution at their mills, they were well prepared for developing corporate climate strategies when climate change emerged on the international agenda. Norske Skog and SCA have accounted for their CO_2 emissions since 1996 and 1998, respectively, much earlier than many other companies in the PPI. They also expressed implicit or explicit support for international efforts to develop a legally binding international climate treaty at a quite early stage. Further, the Carbon Disclosure Project (CDP) in 2010 named Norske Skog and SCA number one in carbon reporting and accounting in Norway and Sweden, respectively. Our expectation that these companies would be among the environmental frontrunners in the sector was confirmed by examination of corporate climate strategies.

A second observation is that their commitment to reduce CO_2 emissions increased gradually from around the year 2000 and onwards, particularly with the adoption of emissions reduction targets in 2007 (Norske Skog) and in 2008 (SCA). The most apparent difference here is the fact that Norske Skog's target is formulated as an absolute reduction target, while SCA's target follows the common practice of reductions related to production level (see Appendix). Absolute reduction targets leave less room for manoeuvring, but may be easier to meet if production levels decrease. Indeed, the fact that Norske Skog has closed down some production units in recent years has contributed to progress towards its target. When production units have been sold, Norske Skog has made downward adjustments to its base-year emissions (Carlberg pers. comm.). These operational restructurings were probably foreseen at the time the target was formulated; that could explain the rationale for adopting an absolute target. Also SCA has been moving in a positive direction, making progress towards its relative target for 2020.

A third observation is that although neither company has adopted any long-term targets beyond 2020, SCA engages in more innovation activities and CO_2-lean investment projects than Norske Skog does. For instance, SCA is part of a joint venture with the Norwegian power company Statkraft for large-scale exploitation of wind power. Engaging in large-scale wind power projects of this kind is an entirely new strategy for SCA and indeed for the industry in general. SCA also reports many small- and medium-scale projects to improve energy efficiency and achieve CO_2-friendly fuel shifts at its mills.

Two key questions emerge from this examination and comparison of the climate strategies of SCA and Norske Skog. First, what is the relationship between the introduction of the EU ETS and the increasing ambitions in the climate strategies propounded by SCAs and by Norske Skog? Second, why does SCA engage in more ambitious climate-related innovation activities and projects than Norske Skog does? To these questions we turn in the next two sections.

5. Linking the ETS to Changes in Corporate Climate Strategies

5.1 Positions on the EU ETS

In the planning and formulation phase of the EU ETS, SCA and Norske Skog were positive towards the idea of a trading scheme, i.e. that climate mitigation is to be conducted where it is most cost-effective (Isaksson pers. comm., Carlberg pers. comm.). Expectation as to the allocation of EUAs were by and large fulfilled. Due to the situation with international competition, the companies expected the EU member states, in their National Allocation Plans (NAPs), to propose overall generous allocations to their domestic manufacturing industries (Isaksson pers. comm.). As regards the political importance of getting the scheme up and running, the NAPs were also approved by the European Commission.[9] For SCA and Norske Skog, like many other pulp and paper companies (see section 3.2), this resulted in a long position during the first trading period. The risk of carbon leakage is still perceived as a shortcoming of the EU ETS, and our two companies would prefer a global emissions trading scheme (Strandqvist pers. comm., Carlberg pers. comm.)

Under the EU ETS, both SCA and Norske Skog have anticipated successive reductions in allocated EUAs. This has often been the case for individual installations, but the aggregate amount of allocated EUAs has nevertheless increased for the two companies from the first to the second trading period (see Figure 5.1 and Figure 5.2). Since the start of the EU ETS, the annual value of SCA's EUA surplus corresponds to somewhere between 2.5 and 6 million euros, depending on the yearly emission-to-cap ratio and the average EUA price. The revenues from selling surplus EUAs – a kind of windfall profit for the PPI – are nevertheless deemed insignificant compared to the increasing costs of electricity experienced over the same period (Isaksson pers. comm.). Group-wide SCA is dependent on a grid supply of almost 7 TWh per year of electricity (SCA 2011). Based on the assumptions that 75 per cent is purchased within the EU ETS area and that the scheme has caused an electricity price increase of 10 euros per MWh, the resultant cost increase for SCA is 50 MEuro per year. Under these conditions

---

9 See Convery and Redmond (2007) for a general discussion on this issue.

the potential revenues from selling surplus EUAs can compensate only some 5–10 per cent of the increase in electricity costs.

Even prior to the introduction of the EU ETS, both SCA and Norske Skog were concerned about the potential effects on electricity prices, and also warned decision-makers about the risk for windfall profits in the power sector (Isaksson pers. comm., Carlberg pers. comm.).[10] SCA views the EU ETS as one underlying factor behind the increases in electricity prices, which clearly have a negative impact on business (Isaksson pers. comm.). Access to abundant and affordable electricity is crucial for SCA; thus, the EU ETS has made it increasingly important to calculate projections about future electricity prices and account for these in relation to new investments and business plans (Isaksson pers. comm.). This 'indirect' effect of the EU ETS overshadows the more 'direct' effect of the EU ETS setting a price tag on CO_2 emissions from internal fossil fuel use (Isaksson pers. comm.). Norske Skog shares this frustration with SCA over the ETS. According to the company, the sales of surplus allowances were far from sufficient to compensate for the rise in electricity prices (Carlberg pers. comm.)

5.2 CO_2 Emissions and Relation to Cap

Assessing CO_2 emissions and relation to cap is important because these measures indicate the degree to which the ETS provides SCA and Norske Skog with incentives for reducing their emissions. Figure 5.1 shows the annual CO_2 emissions from 41 of SCA's installations covered by the EU ETS. The ten largest mills in terms of CO_2 emission levels account for almost 70 per cent of SCA's emissions under the EU ETS. Because some new installations were introduced into the scheme in the second trading period, total verified emissions reached a high of 1.52 Mt CO_2 in 2008 (Sandbag 2011). The large-scale project undertaken at the Witzenhausen mill and overall low production volumes in 2009 are factors that have contributed to the reduced emission during 2009/2010.

Figure 5.1 also shows the total allocation of EUAs received by SCA. The surplus allocation was 10 per cent during the first trading period, increasing further in the second period. The main explanation is the acquisition of several tissue-producing mills in 2007 (Eriksson pers. comm.). As a result, SCA introduced new installations and was granted additional EUAs in the second period. There is also a clear increase in the allocation of EUAs to a mill in Mannheim, Germany under SCA Hygiene. During the second period this mill has held a considerable surplus of EUAs compared to its average annual CO_2 emissions between 2005 and 2010 (Sandbag 2011, CITL 2011). The result for 2010 is that SCA had a total

---

10   According to economic theory, power generators will pass on the opportunity cost of their, to a large extent, freely allocated emission allowances to electricity consumers. The extra cost of fossil-based power generation thus has an impact on wholesale electricity prices in accordance with the carbon intensity of the marginal production unit (Sijm et al. 2006).

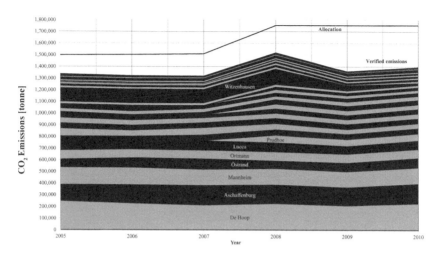

Figure 5.1   Total allocation and verified CO$_2$ emissions for 41 of SCA's installations that are covered by EU ETS

Source: Sandbag (2011).

EUA surplus of 360,000 tonnes CO$_2$ and an emission-to-cap ratio of 80 per cent (Sandbag 2011).

For the upcoming third trading period, SCA expects a decrease in allocated allowances compared to earlier periods (Isaksson pers. comm.). Due to the approach based on best practice benchmarks, many of SCA's mills, with less favourable fuel mixes, will receive fewer allowances than currently needed (Strandqvist pers. comm.). However, from a group-wide perspective SCA may still receive a surplus of EUAs, due to expected allocations to some larger mills that rely heavily on biomass fuels (Fält pers. comm.). Planned and recently made investments that will start generating CO$_2$ emissions reductions will of course contribute to the future situation, like any general investments for increased production. The lime kiln investment at the Östrand chemical pulp mill, for example, is expected to reduce CO$_2$ emissions by 50,000 tonnes/year while also boosting future production capacity. The mill's production benchmark will thus improve considerably, putting the mill in the favourable position of receiving a EUA surplus in the third trading period (DG CLIMA 2011).

Turning to Norske Skog, we see in Figure 5.2 the total allocation and verified CO$_2$ emissions for its seven installations covered by the EU ETS. The largest CO$_2$ emissions from Norske Skog's European mills are from Parenco in the Netherlands and Bruck in Austria, which together account for almost 90 per cent of all CO$_2$ emissions from the company's mills in Europe. The relatively high emissions from these two mills are due to the fact that electricity for the production process is not purchased but produced on-site from natural gas (co-generation of heat and power).

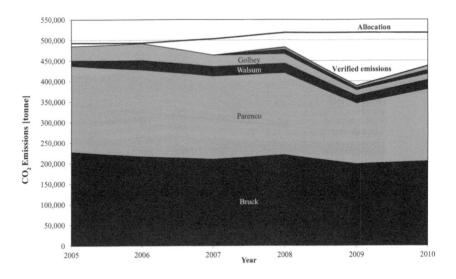

Figure 5.2   Total allocation and verified CO$_2$ emissions for seven of Norske Skog's installations that are covered by EU ETS

Source: Sandbag (2011).

Under the ETS, emissions from the production of electricity are accounted for by these mills – not by power companies, as in the case of purchased electricity. By contrast, the emissions from Norske Skog's three Norwegian mills are miniscule compared to emissions from similar mills in other countries. The explanation is simple and stems from the particular energy and fuel mix used for the Norwegian mills. Although Norske Skog's three Norwegian mills account for more than 30 per cent of the company's total production capacity, indirect emissions from the purchase of energy are very low because this energy is generated primarily from hydropower. Some 62 per cent of the energy used in the Norwegian mills stems from purchased electricity (indirect emissions), whereas 20 per cent of the energy used is generated from process heat, 16 per cent from bioenergy, and only 1 per cent from fossil fuels (direct emissions). As noted, indirect emissions from purchased energy are accounted for by the power sector under the ETS, not Norske Skog or other buyers of energy.

From an initial emission-to-cap ratio of nearly 100 per cent, the allocation of EUAs has increased slightly after the Norwegian mills joined the EU ETS in the second trading period. Over the period 2005–2010 Norske Skog's direct emission decreased by some 10 per cent. The result for 2010 is that Norske Skog had a total EUA surplus of 80,000 tonnes CO$_2$ and an emission-to-cap ratio of 85 per cent (Sandbag 2011).

The starting point for setting performance benchmarks for free allocation of emission allowances in EU ETS post-2012 was the average performance of the

10 per cent most efficient installations in a sector in the years 2007–2008. From a group-wide perspective, Norske Skog could be over-allocated in the third trading period, as in the first and second trading periods, due to the relatively low direct emissions from most of its European mills. Table 5.1 shows the direct and indirect emissions from Norske Skog's European mills measured as tons CO_2 equivalents per tons of paper.[11] Only the direct emissions are reported under the ETS. As the ETS benchmark for these mills will be around 0.3 tons CO_2 equivalents/tonne of paper at the beginning of the third trading period (DG CLIMA 2011), only two mills – Bruck and Parenco – will have to buy emission allowances. All other European mills owned by Norske Skog will have a surplus of allowances (Carlberg pers. comm.). Somewhat paradoxically, the mill with the biggest carbon footprint – Walsum, in Germany – will have a considerable surplus of emission allowances in the third phase of the EU ETS. This mill has a large carbon footprint because it uses electricity from a coal-fired power plant, but emissions from producing this electricity are accounted for by the power plant under the ETS, not by the pulp and paper factory. In sum, we may conclude that Norske Skog's mills are well positioned for the third trading period due to their relatively low emissions compared to other mills. For most mills owned by Norske Skog, the main impact of the ETS will likely continue to be felt through the increase in electricity prices.

Table 5.1 Direct and indirect emissions from Norske Skog's European mills measured as tonnes CO_2 equivalents/tonne of paper

Mill	Bruck (AT)	Follum (NO)	Golbey (FR)	Parenco (NL)	Saugbrugs (NO)	Skogn (NO)	Walsum (DE)
CO_2-e direct	0.55	0.01	0.04	0.72	0.02	0.02	0.06
CO_2-e indirect	0.05	0.02	0.12	0.07	0.02	0.02	1.29

Source: Norske Skog (2011).

To summarize, all of SCA and Norske Skog's European mills received their EUAs for free in the first and second trading periods, based upon the allocation plans of each member state. The mills could then trade allowances internally or externally, and were responsible for their own compliance. Due to generous allocations, there has been no need for the companies to buy any EUAs. According to our interviews, expectations are mixed as regards the effects of the new allocation practices for the upcoming third trading period. The occurrence of mills in 'short' positions will become increasingly common. The net impact may nevertheless be a surplus of EUAs for the total of Norske Skog's and SCA's installations.

---

11 Norske Skog's mills produce primarily newsprint and coated fine paper.

Especially Norske Skog, specialized in publication paper, has demonstrated overall high performance in relation to the proposed benchmark criteria. For SCA, with its more diverse product portfolio and major operations in up to ten EU ETS countries, the future situation is more difficult to predict. In any case, various strategic investments are expected to bring positive outcomes in the course of the upcoming trading period.

5.3 Procedural, Structural and Long-term Changes

SCA established an organizational structure for environmental management, with group-wide and business unit level environmental managers and committees as well as environmental coordinators at each mill, already in the early 1990s. The main environmental challenges of that time were to introduce chlorine-free bleaching methods and to increase the use of recovered fibre, in response to the market demand for such products (Nyström et al. 1997). The climate change issue became increasingly important by the end of the 1990s, as can be seen in the company's first environmental report (SCA 1999). Monitoring and reporting practices for CO_2 emissions – as well as other emissions to air, water and various material flows – were also introduced at that time, through the Resource Management System (RMS) (SCA 1999). This was due to internal driving forces, however, and independent of any expectations about a future emissions trading scheme (Isaksson pers. comm.). Since then, the RMS has been used for the group-wide bottom–up collection of relevant GHG emissions data, but the reporting to national environmental authorities is conducted by each individual installation that holds emission permits. While some of SCA's environmental and social commitments have been altered or replaced over the years, the commitment to mitigate climate change has remained firm since 2001.[12] It was not until 2008, however, that a first target was clearly quantified.

One effect of the EU ETS is that the reporting of emissions has been extended to include more staff hours at the mills. The environmental and accounting departments are especially involved at the site level (Fält pers. comm.). Also the project departments, responsible for technical and economic assessments of larger process changes and investments, have become involved, as they have to consider the price tag on current and future CO_2 emissions as a result of the EU ETS. In every investment decision, the EUAs represent costs or revenues for the mill (Fält pers. comm.). This can be exemplified by the Witzenhausen mill, which opted to outsource its electricity and heat production (see section 4.1). The resultant EUA surplus represents an annual value of about 1.5 million euro (CITL 2011). This, together with other factors, surely contributes to the overall picture, when investments or process changes are considered (Fält pers. comm.). A general

---

12   Since 2006, SCA has had the following environmental and social commitments: reducing CO_2 emissions from fossil fuels, avoiding wood fibre from controversial sources, improved water usage, and compliance with the universal Code of Conduct (SCA 2011).

standpoint held by industry representatives is that energy and climate policies need to provide long-term stable conditions if they are to facilitate investments. Perceived uncertainties may lead to refraining from or postponing investments due to lack of decision support (Fält pers. comm.). As yet there have been no examples where the EU ETS as such has led SCA to refrain from an investment. The same is true for Norske Skog, according to the company Vice-President for the Environment (Carlberg pers. comm.)

At SCA headquarters, it is the division for energy procurement that manages the trading activities introduced by the EU ETS. Work here can involve selling and purchasing EUAs, banking and borrowing, and investing in CDM projects to receive Certified Emission Reductions (CERs). Already in the first trading period, SCA was investing in CDM projects, which allowed CERS to be transferred to the second period. Thereafter, SCA has been able to surrender CERs to offset some 200,000 tons of CO_2 emissions at installations in Germany and Austria (Sandbag 2011). The price difference between EUAs and CERs has provided the rationale for surrendering CERs to the extent possible with regard to the country-specific rules on the maximum share allowed for offsetting (Eriksson pers. comm.)

In 2010, SCA formed the new business unit SCA Energy as a response to the high market demand for biofuels and renewable electricity supply. This unit is organized under SCA Forest Products and will coordinate business activities in the segments of fuel from logging residues, refined biofuels and wind power (Fält pers. comm.). The strategy is a result of SCA's willingness to exploit the following conditions: SCA has large forest assets and logging activities, SCA Timber has several saw mills and SCA BioNorr has pellets production. These activities can generate renewable energy supply for the market as well as for SCA's own mills (e.g. the Östrand mill). In 2010, SCA supplied 3.5 TWh of forest-based fuels, thus emerging as a large player on the biofuel market (SCA 2011). The focus for SCA Energy is on growth and development of new innovative segments like automotive fuels. This is likely to influence the R&D strategies of the company (Fält pers. comm.). Increased interest for electricity generation from wind power and industrial CHP is also coordinated by SCA Energy. Due to its influence on power prices, the EU ETS has contributed to this strategic reorientation that will entail long-term consequences.

As a long-term abatement alternative SCA underlines the importance of its sizeable forest assets. SCA claims that its active forestry practices of clear-cutting and re-plantation imply a net annual growth of one per cent and a carbon sink of 2.6 million tons CO_2 per year (SCA 2011, Strandqvist pers. comm., Isaksson pers. comm.). Though there has been discussion of how a system based on avoided deforestation credits could be introduced under the EU ETS, the Commission does not intend to propose the creation of such a system earlier than 2020 (DG CLIMA 2008).

Norske Skog has introduced new procedures for monitoring, reporting and verification of CO_2 emissions as a direct consequence of the EU ETS (Carlberg pers. comm.). Norske Skog also reports that because of the ETS the company now ensures that investment and project planning include a cost on carbon – whether as a revenue-earning opportunity or as an additional cost. Where this is applicable and cost–effective, the company seeks to generate a maximum number of abatement certificates (emissions reductions) on-site (CDP 2011). Norske Skog reports that its main strategy for complying with the ETS is to reduce emissions internally rather than relying on the purchase of permits, although internal emissions reduction will grow more difficult as compliance targets become increasingly stringent. The company used CERs to offset some 60,000 tons of CO_2 emissions in the years 2008 and 2010 (Sandbag 2011).

According to Norske Skog's Carbon Disclosure Project report, the main impact of the EU ETS has been felt through the increase in electricity prices (CDP 2011). These higher costs could not be passed on to consumers due to the competitive nature of the publications-grade paper market. On the other hand, sale of surplus allowances have presented some mills with opportunities for making a profit. One of the company's European mills has had a significant income from sale of surplus allowances (Carlberg pers. comm.). By contrast, Norske Skog expects the third trading phase (2013–2020) to have a significant financial impact on the industry through higher energy prices and allocation of fewer allowances. Moreover, the sales of surplus ETS allowances could not compensate for the rise in electricity prices during the first two trading periods of the ETS. Although Norske Skog reports that the ETS thus far has had 'minimal financial impacts' on the company (CDP 2011), the increase in electricity prices caused by the ETS has created strong incentives for reducing energy consumption in the production process. Improvements in energy efficiency have thus become a main priority for the Norwegian company (Carlberg pers. comm.)

5.4 Summing up the Effects of the EU ETS

In assessing the overall impacts of the EU ETS on the climate strategies of SCA and Norske Skog, we may say that many activities, like energy efficiency improvements and fuel-switching measures, would have been conducted also without the scheme. Due to the EUA surplus, the scheme has not had any heavy impact on practical investment decisions. With this action-oriented view of climate strategy, the impacts of the EU ETS can be judged as relatively small. However, the EU ETS has served to raise awareness of the climate change issue among company staff and management. The EU ETS is amply covered by the media and in the public debate; it has become a reality that SCA and Norske Skog need to take into account. The scheme requires companies to monitor and report CO_2 emissions and to account for the cost of emissions in their financial procedures. We have seen that although SCA and Norske Skog had been collecting and reporting emissions data prior to the introduction of the ETS, the scheme has resulted in slightly more resources being put into site-level administration and reporting of GHG emissions data. Company project departments have more clearly become involved in integrating carbon prices into investment appraisals.

There are also many examples of synergies between EU directives and member-state implementation of energy and climate policy. The 2004 CHP Directive and the 2009 RES (Renewable Energy Sources) Directive are pushing in the same direction as the EU ETS: towards reduced CO_2 emissions. Biofuel-based and energy-efficient CHP is an attractive solution for the pulp and paper industry (PPI) in countries where there is reliable access to biomass, like Sweden, Finland and Norway. In countries like Germany, Austria, the Netherlands, Italy and the UK natural-gas-based CHP is the main alternative.

The influence of the EU ETS on power prices has contributed to heightened interest in electricity generation from wind power and industrial CHP. This strategic reorientation can be seen, for example, in the 2010 establishment of SCA Energy – a new business unit that coordinates the company's expansion in the market of renewable energy supplies, and serves both external customers and internal demands. Rising electricity prices, due partly to the EU ETS, are also a main driving force for company efforts to develop process changes to reduce specific energy use. Here we should recall that improvements in energy efficiency are also good for business, because energy efficiency gains are one of the few ways in which pulp and paper companies can increase their profits in a competitive market. Efforts to improve energy efficiency are therefore hardly controversial as a climate-policy tool in energy-intensive companies. With the recent developments of energy management system standards on national, European (EN 16001) and international levels (ISO 50001), there appears to be a potential for more structured practices in energy management. Several of SCA's and Norske Skog's mills have implemented and received certification for their energy management systems. From this broader perspective, the EU ETS can be judged as somewhat more influential on corporate climate strategy and related activities. However, our main conclusion is that the direct effect of the scheme with regard to reducing company-internal consumption of fossil fuel has been rather low. Also the companies tend to downplay the role of the EU ETS as a direct impetus for investments.

Regarding future outlook within the pulp and paper sector, changes in the ETS may mean better opportunities for long-term company planning. In the third period, the benchmark criteria are likely to dampen the overall generous allocations experienced so far, making the outlook for individual mills more predictable up until 2020.

6. Company-internal and Company-external Factors

We have observed that SCA has a more proactive strategy than Norske Skog with regard to engagement in innovation activities and CO_2-lean investment projects. How can we explain this divergence? Both company-external and -internal factors seem to have conditioned the effect of EU ETS on the climate strategies of the two companies. To begin with company-external factors: research has shown that companies are influenced by pressures in their home countries, meaning that the domestic context shapes company responses to emerging environmental issues (Rowlands 2000, Levy and Kolk 2002). One interviewee thought that the fact that Norske Skog is based in Norway, where the focus on climate change and other environmental problems was quite intense in the 1980s, was important for the company's early acceptance of the problem of climate change. But he also stressed that Norske Skog has evolved into an international company that applies the same environmental standards wherever it operates, and has one group-wide emissions reduction target (Carlberg pers. comm.)

Turning to SCA, our interviewees could not see how the domestic context of the company would influence its current visions, goals or values concerning climate change. They regard SCA as an international company and also recognize that climate change is on the political agenda all over Europe, subject to widespread attention among such stakeholder groups as customers, legislators and investors. Interviewees also noted the necessity of having a group-wide climate mitigation target that could be adapted to the varying situations in different EU member states and elsewhere. In each country, the mills experience different situations regarding access to feedstock, fuel/electricity supply, and policy context. Consequently, opportunities for reducing emissions vary significantly across regions and countries.

The domestic context clearly matters when it comes to electricity supply and the availability of biomass to replace fossil fuels. In Norway and Sweden, mills can rely on hydropower and CHP from biomass fuels, whereas mills elsewhere in Europe must rely on natural gas for a large share of the electricity and process heat needed for pulp and paper production. Also, since 2003, a Swedish policy scheme of tradable renewable electricity certificates (TREC) has stimulated SCA's mills in Sweden to increase auto-produced electricity (Ericsson et al. 2011). Until recently, there has been no corresponding support scheme in Norway, but in January 2012 Norway and Sweden established a common TREC scheme, so the motivation for Norske Skog's Norwegian mills to increase auto-produced electricity should increase.

Thus we find that the effect of the ETS is conditioned by several factors at the national and local levels, including access to biomass, electricity supply and policy context. This is especially evident for Norske Skog's mills located in Norway, where fossil fuels account for only 1 per cent of total energy demand. The result is very low emissions, as shown in Table 3 in section 5.2. This reflects a priority for using electricity rather than fuel oil for steam generation – which is not a viable option in other countries.

Regarding company-internal factors, SCA and Norske Skog differ as to size and production facilities. SCA's net sales in 2010 were 11.45 billion euros, and the company employed around 45,000 persons worldwide, while Norske Skog's net sales were around 2.6 billion euros and the number of employees was approximately 5,300 worldwide. SCA has a diverse portfolio of production facilities which offers many viable investment alternatives. By contrast, Norske Skog specializes in publication paper, for which market demand is weak. Another important difference is that Norske Skog has sold off the forests it used to own, whereas SCA, as Europe's largest private forest owner, can use its vast forest resources to produce biofuels, generate electricity from biomass sources, and experiment with wind power.

With regard to company 'cultures', it is hardly surprising to expect variation between the two companies that we have selected for closer study. However, it is quite challenging to identify exactly how the cultures differ and, not least, what these differences have meant for differing company responses to the EU ETS. Swedish and Finnish pulp and paper companies are known for their long history of product and process innovations (see e.g. Waluszewski 1990, Smith 1997, Laestadius 1998). According to recent rankings of the top 1000 EU companies by level of R&D investment, the companies Stora Enso, SCA and UPM are the three highest-ranking forestry and paper companies (JRC EC 2011). By contrast, Norske Skog has not stood out as a technological frontrunner in the past, nor is it listed among companies with the highest R&D investments. On the other hand, Norske Skog has been relatively quick to adopt new technology developed in collaboration with equipment manufacturers and the Swedish and Finnish pulp and paper industries. During the 1970s, for example, Norske Skog adopted Swedish technology to reduce air and water pollution. The technical solutions were developed by Swedish companies and delivered by Swedish suppliers (Sæther 2000: 190).

A company's or a mill's production factors are important for the kind of climate-related innovation activities that can be realized. Viable options for larger investments in energy installations depend heavily on the infrastructural and organizational context surrounding the mill in question. The SCA Östrand mill, for example, is located near the company's large forest assets, 2.6 million hectares of land area. At a nearby site, the business unit SCA BioNorr produces refined biofuels out of the byproducts from the surrounding sawmills under SCA Timber. This integration creates a supply chain and a logistic solution that ensures reliable and affordable access to fuel pellets, enabling investments in the biofuel-based lime kiln (Fält pers. comm.). The projects undertaken at the Östrand mill and the Witzenhausen mill (mentioned earlier) show that production factors (type of land area, natural resources, infrastructure etc.) clearly matter for what type of innovative and CO_2-lean investment solution can be accomplished. However, these factor conditions are not entirely inherited or given: they have been exploited and refined by SCA working together with other actors (see Porter 1990; Porter and van der Linde 1995). In the Sundsvall area, SCA has aligned several activities that interplay in something like an industrial cluster, creating a competitive supply chain. In Witzenhausen, on the other hand, SCA has outsourced electricity and heat production and contracted the owner and operator of the CHP plant, a company covering the whole 'waste-to-energy' value chain, to ensure a long-term energy supply. It is only when a long-term stability in the energy system, in combination with other aspects, can be expected that strategic decisions and investments like these are made (Fält pers. comm.). Although the revenues created from selling

EUAs contributed positively to the investment decision, production factors were more important for the final decision. In sum, then, the main explanation for SCA's more proactive strategy seems to be that it is a more diversified company than Norske Skog with more opportunities to experiment with emissions reduction projects in the countries where it operates.

7. Discussion

Regarding the models presented in Chapter 2, we find relatively strong support for the model of *companies as reluctant adapters* in the industry's response to the establishment of the EU ETS. A resisting strategy can be seen especially in CEPI's initial opposition to the EU ETS and in the efforts within the pulp and paper industry (PPI) to question the environmental effectiveness of the system. Most PPI companies responded as predicted by the model of companies as reluctant adapters to the EU ETS; compliance strategies depended on short-term pay-offs. However, the industry has gradually become more accepting of the EU ETS. Although it opposed the introduction of the ETS, the industry has come to accept the scheme as an important part of the regulatory framework for its European operations. This acceptance has partly to do with experiences from the first trading period, particularly the fact that due to over-allocation of emission allowances the economic consequences of the system proved less severe than expected. Whereas the gradual acceptance of the EU ETS can be explained in part by the lack of severe economic consequences from emissions trading, the considerable focus on energy efficiency improvements and fuel shift measures in the industry does not fit with the model of companies as 'reluctant adapters'.

From the model of *companies as innovators*, we expected that the EU ETS would alert and educate companies about the opportunities for better resource use and technological improvements, and raise the likelihood that product and process innovations would be environmentally sound solutions. Although over-allocation of allowances has provided the companies with scant incentives for long-term and far-reaching strategic changes so far, the EU ETS seems to have strengthened on-going abatement efforts. The main impact of the EU ETS has been felt indirectly through the increase in electricity prices, which could not be compensated for through the sale of surplus allowances. In combination with other factors, the EU ETS seems to have given the pulp and paper companies incentives for reducing specific energy use, through increased energy efficiency measures, and for changing energy carriers or fuels, for example from fossil fuels to biofuels or other less carbon-intensive alternatives. In parallel, there has been growing interest in internal energy generation, which often implies energy-efficient co-generation. SCA and a few other major pulp and paper companies are also engaged in partnerships to facilitate investments in new power assets. Our company studies demonstrate that SCA, in particular, has initiated several new projects relating to energy efficiency improvement and fuel switching as ways of reducing CO_2 emissions. These projects show that SCA is working strategically on long-term abatement options, but the EU ETS is only one of many factors that influence energy efficiency and fuel-switching measures.

The EU ETS has also brought about some changes in the monitoring, reporting and verification of CO_2 emissions in the pulp and paper industry. Our examination of SCA and Norske Skog has shown that the reporting of emissions has been extended to include more staff hours at the mills. As noted, the Carbon Disclosure Project has named SCA the best Swedish company and Norske Skog the best Norwegian company in terms of CO_2 emissions accounting. From our examination of accounting and reporting practices of ten major pulp and paper companies, SCA and Norske Skog do appear advanced in this regard. Interestingly, the Finnish–Swedish company Stora Enso, which also belongs to the pulp and paper sector, was named the number one company in the Nordic region (CDP 2010). This indicates that the PPI sector should be particularly advanced in CO_2 emissions accounting, at least from a Nordic perspective.

Although the model of companies as innovators gains some support from our case studies of SCA and Norske Skog, it is very difficult to separate the effects of the EU ETS from other factors that may influence corporate climate strategies. Our case studies have shown that various factors – including the EU ETS, other EU regulations (the RES Directive and CHP Directive), domestic climate and energy policies, and increasing scientific and societal concern over the climate change problem – have exerted influence on the climate strategies of SCA and Norske Skog. These factors have all contributed to increase company attention to the climate change problem, strengthen corporate climate strategies, and stimulate the search for short-term and long-term abatement opportunities. It is impossible to point to specific abatement projects that were initiated solely or chiefly because of the EU ETS. In all cases, various factors have influenced investment decisions and the choice of solutions. As seen in the projects undertaken by SCA at the Östrand and Witzenhausen mills, production factors are important to the type of abatement options that can be accomplished. Large investments in energy installations are capital-intensive and dependent on the infrastructural and organizational context surrounding the mill. As with energy efficiency and fuel-switching measures, the EU ETS is only one out of several factors that influence such investment decisions – and it is not the most important one.

From the model of *companies as socially responsible*, we expected the EU ETS to influence corporate norms about appropriate conduct and voluntary initiatives. According to this model, regulation may 'crowd-out' or 'crowd-in' corporate norms of social responsibility. Our study has shown that all the largest pulp and paper companies covered by the EU ETS seem to recognize the problem of climate change and their responsibility to contribute to problem-solving. We have seen that the largest pulp and paper companies covered by the ETS have gradually adopted more proactive and converging climate strategies. Most of the companies examined have adopted short- and medium-term abatement targets, which they have either achieved or are on track to achieve.

Regarding the two companies in focus in this study, both SCA and Norske Skog had acknowledged the climate change problem and adopted voluntary targets and measures even before the ETS was launched. Since the introduction of the scheme, they have strengthened voluntary targets and measures. From both the sector study and the company study of SCA and Norske Skog, we can thus conclude that there has not been any 'crowding out' of corporate norms of social responsibility following the establishment of EU ETS. Nor do we find any indications of a 'crowding in' of climate protection norms, in the form of significantly strengthened joint innovation efforts to realize a low-carbon economy. The industry has not launched any major innovation programmes for developing climate-friendly technologies within the sector. Indeed, it seems that the pulp and paper industry is more focused on continuous improvements of operations than on innovation and joint efforts to develop breakthrough technologies. The lack of a strong focus on innovation for CO_2-lean solutions is striking in the overall examination of company reports as well. Hence, while the widespread adoption of climate objectives shows that having a climate strategy has become appropriate behaviour in the PPI, the evidence indicates that the EU ETS has neither crowded-out nor crowded-in corporate norms of social responsibility in the industry.

8. Conclusions

This chapter has shown that although the EU ETS has strengthened ongoing abatement efforts in the PPI, emissions trading has had a rather limited effect on innovation activities in the industry thus far. From our in-depth study of SCA and Norske Skog we may conclude that the EU ETS has not had any strong impact on major investment decisions. However, both companies recognize that EUAs represent costs or revenues for their mills in every investment decision. Moreover, the impact of the EU ETS on investment decisions is expected to increase in the third trading period, and the companies are taking this effect into account in their forecasts and long-term planning. Both SCA and Norske Skog have also adjusted their CO_2 monitoring, reporting and verification procedures as a direct consequence of the ETS.

Rising electricity prices are perceived to be the strongest influence of the EU ETS. Rising prices have led to strategic decisions to search for alternatives to being buyers on the deregulated electricity spot market. Electricity-intensive pulp and paper companies are showing greater interest in investing in power assets, either on their own or in different constellations; in making bilateral agreements about long-term power contracts; and in engaging in other forms of energy contracting.

The in-depth company study shows that SCA, in particular, has initiated various new projects relating to energy efficiency improvement and fuel switching as ways of reducing its CO_2 emissions. Since improvements in energy efficiency and fuel-shift measures are often good for business as well, these projects demonstrate that SCA is working strategically on long-term abatement options. While Norske Skog too has adopted a quite progressive climate strategy, it has fewer possibilities, and perhaps less willingness, to experiment with innovative activities. Here it should also be noted that SCA has maintained a low profile when it comes to prospects for the long-term abatement options like black liquor gasification and CCS.

Our analysis shows that some activities, like energy efficiency improvement actions, can also be attributed to other policy programmes or to an autonomous development. In this perspective, the EU ETS can be seen as one factor among others, but one which has as yet had rather limited influence on corporate climate strategies. For the system to have greater influence on company investment decisions, a more stringent cap and a much higher EUA market price are required.

The effect of the EU ETS is conditioned by various factors at the national and regional level, including access to biomass, electricity supply, and policy context. Our case studies also show that company-internal factors influence corporate responses to the EU ETS and help to explain why SCA has initiated more innovation activities and CO_2-lean investment projects than Norske Skog. The effect of the ETS is thus mediated by market factors and factors of production as well as company-internal factors. A closer study of such interactions is called for in future studies of climate-mitigation activities, strategies and innovations in the European pulp and paper industry and other industries. The ultimate performance of the EU ETS depends critically on how the dynamic interactions between the system and other policy programmes and market factors influence corporate climate strategies.

List of Interviews and Personal Communications

Christer Fält, SCA, Environmental Manager SCA Forest Products, 15 April 2011.
Georg Carlberg, Norske Skog, Vice-President Environment, 13 June 2010, 30 June 2011, and 27 October 2011.
Kersti Strandqvist, SCA, Senior Vice-President Corporate Sustainability, 19 April 2011.
Marco Mensink, CEPI, Energy and Environment Director, 28 January 2011.
Patrik Isaksson, SCA, Vice-President Environmental Affairs, 20 April 2011.
Per-Erik Eriksson, SCA, Vice-President Energy, 11 October 2011.
Tomas Wyns, Climate Action Network Europe, EU ETS Policy Officer, 27 January 2011.
Yvon Slingenberg, European Commission, 27 January 2011.

References

Bengtsson, S. 2010. En massa energi från det självförsörjande bruket. *Energimagasinet*, 32(3): 24–8.

Carbon Disclosure Project (CDP). 2010, 2011. Norske Skog. Available at https://www.cdproject.net/en-US/Pages/HomePage.aspx

Carbon Disclosure Project (CDP). 2007, 2008, 2009, 2010. SCA. Available at https://www.cdproject.net/en-US/Pages/HomePage.aspx

CEPI. 2003. *Declaration of Intent on Renewable Energy Sources (RES)*. Brussels: Confederation of European Paper Industries (CEPI).

CEPI. 2007a. *Bio-energy and the European Pulp and Paper Industry – An Impact Assessment. A Summary of the Conclusions of the Report*. Brussels: Confederation of European Paper Industries (CEPI).

CEPI. 2007b. *Framework for the Development of Carbon Footprints for Paper and Board Products*. Brussels: Confederation of European Paper Industries (CEPI).

CEPI. 2011a. *Key Statistics 2010 – European Pulp and Paper Industry*. Brussels: Confederation of European Paper Industries (CEPI).

CEPI. 2011b. *CEPI Sustainability Report 2011*. Brussels: Confederation of European Paper Industries (CEPI).

CEPI. 2011c. *Unfold the Future: The Forest Fibre Industry – 2050 Roadmap to a low-carbon bio-economy*. Brussels: Confederation of European Paper Industries (CEPI).

CITL. 2011. *Verified emissions for 2008–2009–2010 and allocations 2008–2009–2010* (15 April 2011). Accessed 14 May 2011 at http://ec.europa.eu/clima/policies/ets/registries/documentation_en.htm

Convery, F.J. and Redmond, L. 2007. Market and Price Developments in the European Union Emissions Trading Scheme. *Review of Environmental Economics and Policy*, 1(1): 88–111.

DG CLIMA. 2008. 'Questions and answers on deforestation'. Accessed 20 November 2011 at http://ec.europa.eu/clima/policies/forests/deforestation/documentation_en.htm

DG CLIMA. 2011. *Guidance Document n°9 on the harmonized free allocation methodology for the EU-ETS post 2012 – Sector-specific guidance*. European Commission Directorate-General Climate Action, August 2011.

Ecofys. 2009. *Methodology for the free allocation of emission allowances in the EU ETS post 2012 – Sector report for the pulp and paper industry*. Report prepared by Ecofys, Fraunhofer ISI and Öko-Institute. November 2009.

ENDS Daily. 2010. EU consults on low CO_2 future, 29 October.

Ericsson, K., Nilsson, L.J. and Nilsson, M. 2011. New energy strategies in the Swedish pulp and paper industry: The role of national and EU climate and energy policies. *Energy Policy*, 39(3): 1439–49.

Hyvärinen, E. 2005. The downside of European Union emission trading – a view from the pulp and paper industry. *Unasylva*, 56(222): 39–41.

IEA. 2007. *Tracking Industrial Energy Efficiency and CO_2 Emissions – In Support of the G8 Plan of Action*. Paris: International Energy Agency.

IEA. 2008. *Energy Technology Perspectives 2008: Scenarios & Strategies to 2050*. Paris: International Energy Agency.

JRC EC. 2011. *Monitoring Industrial Research: The 2011 EU Industrial R&D Investment Scoreboard.* Luxembourg: Publications Office of the European Union.

Kossoy, A. and P. Ambrosi. 2010. *State and Trends of the Carbon Market 2010.* Washington, DC: The World Bank.

Laestadius, S. 1998. The relevance of science and technology indicators: The case of pulp and paper. *Research Policy*, 2(4): 385–95.

Levy, D. and A. Kolk. 2002. Strategic responses to global climate change: Conflicting pressures on multinationals in the oil industry. *Business and Politics*, 4(3): 275–300.

Möllersten K. 2002. *Opportunities for CO_2 Reductions and CO_2-lean Energy Systems in Pulp and Paper Mills.* Stockholm: Royal Institute of Technology.

Norske Skog. 1991. *Årsrapport 1990.* Lysaker, Norway: Norske Skog.

Norske Skog. 1999. *Årsrapport 1998.* Lysaker, Norway: Norske Skog.

Norske Skog. 2002. *Environmental Report 2001.* Lysaker, Norway: Norske Skog.

Norske Skog. 2010. *Annual Report 2009.* Lysaker, Norway: Norske Skog.

Norske Skog. 2011. *Annual Report 2010.* Lysaker, Norway: Norske Skog.

Nyström, H., Smeder, B. and Mark-Herbert, C. 1997. *Miljöstrategier för produkt- och företagsutveckling i svensk skogsbaserad industri.* Report 110. Uppsala: Swedish University of Agricultural Sciences.

Porter, M.E. 1990. *The Competitive Advantage of Nations.* London: Macmillan.

Porter, M.E. and van der Linde, C. 1995. Towards a New Conception of the Environment-Competitiveness Relationship. *Journal of Economic Perspectives*, 9(4): 97–118.

Rogge, K.S., Schleich, J., Haussmann, P., et al. 2011. The Role of the Regulatory Framework for Innovation Activities: The EU ETS and the German Paper Industry. *International Journal of Technology, Policy and Management*, 11(3–4): 250–73.

Rowlands, I.H. 2000. Beauty and the Beast? BP's and Exxon's Positions on Global Climate Change. *Environment and Planning C*, 18(3): 339–54.

Sandbag. 2011. 'Sandbag data tools'. Accessed 20 November 2011 at http://www.sandbag.org.uk/accounts/login/?next=/data/

SCA. 1999. *SCA Environmental Report 1998.* Stockholm: SCA

SCA. 2002. *SCA Environmental Report 2001.* Stockholm: SCA

SCA. 2009. *SCA Sustainability Report 2008.* Stockholm: SCA

SCA. 2011. *SCA Sustainability Report 2010.* Stockholm: SCA

SEA. 2008. *Kunskapsinventering – kring vindar och vindkraft i skog.* Report ER 2008:21. Eskilstuna: Swedish Energy Agency.

Sijm, J., Neuhoff, K. and Chen, Y. 2006. CO_2 Cost Pass-through and Windfall Profits in the Power Sector. *Climate Policy*, 6(1): 49–72.

Smith, M. 1997. *The U.S. Paper Industry and Sustainable Production: An Argument for Restructuring.* Cambridge, MA: MIT Press.

Sæther, B. 2000. *Miljøarbeid i norsk treforedlingsindustri 1974–1998. NIBRs Pluss-serie.* Oslo: NIBR.

Sæther, B. 2004. From National to Global Agenda: The Expansion of Norske Skog 1962–2003. In A.A. Lehtinen, J. Donner-Amnell, and B. Sæther (eds), *Politics of Forests: Northern Forest-industrial Regimes in the Age of Globalization*. Aldershot: Ashgate.

Vindkraft Norr. 2011. *Översiktskarta*. Accessed 20 October 2011 at http://www. vindkraftnorr.se/vindparkerna_oversiktskarta.asp

Wadsborn, R., Berglin, N. and Richards, T. 2007. *Konvertering av mesaugnar från olje- till biobränsleeldning – drifterfarenheter och modellering*. Stockholm: Värmeforsk.

Waluszewski, A. 1990. *Framväxten av en ny mekanisk massateknik – en utvecklingshistoria*. Uppsala: Acta Universitatis Upsalensis.

Wiberg, R. 2007. *Energiförbrukning i massa- och pappersindustrin 2007*. Stockholm: Skogsindustrierna.

Appendix Table 5.1   Climate strategies of the 10 largest pulp and paper companies in the EU ETS

Company	Short-term climate and energy related targets and measures	Long-term low-carbon solutions	Acknowledgement of problem and responsibility for problem-solving	Stated importance of ETS and EU climate policy
Stora Enso (Finland/ Sweden)	20 per cent reduction in relative fossil CO_2 emissions by 2020, compared to 2006. Increase internal energy generation and power-to-heat ratio. Reduce specific energy consumption (by 1–2 per cent annually).	Currently no target beyond 2020. Innovation on second generation bio-fuels. Shareholder in wind power company.	Yes, since the company was established in 1998.	Accepts the EU ETS, and reports that the scheme is important for the company's climate strategy. Prepares for a tighter phase 3. Takes all costs of carbon into consideration, and investment planning assumes the prolonging of the EU ETS beyond 2020.
UPM (Finland)	15 per cent reduction in relative fossil CO_2 emissions by 2020, compared to 2008. Continuously improve energy efficiency and increase the use of carbon-neutral/biomass energy.	Currently no target beyond 2020. Investments in mills with biomass power plants. Assets in nuclear and hydropower generation. Innovation on second generation bio-fuels.	Yes	Accepts the EU ETS, but is cautious about possible risks arising from uncertain and inconsistent regulation.
Smurfit Kappa Group (Ireland)	20 per cent reduction in relative fossil CO_2 emissions by 2020, compared to 2005. Energy efficiency improvement and increased cogeneration of biofuels (no explicit target formulations).	Currently no target beyond 2020. Investments in biomass technology.	Yes	Accepts and supports the EU ETS, but is cautious about possible risks arising from uncertain and inconsistent regulation affecting the global level playing field. Carbon costs are taken into account, and the EU ETS is an important part of the climate strategy's backdrop.

Company	Short-term climate and energy related targets and measures	Long-term low-carbon solutions	Acknowledgement of problem and responsibility for problem-solving	Stated importance of ETS and EU climate policy
M-real, part of Metsäliitto group (Finland)	No explicit target formulation on CO_2 reductions. A group level climate programme prioritizes energy efficiency and increased share of carbon-neutral energy. A rolling target to reduce 1 per cent of specific energy usage per year, according to latest CDP report.	Currently no long-term target formulations. Investments in biomass technology. Shareholder in a nuclear power plant.	Yes	Accepts and complies with the EU ETS. Carbon costs and climate issues are taken into account, and the rising costs resulting from the ETS is counterbalanced with improvement in energy efficiency.
SCA (Sweden)	20 per cent reduction in relative fossil CO_2 emissions by 2020, compared to 2005. 7.5 per cent reduction in specific energy consumption by 2012, compared to 2005. Increased use of carbon-neutral energy.	Currently no target beyond 2020. Business unit formed to coordinate interests on future energy markets. In a joint venture to facilitate large-scale wind power exploitation.	Yes	Accepts and complies with the EU ETS. Carbon costs and effects on electricity prices are taken into account in investments and planning. EU ETS is one out of several important factors that shape the corporate climate strategy.
Burgo Group (Italy)	Current target to reduce relative CO_2 emissions by 4 per cent, compared to previous year (target updated every year). Increased use of biomass cogeneration.	Currently no long-term target formulations. Research into technologies that minimize energy consumption.	Yes	No data available.

Company	Short-term climate and energy related targets and measures	Long-term low-carbon solutions	Acknowledgement of problem and responsibility for problem-solving	Stated importance of ETS and EU climate policy
Mondi (South Africa/ UK)	15 per cent reduction in relative fossil CO_2 emissions by 2014, compared to 2004. 15 per cent reduction in specific energy consumption by 2014, compared to 2004.	Currently no target beyond 2014. Focus on smart solutions for energy efficiency and increased use of biomass. Research on new products and packaging solutions.	Yes	Accepts and complies with the EU ETS. Cautious about possible risks arising from the ETS, but thinks the scheme will force paper companies to improve their energy management systems.
Sappi (South Africa)	Regional targets. 1 per cent per year reduction of European relative fossil CO_2 emissions. Increased use of renewable energy and biomass cogeneration.	Currently no long-term target formulations. Sustainable packaging solutions, research on alternatives to fossil raw materials.	Yes	Accepts and complies with the EU ETS. Cautious about possible risks arising from the ETS, e.g. related to benchmarks.
SAICA (Spain)	No target or strategic formulations on CO_2 reductions. Recovered paper and biomass cogeneration.	Currently no long-term target formulations.	No data available.	No data available.

Company	Short-term climate and energy related targets and measures	Long-term low-carbon solutions	Acknowledgement of problem and responsibility for problem-solving	Stated importance of ETS and EU climate policy
International Paper (USA)	No explicit target formulation on CO_2 reductions. Fulfilled previous target of 15 per cent reduction in CO_2 emissions by 2010, compared to 2000. Increase internal energy generation by CHP. Improve energy efficiency. Increased bioenergy.	Currently no target beyond 2010.	Yes	Accepts the EU ETS. The regulatory risk posed by the EU ETS so far has been minimal, but this may change in the future.

Source: Company reports and websites, Carbon Disclosure Project, and e-mail correspondence. Companies are ranked in terms of their amount of verified emissions in the EU ETS.

Chapter 6

Cement Industry

Anne Raaum Christensen

The cement industry portrays itself as both a source of and a solution to the climate change problem. A solution because societies will have to adapt to more extreme and unpredictable weather – and such adaptation will be largely about building new infrastructure and housing.[1] More importantly, the cement industry acknowledges that it is part of the climate change problem, accounting for 25 per cent of industrial carbon dioxide (CO_2) emissions and about 5 per cent of global CO_2 emissions (IEA 2009: 2). This makes it the second largest CO_2-emitting industry, after power generation. The sector is now faced with the double challenge of expanding its business while at the same time decreasing its carbon intensity. This chapter asks: How have the climate strategies of European cement companies changed since 2003?[2] To what extent and how has the EU's Emissions Trading Scheme (EU ETS), the cornerstone of EU climate policy, affected these changes?

Studies of cement industry climate strategies and how the EU ETS has affected the sector are limited to date, but there are some exceptions (see e.g. Rehan and Nehdi 2005, Cook 2009). Most studies have looked at how the ETS will affect the competitiveness of the cement industry (e.g. Szabó et al. 2003, 2006, Reinaud 2005, Demailly and Quirion 2006, Ponssard and Walker 2008, Reinaud 2008, Cook 2009). Ellerman et al. (2010) have examined the short-term effects of the scheme, and find no impact of the EU ETS on cement trade during 2005 and 2006. Several NGOs and think-tanks (e.g. Oxfam 2010, Sandbag 2010, Sandbag 2011) have also studied the ETS impact on the cement industry, focusing on the negative effects of over-allocation and windfall profits. Studies of individual companies and their climate strategies are rare: two exceptions are Nordqvist et al. (2003) looking at Lafarge, and Drake et al. (2009) studying HeidelbergCement. On the sector level, the cement industry's own initiative, the Cement Sustainability Initiative (CSI), has been studied as an example of sectoral approaches to climate change (see Fujiwara 2010, Fujiwara et al. 2010). All in all, the literature has focused mostly on the short-term effects of the ETS on competitiveness in the sector as a whole.

---

1   The need to build flood barriers, coastal defences, irrigation systems, bridges and secure buildings will lead to increased demand for concrete, a cement-based material (Cembureau 2009b).

2   Although the main emphasis is on strategy changes after the adoption of the EU ETS in 2003, company changes before 2003 will also be taken into account where they can be seen as anticipatory effects of the EU ETS.

This chapter will complement these findings, examining how specific companies have responded to the climate challenge in general and the mandatory EU ETS in particular. Through company studies we aim to detect the mechanisms through which the ETS works, and gain a fuller understanding of the various drivers of change in corporate climate strategies.

In addition to an aggregate sector analysis, two companies are studied in greater depth in this chapter: Holcim from Switzerland and HeidelbergCement from Germany. Holcim could be expected to be a progressive company, and was selected because of its reputation as a frontrunner in the sector (interviews 2011). Holcim was furthermore chosen due to its position as market leader for clinker substitution, one type of climate strategy available to a cement company. As for HeidelbergCement, it was the first company in the sector to look into Carbon Capture and Storage (CCS) technology in Europe, as well as being an alternative fuels frontrunner. In other words, the two companies have been selected because they have been applying three different and important climate mitigation measures available to the sector: clinker substitution, CCS and alternative fuels use.[3]

Our data indicate that climate strategies in the cement sector have become gradually more proactive in the first decade of the 2000s, but with inter-company differences in strategic responses. It is very difficult to ascertain to what extent these changes can be attributed to the ETS, as the scheme works together with, and is conditioned by, a range of other company-external and company-internal factors. However, some cautious conclusions can be drawn. In the short term, the EU ETS has not been able to provide strong economic incentives for strategy changes, due to significant over-allocation of allowances and low carbon prices. Nonetheless, expectations of a *more stringent* ETS in the near future underpin the companies' increasingly proactive long-term climate strategies. The trading scheme has put carbon on the corporate agenda, and has contributed to a 'mind-change' in the cement companies. We find clear indications that the ETS has directed company attention to previously underexplored or unattended business lines – like energy efficiency measures, clinker substitution and CCS technologies.

We begin by presenting some sector characteristics, focusing on abatement opportunities in cement installations. In section three, the aggregate climate strategy of the cement sector in the EU is presented, with Cembureau's[4] EU ETS position and the Cement Sustainability Initiative (CSI) as starting points. We also present the climate strategies of nine major cement companies, showing that most companies now take responsibility for problem-solving and have taken on voluntary short- and medium-term climate commitments. The main bulk of

---

3   Nonetheless, the two companies have similar climate strategies. Had we chosen other companies (such as CEMEX or Lafarge) we would most likely have found greater variation in climate strategies. Holcim and HeidelbergCement were selected due in part to pragmatic reasons, as it is relatively easy for a Norwegian scholar to obtain information about and from these companies.

4   Cembureau is the organization of the cement industry in Europe.

the chapter is found from section four onwards, where the climate strategies of Holcim and HeidelbergCement are presented and discussed. What are the main similarities and differences in their climate strategies? In section five we discuss possible explanatory factors, focusing on whether and how the EU ETS has affected the observed variation in climate strategies, in the short- and longer term. Section five also brings in other explanatory factors external and internal to the company. Section six offers a discussion of the research questions, followed in the final section by some concluding remarks and reflections.

1. Sector Characteristics

1.1 Cement Industry Overview

World production of cement increased from 1.67 billion tonnes in 2000 to over 3 billion tonnes by 2010, showing an uninterrupted and steadily growing trend (Cembureau 2011: 4). Table 6.1 presents the shares of cement production for the various world regions 2003–2010. China alone accounted for more than half of world production in 2010, with Asia in total accounting for almost 80 per cent (Cembureau 2011: 5). Cembureau countries accounted for about 8.6 per cent of world cement production in 2009, and 7.7 per cent in 2010. The largest operating companies in the world with cement interests in the European Union are CEMEX (Mexico), HeidelbergCement (Germany), Holcim (Switzerland), Italcementi (Italy) and Lafarge (France).

Most cement producers are global actors – with production sites in several countries and continents. As a commodity, however, cement is mostly limited to domestic and regional markets. Transport costs are high compared to the cement price, and since the raw material for cement can be found almost everywhere, this reduces the role of international trade in the sector (Szabó et al. 2006: 73). Despite limited trade, the cement sector – as an integrated clinker-cement producer – still qualifies as a sector vulnerable to carbon leakage under the EU ETS. This is not on the basis of trade intensity, but because of the increased costs directly and indirectly induced by the trading scheme.[5] According to a study by the Boston Consulting Group, at CO_2 prices of 25€ per tonne, approximately 80 per cent of clinker production would be outsourced if no free allowances are allocated (BCG 2008: 1).

The financial situation of the cement sector has fluctuated in recent years. After experiencing growth and a boom around 2006/20007, the cement sector – especially in Europe and the USA – was hit severely by the economic crisis in

---

5   According to Article 15 of Directive 2009/29/EC, for a sector to be deemed exposed to a significant risk of carbon leakage, it must be expected to face a 30 per cent or greater increase in production costs under Phase III of the ET-ETS, or a 5 per cent increase in production costs along with a 10 per cent increase in trade intensity.

Table 6.1 World cement production by region (%)

	2003: (1.94 bT)	2004: (2.11 bT)	2005: (2.27 bT)	2006: (2.54 bT)	2007: (2.77 bT)	2008: (2.83 bT)	2009: (3 bT)	2010: (3.3 bT)
Asia	67.0	67.6	68.1	69.4	70.1	71.2	75.6	77.1
Oceania	0.4	0.5	0.5	0.4	0.4	0.4	0.4	0.3
Europe	14.4	14.0	13.5	13.1	12.1	11.3	9	8
Commonwealth of Independent States (CIS)*	2.8	3.2	3.3	3.4	3.4	3.2	2.5	2.5
Americas	11.3	10.5	10.6	9.7	9.6	9.1	7.9	7.3
Africa	4.1	4.1	4.0	4.0	4.4	4.7	5.6	4.8

Note: * Former Soviet Republics.
Source: Cembureau Activity Reports 2003–2010.

2008/2009 (Cembureau 2010). Cement consumption in Europe declined, and the sector has still not recovered fully. Cement is a capital-intensive and energy-intensive industry, but not a labour-intensive one. A modern plant is usually manned by fewer than 150 people; in the EU today the cement industry represents 58,000 direct jobs.

1.2 The Cement Manufacturing Process and CO$_2$ Emissions

Cement is a binding agent: it sets and hardens independently, and then becomes water-resistant and as durable as rock. The most important application of cement is the production of concrete, where cement is used to bind the aggregate materials together. The most commonly used cement type is Ordinary Portland Cement (OPC), produced by grinding Portland cement *clinker*. The manufacturing process consists of three main steps: raw material preparation, clinker production and cement grinding (IEA 2009).[6] CO$_2$ emissions from cement production are expected to increase in the future, as rapid growth rates are expected in large developing regions of the world (Szabó et al. 2006: 74). The EU stands for around 10 per cent of sector emissions worldwide, which equates to 8 per cent of all emissions capped under Phase I of the EU ETS (Cook 2009: 9). There are three principal sources of CO$_2$ emissions from the cement manufacturing process: process-related emissions, combustion-related emissions and indirect emissions.

Emissions from the manufacturing process stem mainly from the calcination of limestone in the kiln when making clinker. When calcium carbonate is heated, lime and carbon dioxide are produced. This chemical process accounts for some 50 to 60 per cent of total CO$_2$ emissions. The principal option available for reducing these process emissions is the increased blending of cement with other substances than clinker. There are two main alternatives to clinker – blast furnace slag and fly ash. Slag is a byproduct of the steelmaking process; fly ash is waste produced at coal-fired power stations.

Combustion-related emissions stem from the significant reliance on coal and petroleum coke to fuel the kilns for clinker production. This accounts for around

---

6 The cement manufacturing process starts at the quarry, where limestone and other raw materials are extracted through drilling and blasting techniques. To make the intermediate product into clinker, the raw materials are crushed and fed into a rotary kiln for roasting. Clinker production is a highly energy-intensive process, requiring extremely high temperatures – typically between 1200 and 1600 degrees celsius. Conventionally, fossil fuels like coal, petcokes, oil, and natural gas supply this energy. Due to heat and basic oxides, new compounds emerge – small lumps known as 'clinker'. The final phase is handled in a cement-grinding mill. After the clinkers have been cooled down, gypsum and other materials are added. All these constituents are the ground, resulting in a fine and homogeneous powder. The cement is stored in silos until weighing and packing. In the case of OPC, only around 5 per cent gypsum is added, so the clinker content is very high. In Europe, 'blended cements' are used, consisting of a mixture of clinker and other products with cementitious properties.

40 per cent of emissions. Such emissions can be reduced through improvements in thermal and electric efficiency – providing new plants and upgrading old ones. Combustion-related CO_2 emissions can also be reduced by replacing fossil fuels with waste and/or biomass fuels. Waste materials such as tyres, plastics, waste oils and solvents can be combusted, due to the high process temperature and the availability of various gas-cleansing agents. Biomass such as animal meal, recycled wood and sewage sludge are also typical alternative fuels used by the cement industry. Less than 10 per cent of emissions are indirect, and stem from electricity use for raw materials and clinker grinding, transport and cement finishing.

In addition to the abatement opportunities already mentioned, carbon capture and storage (CCS) has been identified as a promising technology (IEA 2009),[7] although it is has not been tested out on an industrial scale in cement production. Two main technologies are of interest here: post-combustion (end-of-pipe) and oxy-firing technologies. Early research and pilot tests are needed to gain practical experience, and ECRA (the European Cement Research Academy) initiated a CCS project in 2007 (Schneider 2010). There is a technical potential for CCS in the cement industry, but currently at very high costs, and so the development of an economic framework will be decisive.

Several studies have looked into the potential for improvements in CO_2 emissions reductions in the cement sector (e.g. Liu et al. 1995, Worrell et al. 2000, Taylor et al. 2006, Damtoft et al. 2008, Deja et al. 2010, Habert et al. 2010, Madlool et al. 2011, and Pardo et al. 2011). The data show a significant decoupling of CO_2 emissions and cement production since 1990 (CSI 2009: 13).[8] According to Cook (2009: 10), this positive development, particularly in Europe, is due to increased blending rates and use of alternative fuels. Europe now has the lowest average CO_2 intensity worldwide. When comparing the effects of potential technological improvements in the sector to global sustainability goals, Habert et al. (2010) find that a technological *turnaround* is needed – affecting both the cement material itself and the way in which it is used. Similar conclusions can be found in the Cement Technology Roadmap (IEA 2009), where it is asserted: 'the cement industry could not be carbon neutral within its existing technology, financing and innovation framework.' In other words, the cement industry will have to conduct or support R&D at a much higher level than today. Since emissions have already been reduced (for instance by replacing old kilns with new and more efficient

---

7   Several works that examine the prospects for CO_2 capture in the cement industry have been published in recent years. These studies generally assess the various technologies that could be used for CO_2 capture in cement plants, their costs and so on (see Barker et al. 2009, Bosoaga et al. 2009, Romeo et al. 2011).

8   We can also note a decrease in average CO_2 intensity in recent years, especially in China and India (due to the replacement of smaller shaft kilns with high-efficiency dry rotary technology), but also in Europe as well. While global cement production increased by 54 per cent from 2000 to 2006, absolute CO_2 emissions increased by an estimated 42 per cent (IEA 2009: 2).

ones), concerns are growing that abatement options may become more limited, and more expensive.

Nonetheless, there are several possibilities for CO_2 reductions in the cement sector. The next question thus becomes: To what extent and how have cement companies' climate strategies changed since the early 2000s? What abatement opportunities are regarded as the most relevant and feasible, and what measures and targets have the sector and major companies put in place? Have the cement companies used the EU ETS as an opportunity to exploit reduction possibilities and increase efficiency – or have they tended to resist change?

2. Changes in Aggregate Climate Strategies

On the sector level, the industry initiated the Cement Sustainability Initiative (CSI) in 1999 (WBCSD 2002). Many major cement companies agreed to voluntary emission-reduction targets as a result of this initiative, so industry efforts to reduce CO_2 emissions predate the beginning of the EU ETS. However, since its adoption, the EU ETS has been at the centre of attention for European cement producers, as the first *mandatory* international climate instrument targeting the sector. Here we will present data indicating changes in climate strategies in a more proactive direction since 2003, as companies have begun to take responsibility for problem-solving, investigating various abatement options. After describing the CSI, we go on to discuss how the sector as a whole, through Cembureau,[9] has responded to the EU ETS. This will provide some first indications of how this instrument was perceived and assumed to affect the industry. Thirdly, the climate strategies of nine individual companies are briefly presented, for a closer look at the level and type of targets and measures chosen to meet the climate challenge.

2.1 The Cement Sustainability Initiative (CSI)

In 1999, three of the largest cement groups – Cimpor, Holcim and Lafarge – approached the World Business Council for Sustainable Development (WBCSD). They wanted to explore what *sustainable development* would mean for the cement industry, and to identify and facilitate actions that companies could take, as a group and individually (WBCSD 2002: 4). Several companies joined, and in February 2000 the Cement Sustainability Initiative (CSI) was launched, involving 10 corporations that together produced one-third of the world's cement. This was a first-of-its-kind project for the cement industry, and stands as one of few examples

---

9   Cembureau is the representative organization of the cement industry in Europe. Its full members are national cement industry associations and cement companies of the European Union, plus Norway, Switzerland and Turkey (Cembureau 2011). The Association acts as spokesperson for the cement industry before the European Union institutions and other public authorities.

of industry-led transnational initiatives linked to the deployment of low-carbon technologies (Fujiwara et al. 2010).

From the start, 'climate protection' was one of the main work areas of CSI.[10] Companies began to realize that climate change would be a prominent global policy issue in the future, with the potential for large economic impact on the sector (Battelle 2002b: v). According to Nordqvist et al. (2003: 76), the cement industry wanted to be prepared for the climate instruments initiated in the aftermath of the Kyoto Protocol (KP): 'It is not explicitly stated in the document of the CSI; still it is not misleading to point out these factors [the ongoing political processes of the UNFCCC and the KP] as driving forces for many Working Group members to participate.'

The first phase of the CSI focused on research and information collection. The Initiative commissioned an independent study of climate change and the cement industry, published in March 2002, which recommended that the industry should establish corporate carbon management programmes, set company-specific and industry-wide CO_2 reduction targets, and initiate process and product innovation (Battelle 2002a). A major programme of scoping, research and stakeholder consultation was conducted, resulting in an 'Agenda for action' in July 2002 (WBCSD 2002). This agenda was presented as a five-year action plan, with climate protection as one of eight key issues. The CSI has provided a rich set of reference materials about the cement industry and sustainable development (ibid.), which has since been used, *inter alia*, by the EU legislators in developing the EU ETS monitoring mechanisms.

CSI also took on the responsibility for developing a CO_2 protocol for the cement industry, to provide some common tools and indicators to monitor emissions. This was finished in 2003, when the CSI created a sector-specific CO_2 Accounting and Reporting Protocol – a universal framework for measuring and reporting CO_2 emissions in the cement industry. Since 2006, all CSI members have been required to have their data audited and verified by an independent third party.

In 2009, the WBCSD/CSI and the International Energy Agency (IEA) published a Cement Technology Roadmap, for carbon emission reductions up to 2050 (IEA 2009). The roadmap outlines existing and potential technologies, their estimated costs, timelines and CO_2 reduction potential. It focuses on efficiency, alternative fuel use, clinker substitution and CCS, and identifies the policy changes and investments necessary for achieving a more low-carbon cement sector. The roadmap is intended to function as an eye-opener for the sector: What can and should be done? What long-term strategies are necessary and possible? To date, the CSI remains one of the largest global sustainability programmes undertaken by a single industry sector (interviews 2011).

---

10   Other main work areas: fuels and raw materials, employee health and safety, local impacts, and internal business processes (WBCSD 2002).

2.2 Cembureau – EU ETS at the Top of the Agenda Since 2003

Cembureau has been actively engaged in ETS discussions since day one, and Cembureau officials still spend much of their time working on this issue. The proposal for a Directive on emissions trading was received with mixed feelings by the cement industry (Cembureau 2002: 5). Although market instruments are preferred over more traditional command-and-control, the prospects of a mandatory scheme created concern, especially as regards what was seen as policy-makers' lack of confidence in voluntary initiatives. Nonetheless, when the Emissions Trading Directive was adopted in 2003, Cembureau asserted that '[the Directive] is acceptable to the European cement industry which is conscious of the efforts that will be required in the future' (Cembureau 2004: 12). The organization started to work on sector-specific monitoring and reporting guidelines under the ETS, which would be in line with the above-mentioned CSI/WBCSD GHG Protocol (Cembureau 2004: 12).

After about one year of experience with the EU ETS, Cembureau stated: 'The European Cement Industry recognizes that climate change is a serious and global problem and is therefore committed to contribute to CO_2 reduction in a fair and effective way ... but the unilateral initiative of the EU to launch its ETS is leading to major problems of competitiveness of the EU cement industry and will not succeed in reducing global CO_2 emissions' (Cembureau 2005b). Cembureau signalled that it preferred a global or sectoral approach, but '(a)s long as this new system is not implemented and the political decision is taken to maintain the ETS for an in-between period', the organization did have some demands – for instance, that allocation should be free of charge, early action should be rewarded, and conversion of credits from JI/CDM should be one-to-one and with no cap. Cembureau has increasingly advocated the need for a sectoral/global approach, not a solely European one.

In 2004/2005, Cembureau was also actively involved in dealing with the indirect impact of the EU ETS on electricity prices (Cembureau 2005a: 9). The European industries of, *inter alia*, cement, ceramics, lime, glass, pulp/paper and steel joined forces within the Alliance of Energy Intensive Industries, arguing that the power sector would receive unjustified windfall profits at the expense of power-intensive industries. In 2006, Cembureau noted: '... the early warning that ETS would trigger higher electricity prices has unfortunately materialized: on average electricity prices have jumped from 14per cent of total cement production cost to 25per cent' (Cembureau 2006: 8).

Together with other energy-intensive industries, Cembureau put considerable effort into lobbying prior to the revision of the EU ETS in 2008 (Key Stakeholders Alliance for ETS Review 2007), and was quite successful in influencing the revised ETS. 'The outcome resulted in the best achievable compromise following all the efforts of Cembureau on an issue which filled the Association's calendar in 2008, bearing in mind that the Commission's proposal at the beginning of the year proved problematic to our sector' (Cembureau 2009a: 19). It was particularly important for the cement industry to be recognized as an energy-intensive industry vulnerable to *carbon leakage*, as this would guarantee a longer period for free allowances on the basis of a benchmark from 2013. Cembureau received support for this position from the Boston Consulting Group, which concluded that the European cement sector could be totally wiped out at a price of €35 per tonne of CO_2 if auctioning were to be applied, with negative impacts on global emissions (BCG 2008). The sector achieved recognition as 'vulnerable' in December 2009, when a list of vulnerable sectors was published (Commission 2009). Now the debate on benchmarking rules could begin. The benchmark should be based on the average emissions of the 10 per cent best-performing installations in 2007/2008. A clear majority of Cembureau members preferred a benchmark based on *clinker* production; Holcim represented the minority, arguing for a benchmark based on cement. This created a period fraught with public battles and counter-lobbying in the sector (interviews 2011). We return to this discussion later; suffice to say here that the Commission finally opted for a clinker benchmark (ENDS 16/12/2010).

In 2010 and 2011, Cembureau argued against any further tightening of EU climate goals. The EU was discussing whether the unilateral EU emissions reduction target should be moved from 20 per cent up to 25–30 per cent. Cembureau, through the Alliance of Energy Intensive Industries (AEII), maintained that a target higher than 20 per cent would be unacceptable and detrimental to competitiveness (AEII 2010, ENDS 07/05/2010). Specific sector perspectives should instead be developed, taking into account the individual technological and economic realities of each sector.

All in all, since the first discussions on the EU ETS, Cembureau has accepted the scheme as a part of regulatory reality, but it has lobbied aggressively to protect the sector's interests against the economic risk that the ETS represents – directly, and indirectly through increased electricity prices. Cembureau's number one task is to make sure that the EU ETS is as 'well-designed' as possible (interviews 2011). It has repeatedly warned that the ETS would threaten competitiveness, arguing that a sectoral, global approach is the only viable solution in the long term. The EU ETS has brought climate change to the top of Cembureau's agenda, mostly as a risk to be handled and a threat to be minimized. Gradually, the industry has come to recognize the strategic challenges involved in climate policies, and is interested in influencing climate regulations. Cembureau is also involved in exploring opportunities and strengthening climate initiatives in the sector, for instance through working with ECRA, the European Cement Research Academy,[11]

---

11   ECRA is a German-based research body, closely linked to FIZ, the Research Institute of the Cement Industry. ECRA was launched at the turn of the century, when cement companies realized they should start to develop a common platform and join forces. A sort of 'spiritual leader' for innovation was needed (interviews 2011). ECRA 'supports and undertakes research projects focusing on issues which due to their general significance for the cement industry as a whole might not be tackled by a single company alone' (ECRA 2011)

funding research projects on CCS and other issues. In December 2009, Cembureau was formally offered a seat on the ECRA Technical Advisory Board.

As to the major cement companies, it is clear that they share Cembureau's concerns of the EU ETS as a risk that must be handled. But, due to expectations of more stringent caps in the future, might the ETS also have led to a search for new abatement opportunities? The next section presents – very briefly – the climate strategies that the major cement companies developed in the period that coincided with the introduction and evolution of the EU ETS (See also Appendix Table 6.2).

2.3 Major Cement Companies and Their Climate Strategies

Studying the climate strategies of major cement companies is quite a challenge. These companies operate in several countries and regions, all of which have diverse regulations related to CO_2 emissions and climate change. It can therefore be hard to detect the overall strategy for a company as a whole.[12] Still, some conclusions can be drawn from the information presented in Appendix Table 6.1.

First of all, the cement industry operating in the EU is dominated by a few major companies, and most of these have signalled more proactive climate strategies based on innovative elements. They acknowledge the climate challenge, and seem to be taking responsibility for problem-solving. This commitment is reflected in CSI membership, systems for reporting and verification of emissions, and voluntary short- and medium-term goals. In general, the major cement companies have all signalled future abatement through the four measures described above: energy efficiency, alternative fuel use, clinker substitution and CCS. However, these measures vary in scale and stage of development, and it can be hard to distinguish 'greenwashing' from actual achievement. Nor is it easy to find information on how much the companies' spend on innovation, and where they direct R&D efforts.

Roughly speaking, long-term innovation in the sector is related either to changing the process or cement product itself (e.g. by developing cement types that use less clinker), or end-of-pipe solutions like CCS. The companies differ when it comes to prioritizing of measures, but in general all the larger players are looking into everything at this point. Despite differences in priorities and responses, most companies cite the EU ETS as an important driver for their climate strategies, together with UN processes, new scientific research and climate-related targets and policies being implemented in other parts of the world. Despite some frustration and resistance, this seems to be a sector gradually coming to terms with a mandatory climate regime – companies formulate targets, invest in reduction measures, and prepare for a future with a more constraining ETS or other type of mandatory climate regime.

---

12 It can also be difficult to distinguish between short- and long-term measures in this sector, as the processes of using alternative fuels and substituting clinker can go on for many years, and involves a mix of abrupt decisions, long-term innovation, changing regulatory and national conditions.

Table 6.2   Key group figures for cement (Holcim)

	2003	2004	2005	2006	2007	2008	2009	2010
Production capacity cement (million tonnes)	145.2	154.1	160.4	197.8	197.8	194.4	202.9	211.5
Cement and grinding plants	—	129	135	152	151	151	154	157
Sales of cement in million tonnes	94.3	102.1	110.6	140.7	149.6	143.4	131.9	136.7

Source: Holcim Annual Reports.

In conclusion, while Cembureau has lobbied fiercely to protect the cement sector's interests, making sure that EU ETS is as 'well-designed' as possible, most cement companies have started to adjust to a new reality with a mandatory climate regime, and are developing more proactive climate strategies. To better understand how the EU ETS affects cement companies, let us turn to the climate strategies of Holcim and of HeidelbergCement.

3. Holcim and HeidelbergCement

This section presents and discusses the climate strategies of two major cement companies more in depth: Switzerland's Holcim and Germany's HeidelbergCement.[13] While Holcim is one of the early movers in the cement sector, HeidelbergCement has gradually become more aware and interested in issues of sustainability and climate change. Holcim is known as a frontrunner in substituting clinker, whereas HeidelbergCement is currently working on what may become a groundbreaking CCS project. As discussed in Chapter 2, a company's climate strategy is revealed both through its objective and goals concerning climate change mitigation, and the actual mobilization of resources to achieve those objectives and goals. Here we look into general climate strategy as communicated by the companies' centralized management, as well as on company-specific aspects of the strategy.

3.1 Holcim

3.1.1 Baseline information
Holcim is a global construction materials company, founded in Switzerland in 1912. It began investing in cement businesses in other countries in the 1920s. Holcim is more globally spread than any other building materials group today, with production sites in around 70 countries across all continents (see Table 6.2). Its core activities include the manufacture and distribution of cement, and the production, processing and distribution of aggregates, ready-mix concretes and asphalt. Holcim was named 'Leader of the Industry' by the Dow Jones Sustainability Index from 2005 to 2008, and has a reputation as one of the most sustainability-conscious companies in the building materials industry (CDP Holcim 2010, interviews 2011).

---

13   Under the EU ETS, HeidelbergCement has 53 installations, while Holcim has 27 (Carbon Market Data). HeidelbergCement was allocated 20,374,924 (year 2005), 23,275,661 (year 2008) and 24,225,029 (year 2010) allowances, while Holcim was allocated 13,682,706, 15,879,003 and 16,322,327 for the same years. The two companies are thus comparable in the ETS context, but we should bear in mind that HeidelbergCement has a larger share of its operations in the EU than Holcim, which has more of a global presence.

Holcim's environmental awareness dates back to the early 1990s, as the company's main shareholder, Thomas Schmidheiny, was interested in issues related to sustainability and corporate social responsibility. In 1999, the Board of Directors discussed climate change and its consequences for the company. The discussion was based on internal anticipation that some sort of climate instrument – most likely an emissions trading scheme – would be adopted at the EU level within few years (interviews 2011). The company wanted to be prepared, and a Holcim CO_2 Task Force was established in 1999. This working group started to develop a Holcim methodology for monitoring, reporting and verification (MRV) of CO_2 emissions. Prior to 1999 only seven of some 40 Holcim companies worldwide had been monitoring CO_2, using six different methodologies. It became clear that a group standard was needed, and the Holcim database was ready in 2001.

Parallel to these company-internal efforts, Holcim also took the initiative on behalf of the cement sector as a whole, and approached the World Business Council for Sustainable Development (WBCSD) in 1999 together with Lafarge and Cimpor. This '... reinforced the environmental and social awareness that the Group has embraced for many years' (Holcim 2001 11). A 'Working Group Cement' (later the Cement Sustainability Initiative, CSI) was set up under the WBCSD, aimed primarily at developing a common cement industry monitoring and reporting protocol on CO_2 emissions. Holcim brought its newly developed MRV methodology to this group, which functioned as important inspiration for the later CSI Protocol.

In November 2002, Holcim published its first sustainability report (Holcim 2002). The report set out – in line with demands from the CSI – voluntary targets for emissions reductions: by the year 2010, the Group intended to reduce specific net CO_2 emissions worldwide by 20 per cent against the benchmark year of 1990. In the same report, Holcim announced that it used four key performance indicators[14] to monitor and report the eco-efficiency of cement production (Holcim 2002: 20–21). It became clear early on that substituting clinker with other materials would be at the core of Holcim's climate strategy. This mirrors Holcim's general business model – the company has historically produced cement with less clinker than other companies (interviews 2011).

Right from the beginning, Holcim was supportive of the EU ETS. Since its first reaction to the Commission's Green Paper in 2000, Holcim has been pleased to be able to participate in a market-based mechanism. Taxation was not regarded as a good option, and voluntary agreements were not considered enough to secure an adequate reaction to the important issue of climate change (interviews 2011). The

---

14   These four key performance indicators: the clinker factor (reducing the amount of clinker needed to make one tonne of cement), specific heat consumption (increasing thermal energy efficiency in the clinker-making process), the thermal substitution rate (increasing the proportion of energy from alternative fuels), and the cement kiln dust rate (reducing the amount of kiln dust discarded per amount of clinker produced).

EU ETS brought carbon to the top of the agenda in Holcim, and the CO_2 price soon became an important and natural element in company decisions and discussions.

The overall impression from this period is that Holcim was an early mover, one of the pioneers in the cement industry regarding sustainability and reducing CO_2 emissions. Holcim took some important early steps to monitor and report on emissions internally, and it was also co-founder and a driving force behind the WBCSD's CSI. By the early 2000s Holcim had established a good environmental reputation, and welcomed the EU's Emissions Trading Scheme.

3.1.2 Main developments since 2003

But has Holcim developed further since the early 2000s – or is it at a 'comfortable standstill', protected by its good reputation and high environmental profile? Here we examine changes in the company's climate strategy from 2003 to 2011 – a period that coincided with the introduction and evolution of the EU ETS.

In 2005 Holcim published its second Sustainability Report, emphasizing climate change even more strongly: 'climate and energy are the most significant environmental issues for Holcim' (Holcim 2005: 20). The company's climate strategy continued to be centred on the four areas for abatement presented above: according to the 2007 report, 'sustainable environmental principles and performance are firmly anchored in our business strategy' (Holcim 2007: 14).

Since 2001, absolute CO_2 emissions have been increasing in Holcim, but the company has been able to produce cement more efficiently, with low CO_2 emissions per tonne of cement produced (see Table 6.3). Due to the low clinker factor, these figures are significantly better than the global industry's general performance (see Table 6.4). While Holcim's global cement production 1990–2007 increased by 103 per cent, the increase in absolute net CO_2 emissions was only 70 per cent (Holcim 2007: 15).

Since 2002, Holcim has had the goal of reducing CO_2 emissions per tonne of cement by 20 per cent, compared to 1990. This target was met already in 2009 (Holcim 2001la: 41), which, according to Holcim, this placed it as the leading company in the industry; with the lowest CO_2 emissions per tonne of cement (CDP Holcim 2010). New medium-range targets were decided on in 2009, when the company announced that it aims to reduce specific net CO_2 emissions by 2015 to 25 per cent below the 1990 baseline. In addition, the objective is to remain 'best performer (net specific CO_2)' in the sector (CDP, Holcim 2010). Thus, the company is gradually tightening its targets, and intends to maintain its frontrunner position.

As part of its short-term climate strategy, Holcim has invested in several projects under the Clean Development Mechanism (CDM). In 2007, the company had more than eight such projects running (Holcim 2007: 12). These have included biomass and other alternative fossil fuel- opportunities in Ecuador and Indonesia, wind energy in India and waste heat recovery for power generation in China, India and Thailand (Holcim 2007: 12).

Table 6.3   Environmental performance data (Holcim)

	Absolute gross emissions (million tonnes CO_2)	Absolute net emissions (million tonnes CO_2)	Specific gross emissions (kg CO_2/tonne cement)	Specific net emissions (kg CO_2/tonne cement)
2001	60.8	58.8	708	684
2003	67.3	65.0	677	654
2004	71.8	69.3	673	650
2005	74.5	71.9	658	635
2007	108.5	105.3	664	644
2008	105.5	102.3	654	634
2009	96.6	93.7	645	626

Source: Holcim Sustainability Reports 2002, 2005, 2009.

Table 6.4   Average percentage of clinker in cement (Holcim)

	Clinker content
1990	81.9
2000	79.9
2005	74.8
2007	72.5
2008	71.8
2009	71.1

Source: Holcim Sustainability Report 2009: 15.

3.1.3 Long-term goals and measures

Holcim has no publicly available long-term goals beyond 2015, and it is thus hard to say how the company pictures itself in 2020 or 2050. In 2007, the company wrote:

> With regard to the manufacture of cement, which remains a resource- and energy-intensive process, innovation is the key to reducing CO_2 emissions. This can be achieved by optimizing products and processes, and investing significantly in research and development such as developing new types of cement and using alternative fuels, within the current technical limitations. (Holcim 2007: 14)

Through product innovation, Holcim is working to develop composite cement in which clinker content is reduced by greater use of materials like slag, fly ash, and natural pozzolans. The company portfolio is gradually switching towards low-CO_2 cements; although figures and strategies are not publicly available, it is claimed that the company has been investing considerably in such product innovation (interviews 2011).

CCS does not yet figure in Holcim's long-term climate strategy, as it is considered too costly to install carbon capture in a cement plant at this point (interviews 2011). However, the company acknowledges that in the future CCS may offer important potentials for carbon reduction. Therefore, Holcim supports the ongoing work of the European Cement Research Academy (ECRA) on CCS technologies, and is interested in getting more information. Also on other issues, Holcim is collaborating with universities, consultancy firms and other companies on innovative solutions. One example can be found in Untervaz, Switzerland, where ABB and Holcim have agreed to deploy a new ABB system to generate electricity at the cement plant there. This system can turn exhaust gas into clean electricity (ABB 2010).

In recent years, the cement industry as a whole has generally emitted less CO_2 than it is entitled to under the EU ETS. Sales of excess emissions certificates can raise large sums of money for the companies. In 2010, Holcim decided to allocate this income to a *fund* to promote energy efficiency, some CHF 100 million a year (Holcim 2011c). The fund is intended 'help ensure the realization of innovative projects across the Group in the field of heat recovery, the utilization of alternative fuels and raw materials, as well as wind power and hydroelectricity', with the aim of saving fossil fuels, reducing CO_2 and boosting energy efficiency (Holcim 2011c). In 2010, five heat-recovery plants (in Vietnam, India, Romania, Lebanon and Switzerland) received grants from the fund; when these are built, they will save Holcim around 200,000 tonnes of CO_2 annually (Holcim 2011).

3.1.4 Summing up

Holcim is among the more proactive companies in the cement sector. This commitment began in the 1990s, and has continued since the adoption of the EU ETS in 2003. Relative emissions and the clinker factor have declined every year, and the company was recognized as 'Industry Leader' by the Dow Jones Sustainability Index between 2005 and 2008. Holcim is a frontrunner when it comes to clinker substitution, and has chosen to focus on 'changing the product' rather than end-of-pipe solutions like CCS. Holcim has in many ways a different understanding of what cement is and how it should be produced compared to other cement companies. Composite/blended cements are a central part of both its business and climate strategies.

On the other hand, it is hard to say how Holcim will actually develop beyond 2020 and up until 2050. Aside from its stated goal of remaining 'best performer' in the sector, there are no publicly available long-term goals, and it is hard to find evidence of *how much* the company has actually invested in innovation and R&D. However, the fund set up to support energy-efficiency projects is expected to lead to future emissions reductions, and the company actively promotes product innovation.

3.2 HeidelbergCement

3.2.1 Baseline information

HeidelbergCement's first cement plant was established in Heidelberg in Germany in 1873. The company was active in the south of Germany until the late 1960s, when it expanded into France and the USA. In the 1990s, it bought up cement producers in Belgium and Scandinavia, and its international expansion started to accelerate. Today, HeidelbergCement operates in Europe, North America, Asia, Australia and Africa, and is one of the world's largest manufacturers of building materials (see Table 6.5 below), with 55,000 employees in more than 2,500 locations. It is the global market leader in aggregates, and a leading player in cement, concrete and other downstream activities.

Table 6.5   Sales of cement and clinker (million tonnes) (HeidelbergCement)

	2004	2005	2006	2007	2008	2009	2010
Cement sales	65.2	68.4	79.7	87.9	89.0	79.3	78.4

Source: HeidelbergCement annual reports.

Prior to 2002, 'climate' and 'CO_2' were not important buzzwords in HeidelbergCement. However, also in the 1990s the company did have policies concerning, for instance, energy efficiency, availability of raw materials and alternative fuels – issues the company today frames as climate-related. Around 1990, when fossil fuel prices were unstable and rising, parts of the company started to use biomass and waste as alternative fuels, which already helped to decrease emissions in the 1990s.

HeidelbergCement joined the World Business Council for Sustainable Development (WBCSD) and became a core member[15] of the CSI in 2002. Climate was 'one of many good reasons' for joining the CSI, and since 2002, the company has assessed CO_2 emissions across the entire Group in accordance with the Carbon Dioxide Protocol of the WBCSD.[16] In 2002, HeidelbergCement made a voluntary commitment to reduce its specific net CO_2 emissions per tonne of cement by 15 per cent by 2010, compared with 1990 levels. Other goals decided on in 2002 were to cover 30 per cent of the energy requirements with secondary fuels[17] by 2010, to increasingly replace natural limestone with secondary raw materials ('clinker substitution'), and to intensify its R&D efforts (HeidelbergCement 2003: 90). We can also note greater awareness of the link between sustainability/climate and energy-intensive cement production in the early 2000s, as the company started to frame well-known issues in a new way: 'By using secondary fuels, we reduce our costs and in addition make a contribution to reducing CO_2 emissions, and thus to environmental protection' (HeidelbergCement 2002: 37).

In other words, HeidelbergCement was one of the first cement companies to react to the new WBCSD initiative; it followed Holcim, Lafarge and other early movers in the sector, and soon became a core member of the CSI. The company announced that it intended to make its contribution to achieving the objectives laid out in the Kyoto Protocol, and had already come up with a voluntary target already in 2002. However, many of these initial efforts were related to 'reframing' of earlier measures (such as energy efficiency and alternative fuels use). Prior to the period in focus here (2003–2011) the issue of climate change did not produce a feeling of *urgency* in the company. Some developments took place, and the company did some abatement through e.g. use of alternative fuels, but climate protection was only 'one of many good reasons' for HeidelbergCement's new sustainability commitment.

3.2.2 Main developments since 2003

During the period under study (2003–2011), HeidelbergCement has fully acknowledged the climate change threat and the responsibility of the cement industry. As stated in the 2004 Annual Report: 'This [The Kyoto Protocol] has

---

15   Core members of the CSI manage the CSI, maintain the CSI Charter, define and fund the CSI work programme, and invite new members. At present (2012), there are 11 core members.

16   Based on this Protocol, a working group at Heidelberg Technology Center collects all CO_2 data of the company's plants. This provides an overview of CO_2 emissions across the Group, as well as a basis for decisions and planning measures (HeidelbergCement 2007: 18). The first external verification of HeidelbergCement's emissions took place in 2007.

17   E.g. fuels from recycled used tyres, plastic waste and old wood, which can be used to replace primary/fossil fuels (HeidelbergCement 2003: 89).

produced a revival in worldwide climate protection. HeidelbergCement aims to be a pacesetter in international climate protection' (HeidelbergCement 2005a: 81). And later: 'Climate change is one of the biggest challenges in today's world. As an energy-intensive industry, we believe that we must take responsibility.' (HeidelbergCement 2007: 18). This is a continuation – and a strengthening – of the company's early involvement in the WBCSD and the CSI.

HeidelbergCement supports international targets and initiatives aimed at combating climate change, as long as these act to create a level playing field and carbon leakage is avoided. Regarding the EU ETS in particular, HeidelbergCement wrote: 'In principle, we welcome emissions trading as economically oriented instrument of climate protection, and at the same time support the competitiveness of the European cement industry' (HeidelbergCement 2005b: 48). Although the company welcomed a market-based climate instrument, it was initially negative towards the new scheme, as the EU ETS was perceived as disorganized and uncertain (interviews 2011). We return to perceptions of the EU ETS in section 5, when we link the ETS to HeidelbergCement's climate strategy.

HeidelbergCement had already achieved the goal of a 15 per cent reduction of specific net CO_2 emissions already in 2007, three years before schedule. The following steps had been taken: modernization of production facilities, increasing replacement of clinker with other materials, investments in CDM/JI-projects and greater use of alternative fuels (HeidelbergCement 2009a: 44). The last point has been the most important measure in HeidelbergCement to date. When it comes to the proportion of alternative fuels in the fuel mix, HeidelbergCement is the frontrunner among the major international cement manufacturers. In 2008, the proportion was 17.5 per cent, with a goal of increasing it to 22 per cent by 2012 (HeidelbergCement 2009b: 18). By 2010, the proportion of alternative fuels in the fuel mix had reached 21.5 per cent (HeidelbergCement 2011a: 105).[18]

In 2008, the company had over-fulfilled its 15 per cent reduction goal, with an 18 per cent reduction in relative CO_2 emissions. In September 2009, the company presented 'The HeidelbergCement Sustainability Ambitions 2020', with the following medium-term goals:

3.2.3 Long-term goals and measures

HeidelbergCement's innovation work can be divided into three areas: products and applications, production and development of cements and concretes with improved CO_2 balance (HeidelbergCement 2011a: 106). An example of production-related innovation is to develop waste-derived fuels and raw materials. This is a central pillar in the long-term climate strategy of the company, which is already the leader in many countries as regards using biomass fuels (interviews 2011). HeidelbergCement focuses on three globally available waste

---

18   By comparison, in Holcim the proportion of thermal energy from alternative fuels in the fuel mix was 12.1 per cent in 2009 (Holcim 2010: 15). In the sector as a whole, the average proportion was 7 per cent in 2006 (ECRA 2009: 14).

streams: sorted household waste, sewage sludge and hazardous waste, and works to provide innovative solutions for co-processing these as alternative fuels (HeidelbergCement 2009b: 11). A central aim of this strategy is to safeguard the availability of alternative raw materials and fuels in the long term. More R&D has also been directed towards new cement types with less clinker, as the company aims at a higher clinker substitution rate of 30 per cent in 2020 (see Table 6.6).[19] However, as regards long-term innovation, it is hard to find evidence of how much is spent on climate-related innovation, with vaguely-worded reports coming from the company itself: 'We are working intensively on the development and implementation of innovative solutions for reducing CO_2 emissions' (HeidelbergCement 2011a: 106). In 2010, HeidelbergCement consolidated its group-wide R&D activities into the new Heidelberg Technology Center (Global HTC), with the aim of optimizing and coordinating construction projects all over the world. 'One of the most important aspects of this process is the development of products with minimal CO_2 emissions' (HeidelbergCement 2011b: 23). Like Holcim, HeidelbergCement has also used some of the revenues accruing from its long position under the EU ETS to invest in low-carbon solutions like energy efficiency, but since the financial crisis hit in 2008, these extra revenues from the ETS have been more of a welcome 'help' than an impetus for investing more (interviews 2011).

Table 6.6   HeidelbergCement 2020 ambitions (%)

Key performance indicators	Ambition 2012	Ambition 2020
Alternative fuels rate/thermal	22	30
Biomass fuel rate/thermal	6	9
Alternative raw materials rate	11	12
Clinker substitution rate	25	30
Net direct CO_2 emissions (kg/tonne of cementitious)	23% reduction in specific net CO_2 emissions by 2015 (compared to 1990 levels)	

Source: HeidelbergCement 2009b.

Despite the problems involved in grasping the 'full innovation picture' in HeidelbergCement, it is still possible to describe one of its innovative projects more in detail. HeidelbergCement is the cement company in the EU that focuses the most on Carbon Capture and Storage (CCS) (see HeidelbergCement 2009b: 12, HeidelbergCement 2011b: 26). A test facility for post-combustion CO_2 capture technologies is planned for the cement plant at Brevik in Norway, under Norcem

---

19 By comparison, Holcim already had a clinker substitution rate of 28.9 per cent in 2009 (Holcim 2010).

(a part of HeidelbergCement Northern Europe). The aim of this project, which is being conducted in cooperation with the European Cement Research Academy (ECRA), is to test CO_2-capture technologies that could be relevant for the cement industry. If built, the entire facility will consist of between one and three test rigs, each with a capacity of 10,000t CO_2 per year. These plans are in accordance with recommendations of the IEA Cement Technology Roadmap: '... early research and pilot tests are needed to gain practical experiences with these new developing technologies' (IEA 2009: 14). In the spring of 2011, it was announced that HeidelbergCement will work with Aker Clean Carbon (ACC) and Alstom in executing a concept study and a pre-engineering study. Alstom will study the application of chilled ammonia process and the carbonate looping process, and the various trials will be given industrial environmental testing by using ACC's Mobile Test Unit. Preliminary work in this project is to be completed in 2012, and Norcem and HeidelbergCement will then decide whether to build the test facility or not. This is a forward-looking project with potential long-term consequences for company emissions – but we cannot exclude the possibility that it is being conducted in order to silence critics and enable the company to say 'well, we have tried'. In any case, this project can supply the cement sector with some valuable results and answers.

3.2.4 Summing up

In the time-frame of this study, interesting and relevant changes have taken place in HeidelbergCement. The company is a sector frontrunner in the use of alternative fuels, and has started to explore the possibilities for CCS. Also prior to the EU ETS, it acknowledged the climate-change problem and the cement industry's responsibility, setting short-term goals in 2002. It is however after 2003 that HeidelbergCement's climate commitment really started to take shape. New goals were presented in 2009 in the '2020' ambition, and the company has long-term innovation projects related to, inter alia, waste fuels, clinker substitution and CCS. Its Group-wide R&D activities were consolidated into the new Heidelberg Technology Center in 2010, and it has invested in several CDM projects. The company has been highly sceptical towards many of the actual design features of the ETS, and fears the scheme may lead to carbon leakage. It is difficult to offer conclusions on how HeidelbergCement's climate strategy will develop in the future, but the company does seem committed to contributing to a low-carbon cement sector in Europe.

3.3 Comparing Holcim and HeidelbergCement

Cement production is a highly energy- and resource-intensive process, so efficient use of fuels and raw materials become a question of cost-savings as well as environmental policy. For several decades, it has been important for both companies studied here to optimize products and process, for instance through innovation related to new cement types and alternative fuels. Here we note that developing a climate strategy has, to some extent, been about framing well-known issues in a new way.

Second, we see that the climate strategies of Holcim and HeidelbergCement have much in common – they have converged in a proactive direction since the early 2000s. They both accept the climate change problem, and claim to be serious in their efforts to take responsibility for problem-solving. In the late 1990s and early 2000s, they 'saw something coming', and took action to prepare for some sort of mandatory climate regulation. However, Holcim became concerned with issues of sustainability earlier than HeidelbergCement, and was one of three companies to approach the WBCSD in 1999. Holcim welcomed the EU ETS, whereas HeidelbergCement was initially more critical towards many of the design characteristics. Both companies like the general idea of a trading system, but are concerned about carbon leakage, and have lobbied strongly for a lenient cap and free allocations. The two companies have systems for monitoring, reporting and verification of emissions, based on the WBCSD Protocol. They argue that, with cement demand rising, absolute emissions will increase, but they are working to decrease their net CO_2 emissions.

And so we ask: How to explain the general change in the direction of higher climate awareness and increasing climate action in both companies? To what extent has the EU ETS influenced these changes? What role has the Cement Sustainability Initiative and other factors played?

The differences between the two companies relate mainly to their preferred choice of abatement measures, as well as Holcim's position as an early mover. Both companies are working on the four general ways to reduce emissions: energy efficiency, alternative fuel use, clinker substitution and CCS, but differ in their priorities. For Holcim, the main strategy is to substitute clinker with other, less CO_2-intensive materials, whereas HeidelbergCement is the first company with plans for a CCS test facility. Both companies are frontrunners in the use of alternative fuels, but HeidelbergCement stands out as the market leader in this respect. What factors can explain these differences? Why was Holcim the first of the two to make a move on climate change? Why is Holcim the leader in substituting clinker, while HeidelbergCement is the first cement company to embark on a CCS project in Europe?

4. Explaining and Analysing Climate Strategies

Before linking changes in climate strategies to the EU ETS, we will briefly look at a factor that predates the EU ETS – the Cement Sustainability Initiative (CSI) – and its significance. The CSI brought together cement companies worldwide on sustainability issues for the first time, in order to develop and promote best practice across the industry in a range of areas – indeed it was the first sector-wide, voluntary initiative of its kind. The cement sector was one of the first energy-intensive sectors to acknowledge the climate change challenge, and the CSI can be interpreted as showing that the sector was prepared to take responsibility for problem-solving (Fujiwara 2010, Fujiwara et al. 2010). These observations point to the CSI as an important part of the explanation for the early climate consciousness in the cement sector. It created a common language and some common tools – not least, a system for monitoring, reporting and verification of emissions. On the other hand, the CSI was a completely *voluntary* initiative, and one that some NGOs derided as involving mostly talking and 'greenwashing' (interviews 2011).

Thus, the *mandatory* EU ETS probably had a greater impact on company climate strategies, stimulating stimulated them to go even further. In fact, part of the CSI backdrop was also the *anticipation* of an ETS. In the next section we explore the direct and indirect impacts of the EU ETS.

4.1 Linking the ETS to Holcim's and HeidelbergCement's Climate Strategies

4.1.1 ETS and short-term strategy

As the discussion of Cembureau's reaction to the EU ETS showed (see section 3), the cement sector welcomed the EU ETS cautiously when it was initiated in the late 1990s. Both Holcim and HeidelbergCement have supported market-based mechanisms, and preferred the ETS to a carbon tax, for instance (interviews 2011). Despite this general support, both companies have questioned several characteristics of the ETS, and when the market started to operate in 2005, HeidelbergCement was critical of (among many things) the system based on historical emissions (grandfathering) and the considerable unpredictability related to future allocations. The system became somewhat more harmonized in the second phase, but, according to HeidelbergCement, allocations continued to be 'messy and unclear' (interviews 2011). Holcim was initially more positive, pleased with the establishment of a carbon market, and the Board approved of it early (interviews 2011). However, Holcim's support for the EU ETS has deteriorated significantly since 2003, as the company now believes the ETS has failed to achieve many of its initial goals (interviews 2011). Holcim is particularly dissatisfied with the benchmarking system decided on for Phase 3, as it rewards companies with higher clinker content. Both companies have been critical towards central features of the EU ETS since the beginning, and even though they still support the system, they also hold that many things must change for it to be a success (interviews 2011). Both companies have advocated the view that as long as the ETS is only a European system, and not a global one, it will be hard to get it to function properly.

The companies' views on the functioning of the ETS should be borne in mind when we now ask: have the *incentives* provided by the ETS affected climate strategies? The macro-characteristics of the EU ETS feature, first and foremost, the cap and the allocation method. Both HeidelbergCement and Holcim have received a surplus of allowances every year (see Table 6.7), which had led the British NGO Sandbag to label them 'carbon fat cats' (Sandbag 2010). It is difficult to distinguish between over-allocation surpluses and actual abatement, but, according to Sandbag (2010c: 24), '… it is clear that the overwhelming majority of emissions reductions

Table 6.7   Verified emissions and emissions-to-cap ratio (as percentage of cap)

	2005		2006		2007		2008		2009	
	Em.	Rat.	Em.	Rat.	Em.	Rat.	Em.	Rat.	Em.	Rat.
HeidelbergCement	16,626, 353	-18.40	17,713, 743	-13.18	20,810, 315	-8.83	21,560, 912	-7.37	18,359, 352	-22.4
Holcim	13,060, 664	-4.55	13,339, 824	-3.72	16,023, 976	3.81	15,046, 900	-5.24	11,495, 470	-27.67

Source: Carbon Market Data.

in sectors holding surplus permits have resulted from declines in production, not investment in abatement'. The over-allocation phenomenon was further spurred by the financial crisis, when the demand for cement dropped dramatically. In fact, the EU ETS appears to have helped subsidize the cement companies when the crisis became serious, since the companies could sell their surplus allowances.

In the ETS, the total number of allowances issued affects the price of carbon, through the carbon market. Previously, the CO_2 emissions of cement companies were not regulated – the carbon they emitted was free. Now CO_2 has a price, and cement companies have begun to see this as an important element in their cost–benefit analyses and investment decisions (interviews 2011). However, with too many allowances on the market, this has led to a rather low carbon price, giving cement companies fewer incentives to change their climate strategies than in a situation marked by scarcity of allowances. When allowances are handed out for free, the opportunity cost (the money to be made from polluting less and selling excess allowances) is the only direct economic incentive remaining. But, as long as the carbon price is low, the opportunity cost is less attractive, making it easier for companies to continue business as usual.[20]

When we look at the effects of the EU ETS, one of many paradoxes is that, although the cement companies are displeased with many ETS characteristics, they have worked very hard to 'achieve' the ETS. As explained in section 3, Cembureau has lobbied fiercely to keep ETS constraint to a minimum, by securing a lenient cap and free allocation, and finally by advocating the adoption of a clinker benchmark. Several NGOs and think-tanks have complained about the cement industry's position in the EU ETS negotiations. The sector has been accused of working to preserve the status quo, of exaggerating the carbon-leakage argument, and of constantly sabotaging a more constraining (hence more environmentally-friendly) ETS (interviews 2011, CAN Europe 2010, Oxfam 2010). The companies, for their part, see their actions as a necessary reply to a system that threatens the level playing field and the very survival of the European cement industry (see e.g. Cembureau 2007:14).

Another economic incentive of the ETS is its indirect effect on electricity prices. Although this effect is hard to measure, there is evidence indicating this as an additional driver behind the changes in the company climate strategies. Cembureau has on several occasions warned of the possibility of 'windfall profits' in the electricity industry and rising electricity prices (Cembureau 2006), advising its members of the 'risk to the competitiveness of the European cement industry from high electricity prices due, inter alia, to the EU ETS' (Cembureau 2007: 13).

20　Several studies conclude that free allowances are working against breakthrough investments. According to a study commissioned by 'Climate Strategies': '... firms within the EU ETS which are just below the thresholds established for free allowances are engaging more strongly in climate change-related product innovation than firms that are just above the threshold' (Martin et al. 2011: 3) This suggests that the current EU ETS practice of generously allocating free allowances to manufacturing sectors leads to less innovation.

Although the cement sector does not rely as heavily on electricity as for instance the paper industry, it is affected when power prices go up. Cembureau (2007: 13) noted that 'Since 2005 electricity has jumped from a stable 14 per cent to over 25 per cent of the total production cost of cement'.[21] Both Holcim and HeidelbergCement see increased electricity prices and climate strategies in connection: greater energy efficiency offers a potential for reducing costs as well as CO_2 emissions. In its 2010 Annual Report, HeidelbergCement notes the importance of minimizing electricity price risks by 'increasing use of alternative fuels and raw materials'. As in that way the company can '... minimize price risk while reducing CO_2 emissions and the proportion of energy-intensive clinker in the end-product' (HeidelbergCement 2011: 92). Holcim seems to have the same view on the answer to higher electricity prices: 'New technologies is the answer to long-term increasing energy prices: To improve the eco-balance and thermal energy costs, the fuel mix is continuously optimized based on the availability of local energy sources' (Holcim 2010: 33). Seen in this way, the extra costs provided by the ETS have to some extent 'worked' – indirectly through the increase in electricity prices.

Nonetheless, on the whole the EU ETS has not represented a heavy cost constraint on the cement companies, in the short term. Quite the contrary: excess allowances and windfall profits have put the cement companies in a rather comfortable position – with few obligations to actually cut CO_2 emissions, and with the EU ETS making a welcomed economic contribution during the economic crisis. The excess of allowances is due to the allocation rules, the financial crisis, and most likely also some actual abatement. We do, however, see some signs of the companies using the windfall profits to promote emissions reductions, with Holcim's fund set up in 2010 as the best example. The ETS has created an incentive for reductions in energy use indirectly, by influencing electricity prices.

4.1.2 ETS and long-term strategy

The main influence of the ETS on the cement industry seems to emerge when we look at changes in long-term strategies. The EU ETS has managed to put a price on carbon, and the companies have learned that governments are prepared to regulate carbon and create a market for pollution permits.

Turning to the future ETS, we see signs of several design features changing, possibly affecting the incentive effect of the scheme. EU ETS Phase III post-2012 will be more challenging for the cement industry; initially there will still be permits for free, but this will be based on a benchmark, punishing the least efficient plants. The prospects of having to pay for allowance costs, as well as a higher carbon price, have to some extent changed the companies' beliefs about the future, and

---

21 These concerns also explain why the European cement industry was officially recognized as vulnerable to carbon leakage, as the increase of its product cost resulting directly and indirectly – through electricity prices – from the ETS is estimated at 45.5 per cent of GVA, which passes the threshold of 30 per cent GVA set by the Directive (Cembureau 2011: 22).

these costs and various forecasts are now taken into account in making long-term decisions. From Phase III onwards a shift will be needed in the sector – and, according to the companies themselves, this is triggering increasingly proactive and innovative strategies already today (interviews 2011). HeidelbergCement's attitude towards the EU ETS has become more deliberate – it is a matter of money, and the future competitiveness of the sector. The company is prepared for a more stringent future; it has acknowledged that the 'dirtiest' cement plants will have to close, while the most CO_2-efficient will survive. HeidelbergCement knows it will have to invest in cleaner technologies, but, given the unpredictability of the ETS, it is hard to say when and how much (interviews 2011). How great a constraint Phase III (and the ETS beyond 2020) will prove to be depends, *inter alia*, on the current discussion of a 20 versus 25/30 present target for the EU. The cement industry, as we saw in section 3, strongly opposes a 30 per cent target – arguing that it is premature and will be detrimental to competitiveness.

Constraints in Phase III will also depend on the benchmark rules adopted, and this was debated fiercely in the EU and in the cement sector in 2010. This debate shows clearly how Holcim and HeidelbergCement's preferred climate strategies affect their views on the ETS design. HeidelbergCement wanted a benchmark based on clinker, whereas Holcim was the only company to advocate a cement benchmark. As noted, Holcim's clinker content is among the lowest in the industry; 'to maximize all available emission-reduction levers, Holcim advocated that industry benchmarks should include the entire production process and products and that these should incentivize both product innovation and consumer choice' (Holcim 2010: 14). Simply put, Holcim wants a benchmark that will reward companies that operate with a low clinker factor. It is reasonable to argue that this strategy of material substitution is the most climate-friendly, as 60 per cent of cement industry emissions stem from the chemical process involved in clinker production. Environmental NGOs have supported Holcim's position in this debate (see e.g. CAN Europe 2010). On the other hand, this focus does not necessarily mean that Holcim is more proactive – it could just as well be related to Holcim's starting point and general business model, as the company has a clear comparative advantage in the area of composite/blended cements.

Despite the continuing efforts of the cement companies to keep ETS constraints to a minimum, the sector has begun preparing for a more stringent system. Expectations of a more ambitious EU climate policy, a decrease in free allowances, allocations based on a cement benchmark, and a higher carbon price underpin company strategies with regard to energy efficiency, alternative fuels use, clinker substitution and CCS. Although they cannot know how great the changes will be or when they will come, the companies expect climate policies to provide stronger and stronger economic incentives for change, and are thus already trying to adjust to the new demands for innovative and low-carbon cement (interviews 2011).

Another way of viewing these effects of the EU ETS is to argue that the trading scheme has worked through a *learning* mechanism so far, and we find evidence to indicate that this has been the case in both companies. The new regulation has *set the companies in motion*, building on the foundation created by the CSI and earlier initiatives related to alternative fuel use and energy efficiency. The EU ETS brought CO_2 reductions to the top of the corporate agenda, and more people in both companies now know more about CO_2 issues and ways of reducing emissions (interviews 2011).

Long-term R&D for product and process innovation has, according to the companies, become increasingly important since the introduction of the EU ETS (interviews 2011). Companies are now investigating how they can change a product they never thought much about changing before. Cement was in many ways a 'perfect' product – limestone can be found almost everywhere, and customers are pleased with the traditional Portland cement. Now, the (expected) cost of CO_2 justifies some extra innovation costs, and a need to look into new ways of thinking about cement and its production. Both Holcim and HeidelbergCement state that they have been working on low-carbon solutions for many years, for various reasons, but add that the EU ETS has spurred this development, making it more pressing and relevant (interviews 2011). The most obvious example is energy efficiency – one of the few ways cement companies can actually increase their profits without expanding. Market and economic forces generally trigger the closure of inefficient facilities, and cement producers will install the latest technologies. These technologies – like new kilns – are also typically the most energy-efficient and are thus climate-friendly. Another example concerns alternative fuels. Companies had already started experimenting with alternative fuels in the late 1980s and early 1990s, to guard against fluctuating coal and oil prices. HeidelbergCement Northern Europe opened a plant for waste incineration as early as in 1995 – long before a carbon price and 'climate protection' had entered the equation. Coal prices were rising, and the company wanted to replace coal with waste-derived fuels. This became a useful and profitable way to dispose of unwanted waste while contributing to cement manufacturing. To date, using waste as an alternative fuel has often meant generating revenues rather than absorbing a cost. The effect of the EU ETS seems to be greater in areas where companies have already identified possibilities for cost-savings and optimization of operations: 'HeidelbergCement optimizes the operational performance of its plants, whether covered or not by the EU ETS' (CDP, HeidelbergCement, 2010). As for clinker substation, this will also lead to a reduction in costs. According to Holcim, the main reason for producing composite cement is that it requires lower capital investments. Smaller clinker kilns are needed, and more cement can be produced with less capital. Whereas this was Holcim's original rationale for working on clinker substation, the rational has now been widened to encompass climate protection (interviews 2011).

4.1.3 Summing up

Prior to the EU ETS, the greenhouse gase emissions of cement companies were unregulated, and companies have opposed important aspects of the system after it was introduced. In many respects, the cement sector stands out as a clear example of climate regulations leading to resistance. In the short term, the incentives provided by the ETS have been very weak, and excess allowances and a low carbon price have made the initial ETS less effective than it could have been. On the other hand, the EU ETS has served to reinforce companies' existing efforts in 'climate-related' areas, and has directed their attention towards totally new opportunities. New, low-carbon cement types are attracting even greater interest than before. Holcim has stepped up its long-established efforts in creating cement with a lower clinker factor, and HeidelbergCement has put greater emphasis on the use of alternative fuels. HeidelbergCement has also started to explore carbon capture technologies, and Holcim has recently created a fund to support innovative energy-efficiency projects within the company.

While the companies have been lobbying against changes in the scheme, and working to preserve the status quo at every crossroad, they are nevertheless *preparing* for changes in a more constraining direction. The companies know that they will have to invest in cleaner technologies. Because of the significant uncertainties in the system, they do not know when or how much – but the ETS is definitely part of the explanation why the cement sector now has a more proactive and active attitude towards measures aimed at decreasing emissions and improving energy efficiency.

5. External and Internal Company Factors

Despite many similarities in their response, we see that Holcim and HeidelbergCement to some extent have chosen different measures for coping with the climate challenge. Why is HeidelbergCement the first company to give CCS a serious try? And why is Holcim a frontrunner in reducing the clinker factor? What can explain Holcim's role as an early mover in the sector? This section looks at some factors beyond the EU ETS that might co-produce changes in climate strategies.

Are there perhaps some external *national factors* that can explain the different choice of measures? HeidelbergCement's CCS engagement started in HeidelbergCement Northern Europe, more specifically at Norcem in Norway. HeidelbergCement has been active in Northern Europe since the acquisition of Scancem in 1999, and the Norwegian company Norcem early became a proactive force within the company as regards climate change (interviews 2011). Norcem has considerable experience in using alternative fuels, and has steadily increased the share of such fuels since the mid-1990s. To continue this early strategy, and also achieve further emissions reductions, Norcem has begun to investigate the possibilities for CCS (interviews 2011). The market for carbon-capture technology in Norway is expected to grow, spurred by active engagement and support from the government. Funds are being allocated by Gassnova through the CLIMIT programme, the objective being to 'accelerate the commercialization of CCS by financial stimulation of research, development and demonstration' (Climit 2011). The project in Brevik is the first industry project in Norway to receive financial support from through the CLIMIT programme – hitherto only projects in the power sector have received such support. The project budget so far is NOK 13.5 million, with CLIMIT support of 50 per cent. According to Norcem, CLIMIT support has been a decisive factor behind the decision to go forward with the project. The Norwegian support scheme is unique in the European context, and has provided HeidelbergCement Northern Europe with opportunities not available to all cement companies. The question now is whether CLIMIT will continue to support the project after the current phase of concept development and pre-engineering is over. HeidelbergCement is dependent on further support to continue the project; new applications will be sent to Gassnova (interviews 2011). Thus, national factors are crucial in explaining why HeidelbergCement has chosen to invest in CCS research (interviews 2011).

National factors can also shed light on why a company's climate change-practices differ in the various countries where it operates. The cement companies studied here have a global presence – Holcim in 70 countries, HeidelbergCement in 40. As country-specific conditions vary considerably, opportunities and requirements for reducing emissions will also vary. Some countries make it 'easier' to be progressive than others – for instance, related to the accessibility of alternative fuels and regulations on waste management. As noted, the fuels traditionally used in cement manufacturing are coal, heavy fuel oil or gas. Waste used as a fuel can substitute these, which has become common practice in the cement industry in many parts of the world. The cement industry argues that this results in a double gain, as the waste would otherwise be disposed at incineration plants – resulting in GHG emissions there. In some countries, the waste management strategies and infrastructure needed to enable the companies to use waste are still being developed; and there are countries that do not allow controlled waste collection and treatment of alternative fuels at all (IEA 2009: 10). Cement manufacturing in The Netherlands uses almost 98 per cent alternative fuels, compared to near to zero in Spain (IEA 2009: 11). Moreover, stakeholders in some countries are concerned about potential health or environmental impacts from the handling and combustion of alternative fuels, making it hard to pursue this strategy if stakeholder relations are not dealt with properly (Holcim 2011b: 2).

Turning to another central measure – clinker substitution – we see that national circumstances and regulations matter here as well. First of all, Holcim, but also HeidelbergCement, have both historically used more clinker substitutes than their competitors. Part of the explanation relates to the locations where they operate: cement plants are usually located near limestone quarries, but not all companies have had the same access to such quarries, which will make it more pressing to find clinker substitutes. Plant location can thus in itself contribute to explain Holcim's frontrunner position when it comes to clinker substitution (interviews 2011).

Second, as also noted, two materials have emerged as the best alternatives to clinker in cement manufacture: blast furnace slag and fly ash. The availability of these by-products restricts their potential impact on CO_2 emission reductions in the cement sector. Fly ash from the coal-powered stations and slag from steel producers are limited in supply – and cement companies need long-term contracts with, for instance, steel manufacturers. However, new steel plants produce less blast furnace slag, so new, more climate-friendly, technologies in the steel sector affect the climate strategies of the cement sector. Predicting the future volumes of iron and steel production, and the future number and capacity of coal-fired power plants is very difficult, and the companies see this as a great challenge: 'The availability of the right quality materials to replace clinker in cement is limited in some markets by logistical factors, and by increasingly competitive markets for appropriate materials' (Holcim 2011b). In other words, the climate strategies of cement companies are clearly influenced (and limited) by international and national policies that target other sectors and technologies.

It is important to include company-internal factors in order to understand differences in climate strategies and ETS response. As noted, Holcim has been an early mover when it comes to acknowledging the climate change problem and the responsibility of the cement sector. The Swiss philanthropist Stephan Schmidheiny represented business and industry interests at the 1992 'Earth Summit' in Rio, and in 1990 he established the World Business Council for Sustainable Development (WBCSD). His brother, Thomas Schmidheiny, was at the time (and up until 2003) the controlling shareholder in Holcim, which helped to bring a greater focus on sustainable development (interviews 2011). Thomas Schmidheiny created an early awareness in Holcim, but it is hard to say how the company is affected now that he is no longer the controlling shareholder. There do not appear to be any figures in HeidelbergCement with the same sort of influence as Holcim's Schmidheiny.

Are HeidelbergCement and Holcim regarded as particularly innovative companies in the cement sector? Is the one company 'more innovative' than the other? Here we can at least argue that HeidelbergCement and Holcim hold somewhat different views on what cement should be, and have therefore directed their innovation efforts in different directions. Holcim's business model centres largely on blended cements, and the company has chosen clinker substitution as an important field of R&D. HeidelbergCement, on the other hand, has a large presence in Northern Europe. Here blended cements are less normal, and due to rough weather conditions, most buyers prefer Ordinary Portland Cement (interviews 2011). This can probably explain why HeidelbergCement has focused its innovation efforts more towards alternative fuels and end-of-pipe solutions like CCS. Limestone, the main material in clinker, is readily accessible to HeidelbergCement, but using slag or fly ash involves depending on other suppliers, and a different strategy than the 'traditional' way of thinking of cement.

In conclusion, we see how different climate strategies are affected by other factors external and internal to a company. Differences in national regulations on CCS and climate financing, waste management, and regulations targeting industries which could provide the cement sector with by-products (like the steel sector), can all shed light on differences in climate strategies – both between the two companies, and within a single company. We also find some evidence to indicate that company management can explain the early progressiveness of Holcim.

6. Discussion

The Emissions Trading Scheme is the EU's main climate policy instrument for putting Europe on a track towards a low-carbon economy. It is essential for this instrument to work as intended, and provide incentives for reducing CO_2 emissions in the installations covered. This chapter has showed that interesting changes are underway in the cement sector, one of the most energy-intensive industries in the world.

Our first main observation is that climate strategies in the cement sector have become gradually more proactive, at the aggregate level as well as for the individual company. The cement sector joined forces in the 'Cement Sustainability Initiative' as early as in 1999, and climate change and the ETS have been at the top of Cembureau's agenda since the early 2000s. The cement sector can in some ways be depicted as one of the energy-intensive sectors that has taken early climate action. Voluntary targets have been put in place, as well as systems for monitoring, reporting and verification of emissions. Cement companies have started by picking the 'low-hanging-fruits' – replacing old kilns, using some alternative fuels and substituting clinker with other materials. More pioneering changes are slowly starting to emerge, with R&D directed towards new cement types and CCS technologies. Companies may prioritize differently among these measures, but both Holcim and HeidelbergCement – and several other companies in the sector – are becoming more aware of climate change mitigation.

It is very difficult to say to what extent these strategic changes can be attributed to the EU ETS, as the scheme works together with, and is conditioned by, a range of other company-external and company-internal factors. Chapter 2 presented several models and propositions, describing how environmental regulations like the EU ETS would affect corporate climate strategies.

According to the first model presented in the Chapter 2, companies will be 'reluctant adapters', and the EU ETS will lead to company resistance. Our second observation in the present chapter is that the cement sector has lobbied fiercely to counter the development of an ambitious climate policy in Europe, which lends some support to the model. The sector has worked hard to keep ETS constraints to a minimum, and has largely succeeded. Recent examples are its position in the benchmark debate and on a possible 25/30 per cent reduction target for EU 2020. The sector wants a sectoral or global agreement, maintaining that, as long as the scheme is solely European, its design flaws will continue.

Model I also expects that choice of ETS compliance response will depend on what pays off in the short term, and that the scheme works mainly through an economic incentive mechanism. Our third observation is that the scheme has not yet been able to provide significant economic incentives for change. Studied in this way, the ETS can only to a very small extent explain the short-term changes in company climate strategies. Although the ETS has probably had an indirect impact through increased electricity prices, the sector has received an abundance of allowances (for free) and large windfall profits. In total, the ETS has provided few economic incentives for extensive and costly business alterations in the period 2003–2010/11 – the exception being Holcim's new fund based on the sale of excess CO_2 emissions certificates. However, cost–benefit analyses are expected to be influenced more by the trading regime after 2012, when the new realities could involve a stricter cap, allocations based on benchmarking and possibly a higher carbon price. These *expected* ETS changes are likely to have created incentives for long-term investments and innovation patterns already.

Thus we see that the first model, which views companies as reluctant adapters, has proven to have some explanatory power. Cement production is highly carbon-intensive, and more stringent regulation is likely to hit cement companies hard. This has led to resistance, but the prospects of rising carbon costs have also spurred the companies into action. The main observation not in line with Model I is that both Holcim and HeidelbergCement have done more than strictly required under the scheme, for the short and the long term – which leads us to explore the second explanatory model of this book.

The fourth general observation is related to the second model, according to which the ETS will lead companies to search for new innovative opportunities. The companies have already undertaken many activities aimed at reducing energy use and emissions, but these have been further spurred – and become more relevant – as a result of the trading scheme. Perhaps the main effect of the ETS is not that the companies have discovered totally 'new opportunities', but that the ETS has contributed to a *reinforced focus* on such solutions. The learning effect of the EU ETS is clearly greater in areas where companies have already identified possibilities for cost-savings and optimization of operations. That being said, we also find some evidence that the scheme has served to direct attention towards formerly totally neglected lines of business, with HeidelbergCement's CCS efforts as the best example. The EU ETS forms part of the backdrop for why cement companies want to know more about CCS technologies – they have recognized that in the end some sort of solution will be needed, and an end-of-pipe system is likely to be part of it. It seems that the (expected) costs of CO_2 justify some extra costs, so we may conclude that there is more 'climate-related' innovation in the cement sector today than there would have been without the EU climate targets and the ETS. The sector is gradually acknowledging that the traditional cement product and production process will have to change. However, here we should underline that the EU ETS is not the only factor behind this change of mind-set: the new business focus on long-term sustainability and climate-friendly production has been induced by several other regulatory and societal factors as well – and by the increasing concern for the global warming challenge among scientists and the wider community.

Fifth, we should explore whether the EU ETS has 'crowded-out' existing norms of responsibility in the companies. Has the EU ETS had a positive or negative effect on corporate norms of responsibility and voluntary initiatives in the cement industry? With cement companies becoming increasingly frustrated with the ETS – arguing that it is poorly designed and distorts competition – this could create hostility towards climate issues. Extrinsic motivation like the ETS could affect intrinsic motivation. Perhaps we see some indications of this in Holcim – a company that was already an early mover and one of the most environmentally aware companies in the 1990s. Now, after the latest benchmark decision, Holcim's support for the EU ETS has deteriorated significantly. When the ETS does not take the entire cement product into account, Holcim believes this will act to punish companies like themselves, and that could reduce the perceived legitimacy of the instrument. On the other hand, when we turn to the CSI, we see a strengthening of voluntary initiatives, not a weakening. External verification of emissions reporting has been put in place (building on the voluntary initiative), companies explore new abatement opportunities every year, and responsibility statements – aggregate and individual – have been strengthened. Importantly, we have *not* observed a 'crowding-out' effect: the ETS does not seem to have reduced voluntary social responsibility activities in the cement sector. Quite the contrary, the ETS has to some extent strengthened such initiatives and statements, although this does not seem to represent any 'deep' change in social norms.[22]

Finally, we have sought to explain the differences in the climate strategies of HeidelbergCement and of Holcim. First of all, the two companies have different business models and ways of looking at the cement product. Company history can to some extent explain why Holcim prioritizes clinker substitution in its climate strategy, whereas HeidelbergCement is a frontrunner in alternative fuels. Second, Norwegian support schemes for CCS can explain why HeidelbergCement has decided to focus on this technology. Third, national policies on waste, alternative fuels and by-products suitable for clinker substitution can all influence a company's choice of climate strategy. And finally, Holcim's role as an early mover seems to be partly explained by the fact that the company management was looking for new opportunities in the 1990s and early 2000s.

Concluding Remarks

The climate-change issue provides the cement sector with opportunities as well as risks. Risks, because cement production is a highly energy-intensive process, and extensive changes will be needed for the sector to reduce its carbon footprint. But the climate challenge can also be seen as providing opportunities to change

---

22 Although it is difficult to determine the effect of such responsibility statements on actual corporate action, the changes in climate strategies do not seem to be the result of a 'deeper norm change'.

old habits, and modernize the technology and diversify the cement product. As mentioned in the introduction, cement is regarded as an important ingredient in climate change *adaptation* and also concrete will be in high demand in the decades to come.

The question is therefore: What is needed to realize a low-carbon development path in the sector? And will the EU ETS be able to contribute to the changes needed? Exploring the effects of the EU ETS on corporate climate strategies in the cement sector has been the overarching aim of this chapter. Our starting point has been the assumption that pioneering innovations and radical new solutions are needed to achieve sufficient emissions reductions in the cement sector. There is much to indicate that the EU ETS will not suffice for this to happen – especially not in the short term. On the other hand, the ETS has brought CO_2 and carbon prices onto the corporate agenda, and it has directed company attention to a broader range of business lines. The effect of the EU ETS is greater in areas where the companies have already identified possibilities for energy- and cost-savings related to energy efficiency, clinker substitution and alternative fuels use. The ETS has also worked as an eye-opener: carbon will no longer be free, and the companies are gradually coming to terms with this new reality. The sector has been set in motion. It is important that the EU's member states reinforce and continue their transition path, and that the 'climate pressure' on companies is maintained.

The EU ETS is also dependent on other national and company-internal factors to co-produce climate-friendly changes. National policies on wastes and alternative fuels have proven central to a low-carbon transition in the cement sector, as have policies affecting by-products from coal-powered plants and steel plants. CCS will require firm governmental support, and, in a sector made up of companies with operations around the world, a *global* carbon scheme of some kind is needed to provide clear and harmonized signals. The sector needs binding policies in China, India and elsewhere – after all, fifty per cent of all cement today is produced in China. Interviewees in this study all emphasized that *Europe is not all.*

This chapter has focused on two 'progressive' cement companies: Holcim and HeidelbergCement. Further studies would do well to study other, less progressive companies, to see why existing policies and regulations are not enough to trigger the innovative turnaround that is needed. It would also be relevant to study the Cement Sustainability Initiative in greater depth, to grasp how this has impacted on corporate norms and behaviour. In our post-Durban world, sectoral approaches to climate change are in demand. The CSI might serve as a starting point and a model for action in other sectors as well.

Interviews

Brevik, Per, Director Alternative Fuels, HeidelbergCement Northern Europe, 8 March 2011, Oslo

Klee, Howard, World Business Council for Sustainable Development (WBCSD), the Cement Sustainability Initiative, 27 April 2011, telephone interview

Kumar, Sanjeev, Senior Associate E3G, 13 May 2011, telephone interview

Loréa, Claude, Technical Director, Cembureau, 13 April 2011, Brussels

Mages, Vincent, Vice President Climate Change Initiatives, Lafarge, 17 May 2011, telephone interview

Schneider, Martin, Managing Director, European Cement Research Academy (ECRA), 10 May 2011, telephone interview

Van der Meer, Rob, Director EU Public Affairs and Group Environmental Sustainability, HeidelbergCement, 13 April 2011, Brussels

Vanderborght, Bruno, Senior Vice President Climate Strategy, Holcim, 9 May 2011, Brussels

Wyns, Tomas, Senior Policy Officer (EU Emissions Trading System), CAN Europe, 8 April 2011, telephone interview.

References

ABB. 2010. Press release: *ABB turns waste heat into clean energy*. Available online: http://www.abb.com/cawp/seitp202/8e34f8163c1796d8852577ed007e5acb.aspx [Accessed January 2012].

AEII (Alliance Energy Intensive Industries). 2010. *Public consultation – Roadmap for a low carbon economy by 2050*. Available online: http://www.eurometaux.eu/LinkClick.aspx?fileticket=4b-2RY-7vE0%3D&tabid=84 [Accessed May 2011].

Barker, D.J., Turner, S.A., Napier-Moore, P.A., et al. 2009. CO_2 Capture in the Cement Industry. *Energy Procedia*, 1: 87–94.

Battelle. 2002a. *Towards a Sustainable Cement Industry*. Geneva: World Business Council for Sustainable Development, March. Available online: http://www.wbcsd.org/web/publications/batelle-full.pdf [Accessed May 2011].

Battelle. 2002b. *Towards a Sustainable Cement Industry, Substudy 8: Climate Change*, written by Ken Humphreys and Maha Mahasenan, March 2002. Available online: http://wbcsdcement.org/pdf/battelle/sub_co2.pdf [Accessed May 2011].

BCG (The Boston Consulting Group). 2008. *Assessment of the impact of the 2013– 2020 ETS Proposal on the European Cement Industry*, Report November 2008.

Bosoaga, A., Masek, O. and Oakey, J.E. 2009. CO_2 Capture Technologies for Cement Industry. *Energy Procedia*, 1: 133–40.

CAN Europe. 2010. *Case of EU industry fleeing climate regime up in smoke. CAN-Europe's Submission to the public consultation on re-assessing the case for carbon leakage under the EU ETS*. April 2010, CAN Europe: Brussels.

CDP (Carbon Disclosure Project) HeidelbergCement. 2010. *CDP 2010 Investor Response, HeidelbergCement*.

CDP (Carbon Disclosure Project) Holcim. 2010. *CDP 2010 Investor Response, Holcim.*

Cembureau.2004. *Activity Report 2003.* May 2004. Brussels: Cembureau.

Cembureau. 2005a. *Activity Report 2004.* Brussels: Cembureau.

Cembureau. 2002. *Annual Report 2001.* May 2002. Brussels: Cembureau.

Cembureau. 2005b. *Climate Change – CO₂ Emissions Trading Points of Convergence within the Cement Industry.* Position Paper, 2 December. Brussels: Cembureau.

Cembureau. 2006. *Activity Report 2005.* Brussels: Cembureau.

Cembureau. 2007. *Activity Report 2006.* May 2007. Brussels: Cembureau.

Cembureau. 2009a. *Activity Report 2008.* May 2009. Brussels: Cembureau.

Cembureau. 2009b. *Building a future, with cement and concrete. Adapting to climate change by planning sustainable construction.* Cembureau Publications. Brussels: Cembureau.

Cembureau. 2010. *Activity Report 2009.* May 2010. Brussels: Cembureau.

Cembureau. 2011. *Activity Report 2010.* May 2011. Brussels: Cembrueau.

Climit. 2011. *Infocenter*, Climit webpage. Available online: http://www.climit.no/infocenter-2 [Accessed 27 September 2011].

Commission. 2009. Commission Decision of 24 December 2009 determining, pursuant to Directive 2003/87/EC of the European Parliament and of the Council, a list of sectors and subsectors which are deemed to be exposed to a significant risk of carbon leakage. 2010/2/EC. Available online: http://eur-lex.europa.eu/LexUriserv/LexUriserv.do?uri=OJ:L:2010:001:0010:0018:EN:PDF

Cook, G. 2009. *Climate Change and the Cement Industry: Assessing emissions and policy responses to carbon prices.* Climate Strategies Working Paper, 23 September 2009.

CSI (Cement Sustainability Initiative). 2009. *Cement Industry Energy and CO₂ Performance: Getting the Numbers Right.* Report June 2009. Available online: http://www.wbcsdcement.org/pdf/csi-gnr-report-with%20label.pdf

Damtoft, J.S., Lukasik, J., Herfort, D., et al. 2008. Sustainable Development and climate change initiatives. *Cement and Concrete Research*, 38: 115–27.

Deja, J., Uliasz-Bochenczyk, A. and Mokrzycki, E. 2010. CO_2 emissions from Polish cement industry. *International Journal of Greenhouse Gas Control*, 4: 583–8.

Demailly, D. and Quirion, P. 2006. CO_2 abatement, competitiveness and leakage in the European cement industry under the EU ETS: Grandfathering versus output-based allocation. *Climate Policy*, 6: 93–113.

Drake, D., Kleindorfer, P. and Van Wassenhove, L. 2009. *HeidelbergCement: Technology Choice under Carbon Regulation.* The Business School for the World (INSEAD) case paper.

ECRA (European Cement Research Academy). 2009. *Development of State of the Art-techniques in Cement Manufacturing: Trying to Look Ahead – CSI/ECRA*

Technology Papers. 4 June 2009. Dusseldorf, Geneva. Available online: http://www.wbcsdcement.org/pdf/technology/Technology%20papers.pdf

ECRA 2011. Research. ECRA homepage. Available online: http://www.ecra-online.org/174/ [Accessed 26 September 2011]

Ellerman, D.A., Convery, F.A. and De Perthuis, C. 2010. *Pricing Carbon: The European Union Emissions Trading Scheme.* Cambridge: Cambridge University Press.

ENDS. 16 December 2010. States agree ETS benchmarks for industry. Available online: http://www.endseurope.com/25280/states-agree-ets-benchmarks-for-industry?referrer=search

ENDS. 7 May 2010. Industry rejects plan for 30% CO_2 cut in Europe'. Available online: http://www.endseurope.com/23897/industry-rejects-plan-for-30-co2-cut-in-europe?referrer=search

Fujiwara, N. 2010. Sectoral Approaches to Climate Change: What can industry contribute? CEPS Special Report, May 2010. Brussels: Centre for European Policy Studies.

Fujiwara, N., Georgiev, A. and Alessi, M. 2010. The merit of sectoral approaches in transitioning towards a global carbon market. CEPS Special Report, May 2010. Brussels: Centre for European Policy Studies.

Habert, G., C. Billard, P. Rossi, et al. 2010. Cement production technology improvement compared to factor 4 objectives. *Cement and Concrete Research*, 40: 820–26.

HeidelbergCement. 2002. *Annual Report 2001*. Heidelberg, Germany. Available online: http://www.heidelbergcement.com/NR/rdonlyres/6FBEF38F-C34D-4A2B-ABC6–62C6FF21B125/0/GB_2001_en_print.pdf

HeidelbergCement. 2003. *Annual Report 2002*. Heidelberg, Germany. Available online: http://www.heidelbergcement.com/NR/rdonlyres/6BACB33B-EE3C-4B72-B700-B912B20131F8/0/GB_2002_en_print.pdf

HeidelbergCement. 2005a. *Annual Report 2004*. Heidelberg, Germany. Available online: http://www.heidelbergcement.com/NR/rdonlyres/CDF92E04-A4B7–4969-B431-C43AE493CCBB/0/GB_2004_en_print.pdf

HeidelbergCement. 2005b. *Group Sustainability Report 2004/2005: Building on Sustainability*. Heidelberg, Germany, May 2005.

HeidelbergCement. 2007. *Today for tomorrow – success and goals for sustainability*. Sustainability Report, Heidelberg, Germany, July 2007. Available online: http://www.heidelbergcement.com/NR/rdonlyres/1C436AC5–2760–4B93-BA49–7B431AA2C7AA/0/Sustainability_report_2007_en.pdf

HeidelbergCement. 2009a. *Annual Report 2008*. Available online: http://www.heidelbergcement.com/NR/rdonlyres/7C9B6F38–3AC4–46C0-BA2C-4F84A7452FBB/0/GB_2008_en_print.pdf

HeidelbergCement. 2009b. *HeidelbergCement Sustainability Ambitions 2020*. Information leaflet, September 2009. Available online: http://www.heidelbergcement.com/NR/rdonlyres/9431D133-C7D8–408A-997E-9B0300B56D57/0/HC_SustainabilityAmbitions2020_Leaflet.pdf

HeidelbergCement. 2011a. *Annual Report 2010*. Heidelberg, Germany.

HeidelbergCement. 2011b. *Foundations: Sustainability Report* 2009/2010. Heidelberg, Germany.

Holcim. 2001. *Annual Report 2000*. Available online: http://www.holcim.com/ uploads/CORP/00_en_annual_report_1.pdf

Holcim. 2002. *Corporate Sustainable Development Report 2002*. Available online: http://www.holcim.com/uploads/CORP/CSDR.pdf

Holcim. 2005. *Corporate Sustainable Development Report 2005*. Available online: http://www.holcim.com/uploads/CORP/CSDR_2005_rev.pdf

Holcim. 2007. *Sustainable development report*. Available online: http://www. holcim.com/uploads/CORP/Holcim_Corporate_SD_Report_2007.pdf

Holcim. 2010. *Corporate Sustainable Development Report 2009*. May 2010.

Holcim. 2011a. *Annual Report 2010*. Available online: http://www.holcim.com/ fileadmin/templates/CORP/doc/AR10/AR_2010_english_final.pdf   Holcim 2011b. – Fact sheet – alternative fuels and raw materials. Available online: http://www.holcim.com/holcimcms/uploads/CORP/Holcim_FactSheet_ AFR_2008.pdf

Holcim. 2011c. Fund to promote energy efficiency. Press release, 10 November 11. Available online: http://www.holcim.com/en/press-and-media/latest-releases/ latest-release/article/fund-to-promote-energy-efficiency.html

IEA (International Energy Agency). 2009. *Cement Technology Roadmap 2009 – Carbon emissions reductions up to 2050*. Available online: http://www.iea.org/ papers/2009/Cement_Roadmap.pdf

Key Stakeholders Alliance for ETS Review. 2007. *Lowering Production is no Benefit for the Environment, says European Industry*. Key Stakeholders Alliance for ETS Review Position Paper, Brussels 21 May 2007.

Liu, F., Ross, M. and Wang, S. 1995. Energy efficiency of China's cement industry. *Energy*, 20: 669–81.

Madlool, N.A., Saidur, R., Hossain, M.S. and Rahim, N.A. 2011. A critical review on energy use and savings in the cement industries. *Renewable and Sustainable Energy Reviews*, 15: 2042–60.

Martin, R., Muûls, M. and Ulrich Wagner, U. 2011. *Climate Change, Investment and Carbon Markets and Prices – Evidence from Manager Interviews*. Carbon Pricing for Low-Carbon Investment Project, Climate Strategies, Climate Policy Initiative, Berlin, January 2011.

Nordqvist, J., Boyd, C. and Klee, H. 2003. Three Big Cs: Climate, Cement and China. *GMI*, 39, Autumn 2003.

Oxfam 2010. *Crying Wolf: Industry lobbying and climate change in Europe*. Oxfam international media briefing, 21 November 07/2010.

Pardo, N., Moya, J.A. and Mercier, A. 2011. Prospective on the energy efficiency and CO_2 emissions in the EU cement industry. *Energy*, 36: 3244–54.

Ponssard, J.-P. and Walker, N. 2008. *EU Emission Trading and the Cement Sector: A Spatial Competition Analysis*. University College Dublin and Ecole Polytechnique, Paris, March.

Rehan, R. and Nehdi M. 2005. Carbon dioxide emissions and climate change: Policy implications for the cement industry. *Environmental Science and Policy*, 8: 105–14.

Reinaud, J. 2005. *Industrial Competitiveness under the European Union Emissions Trading Scheme*. IEA Information Paper, February 2005.

Reinaud, J. 2008. *Issues behind competitiveness and carbon leakage: Focus on heavy industry*. IEA Information Paper, OECD/IEA, Paris.

Romeo, L.M., Catalina, D., Lisbona, P., et al. 2011. Reduction of greenhouse gas emissions by integration of cement plants, power plants, and CO_2 capture systems. *Greenhouse Gases: Science and Technology*, 1: 72–82.

Sandbag. 2010. *Carbon Fat Cats 2010: Company Analysis of the EU ETS*. September 2010. Available online: http://www.sandbag.org.uk/site_media/pdfs/reports/fatcats2009.pdf

Sandbag. 2011. *Carbon Fat Cats 2011: The Companies profiting from the EU Emissions Trading Scheme*. June 2011. Available online: http://www.sandbag.org.uk/site_media/pdfs/reports/Sandbag_2011–06_fatcats.pdf

Schneider, M. 2010. The ECRA CCS Project. Powerpoint presentation WBCSD, CSI-Forum, 13–15 September 2010, Warsaw, Poland.

Szabó, L., Hidalgo, I., Ciscar, J.C., et al. 2003. *Energy consumption and CO_2 emissions from the world cement industry*. Institute for Prospective Technological Studies, Joint Research Centre, Report EUR 20769. Available online: http://www.bvsde.paho.org/bvsaia/fulltext/energy.pdf

Szabó, L., Hidalgo, I., Ciscar, J.C. and Soria, A. 2006. CO_2 emission trading within the European Union and Annex B countries: The cement industry case. *Energy Policy*, 34: 72–87.

Taylor, M., Tam, C. and Gielen, D. 2006. 'Energy Efficiency and CO_2 Emissions from the Global Cement Industry', in International Energy Agency (ed.), *Energy Efficiency and CO_2 Emission Reduction Potentials and Policies in the Cement Industry*. Paris 2006.

WBCSD (World Business Council for Sustainable Development). 2002. *The Cement Sustainability Initiative: Our Agenda for Action*. Geneva, WBCSD, July. Available online: http://www.wbcsdcement.org/pdf/agenda.pdf

Worrell, E., Martin, N. and Pryce, L. 2000. Potentials for energy efficiency improvements in the US cement industry. *Energy*, 25: 1189–214.

Appendix Table 6.1   Climate strategies of major cement companies

Company	Acknowledgement of problem and responsibility for problem-solving	Stated importance of ETS and EU climate policy	Short-term goals and measures	Long-term low-carbon solutions
Lafarge (France)	Yes. CSI member. Signed a global partnership with WWF in 2000.	Supports the EU-ETS. Member of IETA (International Emissions Trading Association)	Cut worldwide net CO_2 emissions per ton of cement by 20 per cent by 2010, compared to 1990 (met in 2009). Reducing CO_2 emissions per ton of cement produced by 33 per cent between 1990 and 2020.	Currently no target beyond 2020. Develops clinker which produces less CO_2. Research on process energy efficiency, design innovative concretes. Improve recycling participates in CCS partnerships.
HeidelbergCement (Germany)	Yes. CSI member.	Supports the EU ETS.	Reduce the specific net* CO_2 emissions per ton of cement by 15 per cent by 2010 compared to 1990 in 2007/08). 23 per cent reduction in specific net CO_2 emissions by 2015.	Currently no target beyond 2015. Carbon Capture and Storage, planning a test facility. Waste and biomass as alternative fuels. Reduce the proportion of clinker.
Holcim (Switzerland)	Yes. CSI member.	Yes, supports the EU ETS, but is cautious about Phase 3. Worked actively against the new benchmark rules. Member of IETA.	Reduce average specific net CO_2 emissions by 20 per cent by 2010 compared by 1900 (met in 2009). Reduce specific net CO_2 emissions by 25 per cent by 2015.	Currently no target beyond 2015. Developing substitutes for clinker. Development of alternative fuels.

Company	Acknowledgement of problem and responsibility for problem-solving	Stated importance of ETS and EU climate policy	Short-term goals and measures	Long-term low-carbon solutions
Cemex (Mexico)	Yes. CSI member.	Yes, supports the principle of using market mechanisms to reduce carbon emissions, such as the EU ETS. Supports the inclusion of importers in the scheme. Member of IETA.	Twenty-five per cent reduction in CO_2 emissions per metric ton of cement product by 2015, from 1990 baseline.	Currently no target beyond 2015. Investing in renewable energy projects. Increasing the use of lower-carbon alternative fuels. CCS research. Carbon footprinting tool introduced in 2010.
Italcementi (Italy)	Yes. CSI member.	Yes, supports the EU ETS. Member of IETA.	Reduce the gross emission factor to 640 kg CO_2 per ton of cementitious product by 2014.	R&D on low-carbon content raw material mixes. Use other constituents in cement rather than clinker Mitigation initiatives down the supply chain, including carbon valorization and sequestration.
Buzzi Unicem (Italy)	Yes, but not a member of the CSI.	Yes, supports the EU ETS.	No specific goals.	Energy efficiency, alternative fuels, substituting clinker.
CRH – Cement Roadstone Holdings (Ireland)	Yes. CSI member.	Yes, supports the EU ETS.	By 2015: (agreed in 2007) 15 per cent reduction in specific CO_2 cement plant emissions compared to 1990 levels. Measures: energy efficiency, alternative fuels, substituting clinker.	Research into new cement/concrete technologies, including carbon capture.

Company	Acknowledgement of problem and responsibility for problem-solving	Stated importance of ETS and EU climate policy	Short-term goals and measures	Long-term low-carbon solutions
CIMPOR (Portugal)	Yes. CSI member.	Yes	Reducing global net specific CO_2 emissions by 15 per cent per ton of cementitious products by 2015, compared with 1990.	The issue of climate change is currently a key driver for many innovation efforts. Producing clinkers and alternative products (e.g. belitic clinker). The production and incorporation of artificial pozzolana in the manufacture of blended cement. Analysing and adopting end-of-line measure (CCS). Eco-fuel project.

Note: * A distinction must be made between net and gross emissions. Gross emissions comprise all direct emissions; net emissions are direct emissions minus the savings that are achieved, for instance, through the use of alternative fuels and which are assessed to be CO_2-neutral (HeidelbergCement SD report 2005b: 28).

Chapter 7

Steel Industry[1]

Jørgen Wettestad and Liv Arntzen Løchen

1. Introduction

What do we know about the development of climate strategies and the impact of the ETS in the economically and politically important sector of steel production? Steelmaking accounts for about eight per cent of total global CO_2 emissions and about three per cent of EU CO_2 emissions (EEA 2009). In July 2011 the European steel federation EUROFER took legal action against the European Commission over benchmarking rules (Euractiv 2011). Given the critical profile of the steel sector in the recent development of industrial benchmarks within the ETS, it might be tempting to see the sector as one of the main industrial stumbling blocks in EU climate policy. Furthermore, the generous distribution of free allowances, leading to a surplus has led at least one critic to denounce steel and other energy-intensive industries as 'carbon fat cats' (Sandbag 2010a). At the same time, important technology development work is being carried out within the European Steel Technology Platform, particularly ULCOS – the Ultra-Low CO_2 Steelmaking programme (ULCOS 2011a). There are certainly puzzling elements in the climate profile of this sector, as further discussed in this chapter.

The main research study on ETS effects to date – by Ellerman and colleagues – indicated that energy-intensive industries in the ETS have in fact undertaken 'a number of measures to reduce emissions per unit of output in both the short term and the longer term' (Ellerman et al. 2010: 192). As to steel, the study briefly noted that emissions intensity decreased slightly more rapidly than the long-term average between 2004 and 2006, but that the change was too small to enable a definitive conclusion that the ETS triggered additional abatement (ibid.: 207). Other relevant studies on the ETS and the steel sector have focused on competitiveness effects and technological options, not on actual strategy responses (see Ecofys 2009, CE Delft 2010, CAN 2010).

Hence, our knowledge of ETS effects on EU steel companies in their abatement efforts and development of climate strategies is generally limited. This chapter seeks to contribute to fill this knowledge gap. Does a closer look at some selected

1   Our thanks to the Fridtjof Nansen Institute project group (Per Ove Eikeland in particular), to Miranda Schreurs and participants at the International Studies Association Conference panel in Montreal 2011, to Alexandra Nikoleris, Lars Nilsson, Jonathan Pinkse and Tomas Wyns for helpful comments and inputs.

companies reveal 'lazy carbon fat cats', where a surplus of allowances has muted any incentives for changing company strategies and practices? Or is such a label grossly misleading? Three main conclusions and messages come out of this chapter, complementing and nuancing existing knowledge. First, in the field of climate change politics, the steel sector shows a kind of 'Janus face'. In the high-profile ETS context, the sector has held a reluctant and sceptical stance from the very start, evident also in recent process of establishing industrial benchmarks. These developments may easily overshadow the other more technical and low-profile face, where a long-standing technology development process has given birth to the promising ULCOS programme.

Second, although steel companies may well have been 'carbon fat cats' for a period, interesting changes have taken place, in the form of more short-term energy efficiency initiatives and more long-term technology development initiatives. This is underpinned by information about aggregate climate strategies and a more in-depth study of the major company ThyssenKrupp and the smaller SSAB (i.e. Swedish Steel AB).

Third, as regards the specific impact of the ETS on companies in this sector, we will point to the impact of the ETS on corporate leaders, their thinking and awareness, as the most important effect as yet. As one steel company interviewee expressed it, there has been a considerable 'mind change' brought about by the ETS, not least induced by the changes adopted in 2009. Still, substantial uncertainty attends the functioning also of a revised ETS unless it can be matched by corresponding global measures.

This chapter is structured as follows: in section two we provide some important characteristics of the steel sector, with an overview of relevant economic, political and technological issues. Section three discusses the evolution of the sector's climate positions and strategies. In section four, we zoom in on the development of the climate strategies of two companies selected for in-depth scrutiny: the major German company ThyssenKrupp and the smaller Swedish company SSAB. Both methodological and pragmatic reasons have led to this choice. First, these companies exhibit several interesting differences in climate strategies and development over time. For instance, ThyssenKrupp has given much more weight to 'external abatement' in the form of involvement in Clean Development Mechanism (CDM) projects than SSAB, whereas the latter has given more weight to the setting and communication of specific emissions reductions targets than the former. Second, as we are based in Norway, including a Swedish company in our sample eases some of the problems involved in getting close to the companies and understanding the broader societal contexts that influence them. In section five we discuss the direct and indirect influence of the ETS. In section six we ask whether other central factors – other relevant EU and national policies, market developments, company characteristics – may have interacted with the ETS or even caused the development noticed in the climate strategies of these companies. Section seven discusses the findings up against the three core models of this book, and section eight offers some concluding reflections.

2. The European Steel Sector: Main Characteristics

2.1 What is Steel and How is it Produced? Two Key 'Routes'

Steel is among the most robust and sustainable of materials, and has become a basic component of today's modern world. Bridges, vehicles, building constructions, medical devices and household equipment – all are made with steel. In addition, as steel is 100 per cent recyclable, its production can be seen as contributing to the long-term conservation of fundamental resources for future generations (Ecorys 2008, Eurofer 2010).

The steel industry includes crude steel production and the further processing (casting) of crude steel into various semi-finished products. As the making of crude steel is the most emissions-intensive part of the industry, this dimension is in focus here. The production of crude steel is in principle carried out via two processes, 'Basic Oxygen Furnace' (BOF) steel and 'Electric Arc Furnace' (EAF) steel. These processes differ in terms of the metallurgical process, energy input and process emissions, as well as in the quality and application purpose of the end-products. Using iron ore as its starting base, BOF crude steel is made by producing hot metal in a blast furnace, and then converting it to crude steel in a basic oxygen furnace. This steel process now accounts for approximately 75 per cent of world steel production, and around 60 per cent of EU 25 production (World Steel Association 2008, Ellerman, Convery and Perthuis 2010: 203).

The production of BOF steel requires two preceding processes: coke-making and sintering. The former involves converting coal to coke by heating coal in the absence of oxygen so as to remove the volatile components and other substances like tars, present in coke oven gas. In the sintering process, iron ores of varying grain size are agglomerated together with additives to create a material feed for the blast furnace with improved permeability and reducibility. Both coke-making and blast-furnace processes produce gas which can be used 'in-house' in power production to cover electricity needs, or sold to power companies (Ellerman, Convery and Perthuis 2010: 205).

EAF steel makes up roughly 25 per cent of world steel production, but stands for 40.5 per cent of steel production in the EU. It is produced by smelting scrap or directly reduced iron in an electric arc furnace. Using scrap is the least costly method of making steel. Steel scrap from various sources – households, vehicles, old bridges – is placed in an electric arc furnace. The intense heat produced by carbon electrodes and chemical reactions melts the scrap, converting it into molten steel.

2.2 CO_2 Emissions From Steel: Substantial – But Decreasing

Steelmaking is one of the most energy-intensive industrial production processes. According to the IEA (2007), the iron and steel industry accounts for roughly 19 per cent of final industrial energy use, as well as about a quarter of direct CO_2

emissions from the industry sector. The main reason for the relatively high CO_2 emissions is the large share of coal in the energy mix. Most emissions come from primary BOF steelmaking: the secondary EAF steelmaking process is roughly 4.5 times less emissions-intensive than the BOF process (de Bruyn, Markowska and Nelissen 2010).

Over time, emission intensity (emissions per unit of output) within the European steel industry has decreased steadily and significantly. According to the European Confederation of Iron and Steel Industries (EUROFER), emissions per tonne of steel produced have been more than halved since 1975. The main part of the decrease witnessed after 1990 (around 54 per cent) has been due to the shift in steelmaking methods, from BOF to EAF (Ellerman, Convery and Perthuis 2010: 206). In 2008, steel sector ETS emissions stood at 202 Mt, falling to 142 Mt in 2009 (Point Carbon 2010a). These recent developments are also connected with the financial crisis and falling levels of production.

2.3 Limited Abatement Opportunities: Need for 'Breakthrough Technologies'?

As noted, technological advancements during the last 25 years have caused a substantial reduction in CO_2 emissions from the steel industry. These include enhanced energy efficiency in the steelmaking process, improved use of by-products, improved recycling of steel products, as well as better and more suited environmental protection techniques. At least six principal categories of CO_2 abatement opportunities for the steel sector can be identified (see Table 7.1 below).

Table 7.1   Categories of possible abatement options in the steel sector

1. Closure of inefficient, highly polluting plant
2. Improving energy efficiency and carbon efficiency at existing, non-obsolete plant
3. Ensuring that new plant is built using best available technology
4. Increasing the use of recycled scrap
5. Adopting Carbon Capture and Storage (CCS)
6. Developing and implementing breakthrough technologies

Source: Wooders et al. (2009: 4).

With regard to energy efficiency improvement, on the one hand, the claimed substantial progress in reducing emissions prior to 2005 may mean that many easy, low-cost, measures have been implemented. For instance, already in EUROFER's ETS position paper from 2000 it was stated:

> Steelmaking processes have been developed and refined over a very long time, resulting in energy efficiency close to the theoretical limit. This implies

a low potential for further improvements and high marginal costs for any such improvements. Major reductions in process emissions will require step changes in technology, development of which has a timescale of 20–30 years and requires large investments. (EUROFER 2000b: 1)

However, several other analyses indicate that there is a substantial energy efficiency potential still untapped, also in industries such as steel (e.g. Henningsen 2011: 138).

As to the use of best available technology, most of the CO_2 currently being generated by the steel industry results from the chemical interaction between carbon and iron ore in blast furnaces, where molten iron is converted to steel. The other option is to use scrap/recycled steel. The energy performance of the two processes varies greatly: according to a study by Lindtke and colleagues, 'electric arc steel furnaces use one-tenth of the fuel, one-eighth of the water, one fifth of the air and less than one-fortieth of other materials compared with traditional basic-oxygen blast furnace steel plants' (quoted in von Weizsäcker et al. 2009: 144). On average, the energy intensity of steel produced by using BOF is more than twice that of EAF. According to von Weizsäcker et al. (2009), both BOF and EAF processes have improvement potential. Still, taking all factors into account, the EAF process offers the greatest opportunity for energy productivity improvements. The problem is that most steel products remain in use for decades before being recycled, so there is not enough recycled steel to meet the growing demand (World Steel Association 2008, Kundak, Lazic and Crnko 2009).

All in all, as steelmaking remains highly carbon-intensive, recent assessments claim that there is a maximum 10 per cent potential for further CO_2 cuts to be gained from refining existing technologies (ENDS Report 2011:28). Hence, as noted by Kundak, Lazic and Crnko (2009: 195), 'there is no way of reducing CO_2 levels to where the scientists say these should be by 2050, unless radical new ways of making steel, the so-called breakthrough technologies, are identified, developed, and introduced'. Carbon Capture and Storage (CCS) is an important element here. CCS can be characterized as 'a suite of technological processes which involve capturing carbon dioxide (CO_2) from the gases discarded by industry and transporting and injecting it into geological formations' (European Commission 2008a). The steel sector and steel companies are variously involved in the European process of developing CCS. Not least, CCS is a central part of what must be characterized as the key European initiative in terms of achieving 'breakthrough technological improvements': the Ultra Low CO_2 Steelmaking (ULCOS) programme, further described in section 3.1.

2.4 Economic Context: The Second Biggest Global Producer

Around 40 per cent of global steel production is traded. The EU is currently the world's third largest exporter but also its largest importer (with China as the top supplier, followed by Russia and Ukraine) (CAN 2010: 4). The European steel industry is well established in most the EU member states and is an important employer. The steel sector has a turnover of approximately €190 billion, employs 360,000 people, and produces on average 200 million tons of crude steel per year. This represents 16 per cent of world output, which makes the European steel sector the second biggest global producer after China. However, there is a trend here that should be noted: the EU has gradually been losing ground to China and India, within the EU market and within export markets.[2] Still, due to higher global demand and the 'spectacular increase' in world steel prices in recent years, the total EU steel turnover has increased, and productivity and profitability show a positive development (Ecorys et al. 2008: iii). There are clearly counterbalancing trends at work.

Nevertheless, from 2008 onwards, European steel production was hit by the economic downturn. In 2009, steel output went down some 30 per cent, resulting in lower emissions and less need for allowances. The 142 Mt emitted by ETS steel companies was far below the sector's EU-wide cap of 254 Mt CO_2. This meant a surplus for companies of 112 million allowances, worth around 1.45 billion euros. As explained later in this chapter, the related possibility of banking of substantial number of allowances to the ETS post-2012 has important implications for the dynamics in this phase (see Sandbag 2010a, Sandbag 2010b). In 2010, steel production picked up again, with output rising some 33 per cent in the first nine months of 2010 compared with the previous year (Point Carbon 2010b).

According to Ecorys et al. (2008), the main competitive strength of the European steel sector is its high-quality products, product innovation and technological development, as well as efficiency and skilled manpower. The main challenges facing the European steel industry are the cost and availability of inputs (raw materials, energy), the obligation to reduce emissions, the need to attract and keep a skilled workforce, as well as the competition from third-country producers (ibid.). The industry has to operate in accordance with the EU ETS regulations, whereas key competitors in countries such as Brazil, India and China are operating in far less carbon-constrained contexts. Steel is a global business, so there is a need for global standards and regulations. Manufacture of basic iron and steel and of ferro-alloys was included in the EU list of sectors deemed to be exposed to a significant risk of carbon leakage, adopted in late 2009 (Official Journal of the European Union 2010). Steel was also one of three sectors identified by consultants as being at risk, together with cement and aluminium (see e.g. Droege 2009).

---

2   The EU's overall share of world crude steel output was 24.3 per cent in 1997. See Ecorys et al. 2008: iii.

3. The Evolution of Climate Positions and Strategies

3.1 The Early 2000s: ETS Reluctance – But also Technology Initiatives

At the sector level, a key actor is the European Confederation of Iron and Steel Industries (EUROFER), founded in 1976. It membership includes steel companies as well as national steel federations throughout the EU. The objective of the EUROFER is cooperation among companies and the national federations on issues concerning the development of the European steel industry. Another objective is to represent its members vis-à-vis third parties, like European institutions and other international actors. In 2000, EUROFER published a position paper on climate change, noting that the steel industry operates in 'a highly competitive environment on a global market', and the industry's tradition of developing new processes to reduce energy and raw material production (EUROFER 2000a: 2). Implicitly, the paper accepted climate change as a real environmental challenge, and acknowledged the steel industry's responsibility to contribute to responding to the challenge, due to the industry's CO_2 emissions. But the paper also underscored that previous technological development had brought steelmaking processes 'close to the theoretical minimum' as regards energy use: any significant further emissions reductions would be possible only in a long-term perspective and would depend on 'new break-through processes' (ibid.: 4).

When the ETS was initiated, EUROFER and the European iron and steel industry responded to the European Commission's first discussion of main institutional choices and possibilities (ETS Green Paper, see European Commission 2000). EUROFER welcomed an ETS reluctantly, seeing it as a system which 'could provide a flexible means for companies to achieve their targets', but in fact preferring a clearly decentralized system that would leave main powers and discretion as to the allocation of allowances firmly in the hands of the member states. Furthermore, in order to protect the global competitiveness of the industry, instead of capping emissions, allowances should be distributed according to a Baseline and Credit logic, based on benchmarks and national environmental agreements (EUROFER 2000b). With regard to this latter element, EUROFER and other energy-intensive industries lost out initially, and the ET Directive was adopted as a 'cap-and-trade' system in 2003.

In spring 2004 a broader steel sector green technology drive was launched, with the EU Steel Technology Platform (ESTEP) as the main organizational platform. ESTEP responded to the call from the European Council in 2002 to boost research (European Commission 2004: 3).[3] ESTEP is one of some 40 technology platforms initiated by DG Research; it brings together key European stakeholders

---

3 Industrial actors participating in the launching of ESTEP included Arcelor, Acerinox, Badische Stahlwerke, Böhler-Uddeholm, Celsa Group, Corus, Federacciai, Riva, ISPAT Europe Group, Megasa, Outukumpu Oyj, Rautaruukki Oyj, Salzgitter, Steel Institute VDEh, SSAB, Thyssenkrupp and Voestalpine. See European Commission (2004: 1).

such as enterprises, research institutes and organizations of steel users, to discuss and promote green and low-carbon technologies (ESTEP 2011). ESTEP's 'Vision 2030' paper published in 2004 noted growing consensus on the anthropogenic causes of climate change and that emissions from the steel sector constituted an important share of European emissions (ESTEP 2004: 26).

As a part of ESTEP, the mentioned Ultra Low CO_2 Steelmaking (ULCOS) programme was also established in 2004. ULCOS was initially jointly funded by the European Union and the European steel industry (50–50) and can be seen as a steel sector climate 'flagship'. The goal of ULCOS is to 'reduce the carbon dioxide (CO_2) emissions of today's best routes by at least 50 percent' (ULCOS 2011). ULCOS is a consortium of 48 European companies and organizations from 15 European countries. All major EU steel companies, energy and engineering partners, as well as research institutes and universities, have joined the consortium. The first phase ran from 2004 to 2010, with a budget of 45 million euros.

In order to put ESTEP and ULCOS into a proper historical context, we should note that the EU and the steel sector have a long shared history,[4] dating back to the 1950s and the establishment of the European Coal and Steel Community (ECSC), which was to become the EU. So the steel sector is a sector where industrial policy has been around for a long time at the Community level. Moreover, EU funding for ULCOS comes from the European Coal and Steel Fund, which is the remainder of the ECSC. Given the timing of the establishment of ESTEP and ULCOS it is reasonable to assume that the then-recent adoption of the ET Directive played a certain stimulating role. After all, it was a process that had attracted substantial attention and political energy from state and non-state actors alike in the early 2000s. However, by 2004 the ETS had not started functioning at all. For instance, the press release from the Commission issued at the launch of ESTEP gave prime emphasis to the general competitive advantages related to keeping the European steel sector in the forefront globally (European Commission 2004: 2).

At the same time, and thus even before the ETS had been officially launched, the steel industry and other energy-intensive industries started to draw attention to the possibility that power companies might reap huge windfall profits related to the introduction of the ETS. These companies would receive allowances for free, and could simply cash in as electricity prices rose – at the expense of energy-intensive industries (Wettestad 2009). From 2006 on, this concern about windfall profits and internal distribution anomalies was increasingly complemented by concern over loss of global competitiveness and possible 'carbon leakage'. To cut a long story short: there is increasing evidence that national policy-makers responded to the diverse and vocal concerns of the energy-intensive industries by granting them generous amounts of allowances. This formed some of the main backdrop when the discussions on a revised ETS for the period 2013–2020 started in 2006.

---

4   Special thanks to Per Ove Eikeland and Tomas Wyns for pointing out these historical long lines.

3.2 Recent Developments: ETS Opposition – But also Continued Technology Development

In the ETS revision process, a EUROFER position paper welcomed the EU taking on a leading role in combating climate change through the ETS (EUROFER 2007). However, EUROFER argued that the ETS itself was fundamentally flawed. The main criticism was that the ETS restricts Europe's ability to compete globally and that the system applies to Europe only. A fundamental review of the ETS system would be necessary. EUROFER's main suggestion, with clear links to its positions back in 2000, was to adopt a new Baseline and Credit approach for the steel sector. Such an approach would encourage investment in innovative technologies and promote – not hinder – global competition. EUROFER wanted a system that focused on plant efficiency and not absolute output levels. Such a new system could create a globally acceptable model that could ensure long-term cuts in carbon emissions from the steel industry (EUROFER 2007).

On the whole, EUROFER and the other EU energy-intensive industries had quite mixed success in shaping the revised ETS 2013–2020. On the one hand, these industries did not manage to prevent an overall shift of the ETS towards a system based on auctioning of allowances. On the other hand, the industries did manage to fight back plans for full or at least major auctioning of allowances to the energy-intensive industries from 2013 on. Already the Commission proposal launched in January 2008 suggested that the initial amount of free allowances should be 80 per cent, decreasing gradually up to 2020. Furthermore, the proposal acknowledged that sectors particularly exposed to carbon leakage should receive free allowances all the way to 2020 (European Commission 2008b).

According to the final Directive agreed in December 2008 and formally adopted in 2009, transitional free allocations would be based on harmonized benchmarking rules, taking into account the most energy-efficient techniques, substitutes and alternative production processes (European Council 2009, Skjærseth and Wettestad 2010a, 2010 b). Article 10a of the ETS Directive was intended to prevent 'carbon leakage' from taking place, and sectors determined to be at risk were to be eligible to receive allowances free of charge at the level of the benchmarks. The system with benchmarks would take the average performance of the 10 per cent most efficient installations in a sector as the best performance eligible for 100 per cent free allowances. The list of vulnerable sectors, published in late autumn 2009, showed 164 sectors, including iron and steel.

In early 2010 EUROFER argued against any EU move to a more ambitious climate target. Gordon Moffat, Director General of EUROFER, argued that such higher targets would further damage the industry's competitiveness. He went on to state that 'steel already has to reduce its emissions in 2020 compared to 1990 by over 40 per cent due to the ETS' (Sandbag 2010b: 11). There was certainly a lot of money at stake, as the Commission reckoned that the six billion or so allowances to be allocated for the 2013–2020 period would add up to some 100 billion euros (Euractiv 2010).

Meetings with stakeholders continued in 2010, and inter-service consultation within the Commission started in early September. Analysts *ENDS Daily* noted that 'deriving benchmark values has been an *intense technical and political exercise*' (ENDS Daily 2010a, emphasis added). Final benchmarks were then formally adopted in April 2011. Soon after, EUROFER announced its intention to legally challenge the benchmarking decision. In July 2011, EUROFER released a press statement notifying that action had been initiated at the European Court of Justice (ECJ) for the annulment of the European Commission Decision, claiming that the benchmarks for primary steelmaking were not based on the '10 per cent' rule (EUROFER 2011). The complaint was then dismissed by the Court in June 2012. The Court stated that the matter needed to be taken up at the member state level (Bloomberg 2012).

However, although somehow easily overshadowed by the steel sector's critical and high-profile ETS inputs, important technology development work has continued. As noted, the first phase of ULCOS ran from 2004 to 2010. The next phase of ULCOS, to run until 2015, includes the development of large-scale demonstration plants, at an estimated cost of at least 300 million euro (Rodgers 2008). ULCOS is testing mainly four technologies: top gas recycling blast furnace with CCS, Hisarna smelter technology with CCS, ULCORED (including direct-reduced iron) with CCS, and electrolysis (reducing the need for coke ovens and blast furnaces) (ULCOS 2011b).

One noteworthy project is the Tata company's Hisarna project in Ijkmuiden, the Netherlands. This project involves a re-design of blast furnaces in order to optimize CO_2 production along with carbon capture and storage (or CCS) (CAN 2010, ENDS Report 2011). Such a 'coke-free' blast furnace would function more effectively, *inter alia* obviating the need for sintering, as well as opening up for the use of charcoal from biomass instead of coal. In addition, the use of oxygen creates a waste gas with high CO_2 density, well suited for CCS. It is estimated that this technology could reduce CO_2 emissions by 20 per cent compared with a conventional blast furnace. If combined with CCS, an 80 per cent reduction is envisaged. ULCOS has financed a pilot plan in Ijmuiden (without CCS) which started operating in May 2011 (Birat 2011). The next step involves a larger and considerably more costly demonstration plant also including CCS, to be constructed around 2017–2018. Overall, this is a technology seen as promising with regard to comparative capital investment costs, but it can be considered 'fully proven' only in a 10-year perspective (CAN 2010, ENDS Report 2011: 29).

In addition to the developments within ULCOS, the ESTEP platform has initiated several processes on enhancing energy efficiency: a technology map (related to the EU's Strategic Energy Technology Plan), a working group co-piloted with EUROFER, and two energy 'roadmaps' (ESTEP 2011).

3.3 The Climate Strategies of Big European Steel Companies: Brief Overview

When we look at the climate strategy rhetoric of the eight biggest steel companies in Europe, including ArcelorMittal, Corus/TataSteel, ThyssenKrupp (analysed in greater depth in section 4), Voestalpine, and Salzgitter, we can note interesting elements. As roughly summarized in Appendix Table 7.1, these companies are generally developing measures to increase energy efficiency. For instance, Corus/Tata Steel has entered into a specific energy-efficiency agreement with the Dutch government, and Voestalpine emphasizes several energy-efficiency measures that are being undertaken. However, energy-efficiency measures need not necessarily be motivated by climate change concerns: more general economic, competitive concerns may also serve as causal drivers. Furthermore, as addressed in sections 2.3 and 2.5, there are several interesting R&D and technology initiatives. With steel giants like ArcelorMittal and Corus/Tata, the flagship in question is, not surprisingly, the ULCOS project. Not least ArcelorMittal was a key industrial actor in the establishment of ULCOS.

As to acknowledging the problem and responsibility for problem-solving, ArcelorMittal states that it 'recognizes that it has a significant responsibility to tackle the global climate challenge' and to take 'a leading role in the industry's efforts to develop breakthrough technologies'. Corus/TataSteel recognizes that climate change is a global issue and that the company can contribute to reduce the problem of greenhouse gases. But these and other big steel companies also stress the need for a global approach and agreement in the case of climate change. Hence, as with most societal actors in the EU, it is reasonable to assume that also steel companies have gradually become more concerned about climate change, although we lack a detailed baseline. As long as a more stringent and comprehensive global climate agreement is not in place, the practical willingness to undertake costly abatement measures specifically motivated by climate change concerns will remain moderate. But the ULCOS project shows that willingness is not totally lacking: steel companies provide 60 per cent of the funding for the second phase of ULCOS, up 10 per cent from the first phase.

3.4 Summing Up: A Sector with a Climate Policy Janus Face

In the early 2000s the EU steel sector expressed rhetorical acceptance of the climate change problem and a general responsibility to contribute in responding to it. However, the sector responded rather cautiously and reluctantly to the idea of an ETS. A central reason for this was the fundamentally global character of the sector, and the fear that an ETS would put other and more regulatory burdens on EU industries than in other regions of the world. Like many other energy-intensive industries, the steel sector preferred a different, Baseline and Credit design based on benchmarking, and not the cap-and-trade design which came to be dominant and was adopted in 2003. Despite being given all allowances for free, the steel sector and the ETS were off to a somewhat shaky start.

However, the industry also became involved in a long-term R&D infrastructure aimed at finding new innovative low-carbon solutions for the longer term, in collaboration with the European Commission. This must be seen in light of the long history of the industry in joint R&D development under the all-European regulatory framework established within the European Coal and Steel Community. Hence, from the mid-2000s on, the steel sector stood forward with a sort of climate-policy Janus face: sceptical and reluctant in the context of the ETS, but more progressive as regards involvement in long-term technology development.

Subsequent developments have not changed this picture radically. Concern among energy-intensive industries over increases in electricity prices and windfall profits reaped by the power producers has been followed by increasing concern over global competitive effects and 'carbon leakage', voiced by the steel sector not least. The energy-intensive industries managed to get acceptance for continued free allowances and benchmarking as a central part of the ETS post-2012. But the steel sector has been highly dissatisfied with the Commission's development of specific benchmarks, which led the sector to initiate a legal case against the Commission in July 2011.

Scrutiny of the climate strategies of the eight major steel companies in the EU show a rhetorical acceptance of the climate change challenge, and also that the steel sector acknowledges its responsibility for contributing to problem-solving. Concerning more short-term measures, we observe that industrial companies are reluctant to quantify emissions reductions goals, and continue to insist that any short-term reduction would be impossible. Getting a more stringent and comprehensive global climate agreement in place, and hence a more level global economic playing field, is heavily emphasized. Still, some relevant measures have been implemented (like improved energy efficiency), and, importantly, the development of possible new 'break-through' technologies involving CCS is attracting substantial attention and resources within the ULCOS project.

Against this backdrop, let us turn to a more detailed comparative discussion of two selected EU steel companies: the big German company ThyssenKrupp and the much smaller Swedish company SSAB.

4. A Closer Look: ThyssenKrupp and SSAB

In order to say something about the possible effects of the ETS, we need first to sum up the baseline: what climate strategies looked like in the early 2000s, prior to the adoption of the ETS and its start-up. Then we trace developments over time. This section winds up with a summary discussion of main similarities and differences in the development of climate strategies. In addition to official documents and information published by the companies and news articles and research reports, the analysis here builds on interviews and e-mail communications with representatives of companies, European industry associations, the European Commission, and environmental organizations.

4.1 German Steel Giant: ThyssenKrupp

The big industrial conglomerate ThyssenKrupp is the result of the merger in March 1999 of two of Germany's industrial giants, Thyssen AG and Fried. Krupp AG Hoesch-Krupp. The roots of the company embody more or less the entire history of heavy industry in North Rhine-Westphalia. The Thyssen and Krupp families were heavily engaged in the coal, iron, and steel industry in Germany with their production of cars, ships and weapons. Today the company consists of around 670 companies worldwide and is one of the world's largest steel producers. ThyssenKrupp produces components and systems for the automotive industry, lifts, escalators, material trading and industrial services. It has a special sub-division – Uhde Services – focused on technology development for metallurgical plants, power plants, and petroleum processing, which has been part of the ThyssenKrupp group since 1987 (ThyssenKrupp Uhde 2012). After a re-organization in 2009, ThyssenKrupp is structured into two major divisions: Materials and Technologies.[5] As of September 2011, the company had around 180,000 employees in 80 countries, and assets/revenues of some euro 43.50 billion (ThyssenKrupp 2012).

4.1.1 Baseline information about ThyssenKrupp and climate change

What do we know about ThyssenKrupp and the climate change issue prior to the ETS? Finding relevant older documents here is certainly challenging. From EUROFER publications, as indicated above, we know that the steel industry was discussing climate change and climate policy issues already back in the 1990s. This has also been confirmed by steel industry representatives (interview and e-mail communication, April 2012). As an important member of EUROFER, the positions of ThyssenKrupp on these issues can reasonably be assumed not to have differed much from those of EUROFER.

The company's initial positions and thinking on climate change were part of a broader policy on 'sustainability'. There was no specific 'climate strategy', but rather what the company refers to as a 'cost-carbon strategy' (ThyssenKrupp interview, February 2011). In the early 2000s, and prior to the ETS, the issue of climate change was not generating any feelings of real urgency in the company (ibid.). As to the company's exposure to relevant climate change regulation, we know that a main initial policy instrument targeting German industry involved various voluntary agreements (VAs) between the government and industry, of which there were about 130 by 2001. The BDI (Federation of German Industries) and several sector-wide industry umbrella groups agreed to VAs in 1995, 1996 and 2000 (Wurzel et al. 2003, Watanebe 2011). The steel industry had a VA target of 16–17 per cent reductions of CO_2 emissions per ton of rolled steel by 2005, against

---

5 The Materials division concentrates on carbon steel, stainless steel and material services, while the Technology Division concentrates on lifts, plant and components technology, and marine systems.

a 1990 baseline (OECD/IEA 2003:21). But this did not mean specific emissions caps for individual companies like ThyssenKrupp.

ThyssenKrupp's CO_2 emissions were not measured but more generally calculated. The company's location had resulted in its being more in the regulatory limelight as to the more direct environmental effects and problems felt in the local surroundings. Local air pollution (dust and SO_2) was an important issue, and had made the company the subject of substantial environmental regulation also prior to the ETS.

4.1.2 Main developments

Soon after the ETS was adopted, the involved Group companies installed monitoring systems in order to better measure and calculate their CO_2 emissions. Furthermore, ThyssenKrupp put an operational emissions trading strategy in place, aimed at covering possible 'short' positions (i.e. with less allowances than needed) at the lowest possible cost and at controllable risks. The company established a market access which included membership in the European Energy Exchange (EEX) and in the Intercontinental Exchange (ICE) and European Climate Exchange (ECX). The Raw Materials and Energy department of ThyssenKrupp Steel Europe AG was assigned responsibility for emissions trading (globally) and energy trading (power and gas in Germany) (ThyssenKrupp 2011b).

With regard to more short-term measures, the company has stepped up its work on improving energy efficiency, but also on waste reduction and improvements in recycling. An important project was the Groupwide project ECI (Energy, Climate and Innovation), which examined over 400 measures for saving energy and reducing emissions up to mid-2008 – concluding, however, that most measures offered inadequate returns (ThyssenKrupp 2009). Still, the company completed a programme in 2010 that claimed to save almost 45 million euros in energy, with a subsequent programme estimated to reduce CO_2 emissions by 78,000 tons by 2014 (ThyssenKrupp 2011, 2012). The company particularly emphasizes its success at the Duisburg plant, where an integrated steel mill now converts process gases from blast furnaces, steel melt shops and the coking plant into electricity in the company's power plants and which is used in its production facilities. In addition, the 'shaft furnace' technology installed in 2005 has led to recycling of waste products (ThyssenKrupp 2009).

As discussed in section 2, a dominant perception within the steel industry is that the production processes used are already operating at a level with minimum CO_2 emissions. According to ThyssenKrupp, the technical limits were reached even before the introduction of the ETS. Considerable further reductions are not seen as possible, given the technologies available today. In ThyssenKrupp this has resulted in weight being given to 'EU-external' abatement projects as well as involvement in activities with a more long-term perspective. As regards 'external abatement', ThyssenKrupp established a policy for complying with ETS requirements by involvement in CDM and JI projects, and generally for maximizing the business possibilities from the use of allowances from such projects. The company has become involved in more than 100 different CDM and JI projects, either through CO_2 funds, direct participation or as technology provider (ThyssenKrupp 2010b, 2011a). The Sandbag organization places ThyssenKrupp at the top of the top 10 'carbon fat cats' also as regards the use of external credits/offsets (Sandbag 2011: 6).[6]

ThyssenKrupp emphasizes the contribution of its products and services to carbon emissions reductions for its customers. Products and processes here include the InCar project, with innovations that have led, *inter alia*, to reductions in the weight of vehicles and hence reduced emissions, more energy-efficient lifts, and the EnviNOx process, which breaks down nitrous oxide and helps to prevent emissions in fertilizer production (ThyssenKrupp 2011).

With regard to measures and activities with a more long-term perspective, the company has established as a main strategic priority to reduce the emissions of the steelmaking process via the blast furnace route. R&D efforts to explore new paths have been intensified. The company increased its R&D budget gradually up to 2007/2008: see Table 7.2.[7]

Also as discussed in section 2, developing the Carbon Capture and Storage (CCS) technology is regarded as a necessary move for securing significant reductions in CO_2. As a natural move, a main R&D activity emphasized by ThyssenKrupp has been the earlier-mentioned ULCOS (Ultra Low CO_2 Steelmaking) consortium. The company was one of the founding fathers of this project back in 2004, part of a core group of six companies (ThyssenKrupp 2005).[8] In the years immediately following the launching of the ETS, ThyssenKrupp signalled what seemed like a steel-sector 'standard' interest in CCS, for instance joining a German CCS information network in 2008, together with E.On, Vattenfall and Alstom, among others (IZ Klima 2008).

Concerning external communications and climate policy reporting, the company initially chose not to establish specific emissions reduction targets, giving priority to 'explaining actions' instead (ThyssenKrupp interview, February 2011). With regard to a public communication process such as the Carbon Disclosure Project, ThyssenKrupp answered questionnaires from 2006 on, but did not accept public disclosure at first.

From around 2008/2009, several changes can be noted. As part of a re-organization in 2009, a ThyssenKrupp Group environment and climate officer was appointed, to 'set a framework for a Groupwide environmental and climate policy, network existing knowledge and coordinate transnational environmental and transnational climate protection activities' (ThyssenKrupp 2011b). The Executive

---

6   Sandbag indicates a 7,375,300 million figure for CERS/ERUs for ThyssenKrupp, compared for instance to 0 for ArcelorMittal and 5,335, 000 for Salzgitter (Sandbag 2011: 6).

7   The increase over time is not very large, and it is not clear if the increase has anything to do with climate change concerns.

8   The others were Arcelor, Corus, Riva, Voestalpine, Saarstahl and Dillinger Huttenwerke and LKAB.

Table 7.2 Research and development spending (in million €)

	2001/2002	2002/2003	2003/2004	2004/2005	2005/2006	2006/2007	2007/2008	2008/2009	2009/2010
Basic R&D	191	183	191	186	241	257	316	284	249
Customer-related development*	156	156	182	266	230	294	224	193	215
Technical quality assurance	294	290	275	281	272	264	301	258	213
Total	**641**	**629**	**648**	**733**	**743**	**815**	**841**	**735**	**677**

Note: * Including outside R&D funds and public funding.
Source: Annual Reports.

Board of ThyssenKrupp defined responsible environmental and climate protection as an important corporate objective. To implement this objective, a Group-wide environmental and climate management system was set up in 2009, coordinated by the department of Corporate Sustainability, Environment and Politics. This system is focused largely on the production activities which are covered by the ETS. In dealing with climate protection, ThyssenKrupp is mainly concerned with CO_2 reduction.

A change can also be noted with regard to external communication and climate policy reporting. From 2010 on, ThyssenKrupp accepted public disclosure of its CDP reporting. Also in 2010, ThyssenKrupp was given a 70 point score on the CDP Leadership Index (CDP 2010a). The company joined the UN Global Compact initiative in 2011. The more overall impression is that, from 2008/2009 on, the company has been giving increasing weight to communicating its climate change measures and profile, as reflected for instance in the description of environmental and climate protection in the 2009–2010 Online Annual Report and CEO Hiesinger's 'Group Policy Statement on Environment and Climate' from 2010 (ThyssenKrupp 2011b, 2012). This is probably related to the above-mentioned appointment of a specific climate officer in 2009.

But there is also a change that points in a less 'climate-friendly' direction. Company representatives have recently expressed sobering and not very enthusiastic views on CCS technology. According to environmental adviser Gunnar Still, a cost–benefit analysis of CCS for the steel industry today leads to the conclusion that 'it is unaffordably expensive in view of competitiveness and so cannot "solve" the problem of CO_2. There are also 'gaps in acceptance in society' … without a special solution for steel as long as no worldwide regime for CO_2-ETS exists, compulsory CCS will mean the end of steel production in Central Europe' (Still 2011, also ThyssenKrupp interview, February 2011).[9]

4.2 The Smaller Swedish Company: SSAB

Following a decision of the Swedish Parliament in 1977, SSAB was formed in 1978 through the merger of three Swedish iron mills, with the Swedish state as owner.[10] Since then SSAB has developed into one of the most profitable steel producers in the world, focusing on niche products with high-strength steel qualities. When it was established, SSAB had eight blast furnaces; by 2011, the number of furnaces in Sweden was down to three, but the company is producing significantly greater quantities of steel than before. In July 2007, SSAB acquired

---

9   Well-informed analysts claim that other German steel companies such as Salzgitter and ArcelorMittal Eisenhüttenstadt have not been so swayed by cost considerations and implemented more 'semi-radical' innovation projects than ThyssenKrupp, even at a time when CCS was much less controversial in Germany.

10   Domnarvets Jernverk in Borlänge, Norrbottens Järnverk in Luleå and Oxelösunds Järnverk in Oxelösund.

the North American steel producer IPSCO, one of the largest Swedish corporate acquisitions of modern times. Sales in 2010 amounted to some 40 billion Swedish kroner and the company employed around 8500 worldwide. SSAB presents itself as a global leader in value-added, high-strength steel (SSAB 2012).

4.2.1 Baseline information about SSAB and climate change

What do we know then about SSAB and the climate change issue prior to the ETS? Company representatives maintain that SSAB recognized climate change as a challenge requiring political action. But in those early days, the company had no specific vision or policy paper setting out its main priorities for dealing with such issues. Concerning the collection and reporting of greenhouse gas emissions, SSAB reported on CO_2 emissions prior to the EU ETS, based on Swedish requirements. However, the company did not have specific personnel or a specific unit for dealing with climate change issues. The overall impression is that SSAB took a rather 'indifferent' stance – but indifference more in terms of the lack of policies and visions than as an outright denial of climate change challenges and problems.

4.2.2 Main developments

The introduction of the ETS meant the introduction of significant new requirements for monitoring, measuring and reporting on emissions. For instance, the mines supplying the coal for steel production had no measurement systems previously; in order to catch the 'coal input factor', proper systems now had to be introduced. SSAB has three separate sites which must report, and these sites differ somewhat. According to company representatives, meeting this (differentiated) reporting challenge required considerable effort for a rather small company.

In 2004, SSAB took part in establishing the above-mentioned ULCOS programme, and in 2005 it joined the Swedish 'Steel Eco-Cycle' research programme. This was a four-year environmental research programme jointly financed by the Swedish steel industry and MISTRA (the Foundation for Strategic Environment Research), aimed at 'developing safe, resource-efficient and recyclable products which satisfy the more stringent demands of society' (SSAB 2009a: 35).

Company representatives say 2007/2008 marked a turning point within the company as to climate strategy and attention. In 2007, the post of 'Director of Environmental Affairs' was established. The portfolio includes more than only climate change, but the current incumbent has explained that dealing with climate change issues constitutes a considerable part of the workload. No new unit specifically devoted to climate change issues was established: instead, new tasks were incorporated in existing (local) units. One company representative at the Stockholm office deals with the more technical details of actual trading of allowances – but that is merely one minor element in the task portfolio.

SSAB did not provide information to the Carbon Disclosure Project in 2007,[11] it started answering the questionnaires in 2008 and, similar to ThyssenKrupp, accepted public disclosure in 2010. In 2010, SSAB was accorded a 78 point score on the CDP Leadership Index (CDP 2010b). SSAB joined the UN Global Compact initiative in 2010. The company is also part of the World Steel Association CO_2 data collection scheme established in 2007 (World Steel Association 2012). Thus we may conclude that, from 2008 on, SSAB became significantly more active in its external communication on climate change issues. As a 'crowning jewel', in 2009 the company published a separate White Paper on 'SSAB and CO_2 emissions' (SSAB 2009a), in which it was clearly stated: 'SSAB aims to be one of the very best companies in the steel industry.' SSAB's emissions were described as being among the lowest in the world, and performance was claimed to be better than comparable plants in Germany (ibid.: 6).

As regards acceptance of the climate change problem, the company's 'Steel Book' stated: 'We are all worried by the gradual warming of our atmosphere. The UN Intergovernmental Panel on Climate Change (IPCC) has shown that human activities on Earth influence the climate to such an extent that we must be prepared for a change in our living conditions' (SSAB 2009b: 6). Still, company representatives are hesitant about claiming that SSAB actually has a specific climate *strategy* (SSAB interview, August 2010).

As a main short-term goal, the company had established a goal of reducing CO_2 emissions by 2 per cent per tonne of produced steel by 2012, to be achieved by 'rationalization of our production processes' (SSAB 2009a: 3). The White Paper mentioned 11 additional short-term measures, including the installation of more efficient burners for heating steel, various energy-efficiency measures, and investigating the possibilities for replacing coal with gas and bio-coal. The company also emphasized the use of rail and ship in its transport and the fact that it had been awarded a 'Green Cargo climate certificate' for several years (ibid.: 10–11).

With regard to more long-term low-carbon solutions, SSAB underscored that 'blast furnace technology is now very close to the theoretical limit for what can be achieved as regards the reduction of carbon dioxide emissions' (ibid.: 7). The stated main long-term goal in the White Paper was hence to 'create the conditions for new steel production technologies with a significant reduction in CO_2 emissions'. The most important instrument referred to in this connection was the ULCOS project. SSAB has gradually stepped up its involvement in ULCOS, becoming a member of the steering group in 2009 (SSAB 2009a). The company participates in the Hisarna project (see section 3.2) (SSAB 2010:26). The 'Steel Eco-Cycle' project, which started in 2005, was initially planned as a four-year project, but in 2009 it was extended by another four years. SSAB is one of the industry financers, and in the second phase SSAB participates in four sub-projects (SSAB 2009b: 23).

---

11 According to the company, the task of responding to the questionnaire was at first seen as daunting and complex. Interview, August 2010.

Concerning the development of Carbon Capture and Storage, in addition to ULCOS, SSAB is involved in three projects and research programmes, studying the conditions for and the effects of CCS (SSAB 2009b, 2010b, 2011). All these activities started in 2009 or 2010. One research programme is national, initiated by the Swedish Energy Agency (SSAB 2009b: 22). A second project is focused on studying the potential for CCS in the Baltic region and is a collaboration between Baltic companies and the Swedish Energy Agency. The third project was initiated in 2010; it aims to calculate the effects of implementing CCS in an integrated steel plant. This project is co-funded by companies, the Swedish Energy Agency and the International Energy Agency (SSAB 2010b). SSAB sees itself as a leading force in all three projects. According to company representatives, CCS is the most important technological option for bringing down CO_2 emissions radically.

4.3 Summary and Assessment

Comparing differing and complex companies and distinguishing between fancy rhetoric and practical effects pose formidable challenges as to methodology. That said, we may proceed with an attempt. Prior to the adoption of the ETS, the ThyssenKrupp company projected a generally indifferent stance to the issue of climate change, with no specific climate strategy and no sense of urgency. Since the establishment of the ETS, interesting changes have certainly taken place. Better monitoring and reporting of CO_2 emissions has been established, a 'climate protection management system' has been introduced, relevant organizational changes have been made for handling emissions trading, and, with an environment and climate officer appointed in 2009, several measures for stimulating energy efficiency have been stepped up and new ones established. ThyssenKrupp has been one of the founding fathers in the establishment and development of the European steel sector's ULCOS programme, seeking to develop new steelmaking methods and technologies which may halve CO_2 emissions.

But still, the main impression is that ThyssenKrupp has not really radically changed its way of dealing with the issue of climate change. There has been a reluctance to make public its short- and long-term emissions reductions goals – however, with more emphasis on external communication recently. Perhaps symptomatically, the company has introduced a system for 'managing' climate policy challenges, not a public strategy for actively and forcefully setting the company on a more low-carbon course. It fits in this picture that the main emphasis has been on the possibilities for taking advantage of company-external CDM credits. Furthermore, we have not found any significantly greater emphasis on innovation with long-term implications, although ThyssenKrupp's prominent, active and continued involvement in ULCOS is surely a 'jewel in the climate crown' for the company.

With SSAB, the picture is in many ways similar, but there are also several differences. Prior to the adoption of the ETS, the company did not project a very active or explicit profile on the issue of climate change. SSAB had no specific climate strategy, and must be characterized as rather indifferent as regards climate change. In the years following the establishment of the ETS, several interesting changes have taken place. Better monitoring and reporting of CO_2 emissions has been established, various energy efficiency measures have been introduced, and organizational changes have been made to enable better handling of climate change and emissions trading. A specific environmental director position was established in 2007, giving considerable attention to climate change. From 2008 on, the company has formulated what may in practice be seen as an explicit climate strategy, communicated both in a specific climate white paper and in sustainability reports.

Compared to ThyssenKrupp, SSAB has placed far less emphasis on 'external abatement' such as the CDM. This is the most striking difference between the two companies. It seems also that SSAB started to formulate and publicly communicate clear and explicit measures, goals and ambitions slightly earlier than ThyssenKrupp. As to long-term ambitions, the comparative picture is mixed. On the one hand, although both companies give significant weight to the ULCOS project, ThyssenKrupp is a more central partner – perhaps naturally, due to its greater resources. On the other hand, SSAB appears to maintain a more enthusiastic CCS position than the increasingly pessimistic ThyssenKrupp.

What does this mean in terms of points of departure for our subsequent analysis? First, we can note a rather striking similarity between the companies: although there has been a common problem of establishing a clear baseline, company strategies have certainly changed towards greater proactivity. The first explanatory challenge to be addressed below then becomes: is the ETS the main factor that has caused these changes? We have identified certain differences between the companies. The main one is evident: much greater involvement in CDM and 'external abatement' in ThyssenKrupp than in SSAB. A second difference is less striking: a more stable positive attitude towards CCS in SSAB than in ThyssenKrupp. A third difference is also less striking: apparently greater and earlier weight given to external communication of climate-policy ambitions and targets in SSAB than in ThyssenKrupp. Discussing the background for these differences is the second explanatory challenge in this chapter.

5. Direct and Indirect Effects of the ETS

In order to shed light on similarities in the development of climate strategies, we should note some common design elements of the ETS policy instrument that may have influenced both companies studied here, in a roughly similar way. The ETS has been called 'the new grand experiment'. It was certainly a new instrument in the EU policy portfolio. Although European industry associations and companies had been involved in stakeholder discussions about the emerging design of the instrument from the 2000 ETS Green Paper and onwards, most actors entered the world of actual emissions trading with some trepidation. How would it work?

What would it mean for them? Not least: was it something that would take hold and be a stable factor, influencing future investments – or would it turn out to be just a passing policy fad in the EU? These and other similar questions were bothering many actors in the first half of the decade – including the companies studied here, as evident from company interviews August 2010 and February 2011.

A related matter is that, before the ETS, the GHG emissions of companies were unregulated or regulated only rather loosely (see below), and many companies did not collect and report data about their emissions in a systematic way. So it makes sense that a key early ETS influence reported by our selected companies concerns the new procedures for data collection and reporting. Both ThyssenKrupp and SSAB report that as a direct result of the EU ETS the companies changed monitoring and reporting systems in order to measure and calculate their CO_2 emissions. Particularly for the relatively small SSAB, this was described as a challenging venture.

A second important ETS design characteristic concerns cap-setting and allocations. Due to the initial decentralized character of the ETS, with main decisions about cap-setting and allocation very much left to the member states, national governments quite naturally feared that a 'strict' treatment of their own industries – while the industries in other states were treated far more generously – might well put their industries in an unfavourable competitive situation. So it is understandable that governments generally opted for a generous strategy. From 2004 on, published assessments indicated that member states had been handing out allowances liberally, not least to the energy-intensive industries (see e.g. Ecofys 2004).

Evidence about the companies studied here backs this up further. As noted with regard to ThyssenKrupp, in the ETS phase I (2005–2007), the company had a quite steady allowance surplus. In the ETS phase II (2008–12), the surplus increased further: in 2008 and 2009 the company was listed by Carbon Market Data among the three companies with the highest carbon allowance surplus (Carbon Market Data 2009).

Table 7.3   ThyssenKrupp allocated allowances and verified emissions

	2005	**2006**	**2007**	**2008**	**2009**
ThyssenKrupp	*Allowances* 21,248,943 *Ver. Emiss.* 18,423,050	*Allowances* 21,248,943 *Ver. Emiss.* 18,385,969	*Allowances* 21,263,273 *Ver. Emiss.* 19,290,933	*Allowances* 25,648,952 *Ver. Emiss.* 20,000,349	*Allowances* 25,648,952 *Ver. Emiss.* 14,095,937

Source: Carbon Market Data, 2009.

Verified emissions fell by around a quarter to a historic low in 2009, due to the financial crisis (see below). However, from 2010, both production and emissions from steel companies started to increase again.

Also SSAB has had a steady surplus of allowances. In percentage terms, the SSAB surplus has in fact been higher than that of ThyssenKrupp.

Table 7.4   SSAB allocated allowances and verified emissions

	2005	2006	2007	2008	2009
SSAB	*Allowances* 6,627,333 *Ver. Emiss.* 3,670,338	*Allowances* 6,627,333 *Ver. Emiss.* 3,423,787	*Allowances* 6,627,333 *Ver. Emiss.* 3,678,992	*Allowances* 7,164,051 *Ver. Emiss.* 3,913,416	*Allowances* 7,259,447 *Ver. Emiss.* 2,048,297

Source: Carbon Market Data, 2009.

This overall generosity and surplus of allowances has meant fewer incentives to change steel company practices than would have been the case if allowances were scarcer. According to ThyssenKrupp representatives, 'the ETS has not induced any (additional) abatement at all so far'. A similar view was expressed by SSAB representative: 'we have simply been given what we need so far (and more than that)' (interviews, August 2010 and February 2011). A further dimension of the allocation issue concerns the choice between handing out allowances for free and selling them at auctions. In the ETS, it was decided to initially hand out allowances for free. This also contributed to the moderate incentive effect of the ETS in Phases I and II.

We now turn to a third and related important characteristic of the ETS: the carbon price. An emissions trading system affects targeted actors both directly through the cap and the allocations, and more indirectly and 'dynamically' through the carbon price. A relatively stable and high price will tend to provide greater impetus for companies to change their strategies and practices than will a volatile and rather low price. In the case of the EU ETS, the latter scenario has been dominant, at least initially. After the price climbed to a relatively high level in 2005 (around 30 euros), it quickly halved in May 2006, when the verified 2005 emissions figures confirmed suspicions: more allowances than needed had been handed out in the ETS Phase I. In 2007, the price fell to near zero.

Tighter allocations were decided for Phase II. For a while, the price stabilized at around 15 euros, but dropped below 10 euros in late 2011. Our company interviewees explicitly cited the volatile and at times low carbon price as a factor explaining their companies' hesitancy in changing practices. The volatility and 2006 price crash created further uncertainty as to the potential of ETS instrument. Would it last? Would it be effective? The somewhat higher and more stable price in Phase II can help to explain why many companies took new initiatives in 2008/2009, including ThyssenKrupp and SSAB. Moreover, both companies studied here have underscored that they now see the future carbon price as a central factor, both for 'good and bad'. 'Too high' a price will drive companies out of Europe.

Thus far, apart from the monitoring and reporting requirements, we have not really found common ETS factors that can convincingly shed light on the changes noted. In fact, rather the opposite: the factors discussed so far shed much more light on why it is not so far-fetched to describe the changes noted within the companies as slow, tentative and relatively moderate. However, there is one key ETS design development which needs to be acknowledged: the considerable changes to the ETS in Phase III (2013–20) which were prepared and adopted in 2007 and 2008. Compared to the 'old, decentralized ETS', the revamped ETS from 2013 onwards will be governed quite differently, in a far more centralized way (see Directive 2009/29: European Council (2009)). Furthermore, around 20 per cent of the allowances of energy-intensive industries are to be auctioned, to be tightened over time.

Not surprisingly, these thoroughly-discussed ETS changes induced changes also in the companies in focus here. Both ThyssenKrupp and SSAB offer good examples of this dynamic. Company representatives explicitly point to 2007/2008 as an important turning point when organizational changes were enacted. We can also note a clear increase in the companies' relevant publications and activities in the wake of this turning point. But ThyssenKrupp in particular also emphasizes the considerable remaining uncertainty surrounding the actual content of these changes and not least their possible effects (note benchmarking process, for instance) (interview, February 2011).

There are also several more indirect ways that the ETS may have influenced these companies. In addition to the market for products, as *energy-intensive* industries, these actors are particularly concerned about the electricity market and energy prices. The introduction of the ETS from 2005 on was accompanied by a clear increase in electricity prices. As these developments were seen as related, this did not make the ETS more popular among the energy-intensive industries. It also sparked heated debate about power producers reaping 'windfall' profits (see Sijm et al. 2006, Point Carbon 2008, Sijm et al. 2008). The steel sector was among those seen as most affected by the rise in electricity prices, along with cement (Lund 2007). This perceived interaction can be seen as a factor that sheds further light on the widespread reluctance towards the ETS among these industries, steel included.

However, the link between the ETS and electricity prices received less attention after the carbon price crash in the spring of 2006. A relevant question in the company-comparative context then becomes: did this dynamic play out significantly different in Germany and in Sweden? This is a complicated question. Some have indicated that Swedish energy-intensive industries in the pulp and paper sector were affected by increases in electricity prices linked to the introduction of the ETS, along with electricity market reform (see Ericsson et al. 2011). However, our steel company interviewees did not emphasize this aspect of the ETS impact discussion particularly, and we find no discernible difference in the responses and information provided by our two companies in this regard.

What can be seen as another potential indirect effect of the ETS is the link to the evolving EU CCS policy. A central strategy change noted in both companies involves paying more attention to the CCS option (not least through the ULCOS project). Here we should note that EU CCS policy started out as a quite independent policy area, with only very general and diffuse links to the ETS (Boasson and Wettestad 2011, 2013). For instance, CCS was not mentioned at all in the 2003 ET Directive. Only from 2005 onwards did the CCS policy really start to develop (Boasson and Wettestad 2013, see also Claes and Frisvold 2009). A key instrument for industry involvement and contact was the establishment of the European Technology Platform (ETP) for Zero Emission Fossil Fuel Power Plants (hereafter: ZEP). No steel company is among the 27 companies on the ZEP Advisory Council. Neither ThyssenKrupp nor SSAB participated in ZEP meetings, as far as we know.

However, the ETS and CCS policies became more closely linked in the decision-making process on the EU climate and energy package in 2008. According to the preamble of the revised Directive adopted in December 2008, 'the main long-term incentive for the capture and storage of CO_2 and new renewable technologies is that allowances will not need to be surrendered for CO_2 emissions which are permanently stored or avoided' (point 20 in 2009 Directive). The Directive also makes a reference to allowances set aside from the New Entrants Reserve (NER).[12] Based on an idea allegedly stemming from E3G and Shell representatives, and introduced and pushed in the decision-making process particularly by MEP Chris Davies, a formal link was established between the revised ETS and the building of EU CCS demonstration projects (Boasson and Wettestad 2011, 2013). More specifically, 300 million allowances in the NER were set aside to contribute to the funding of eight CCS and 32 renewables projects (the 'NER 300 fund'). Although our two companies have not applied for these funds, the companies indicate that this process has contributed to raising their awareness about CCS activities. In February 2011, it was reported that steel giant ArcelorMittal would apply for NER CCS funds for a plant in France (Point Carbon 2011b).

But the ETS is clearly not the only factor at work here. In order to illuminate the differences between the companies, as well as the similarities, we need additional explanatory factors.

---

12 In addition, Article 10 mentions CCS as one of the activities to benefit from the recommended setting aside of 50 per cent of member-state auctioning revenues for purposes related to climate change. See particularly point (e).

6. Company-internal and Company-external Factors

We have noted what appears to be greater attention to strengthening energy efficiency in both companies. At the EU and the national levels, there has been attention to energy efficiency, dating back to the SAVE programme adopted in the 1990s (see Wettestad 2001). Although EU energy efficiency policy so far has had a fairly programmatic and non-binding character, and been characterized as not particularly effective (see e.g. Henningsen 2011), it has clearly been an element in the broader package of climate and energy policy measures which has formed the regulatory environment for steel companies in the EU. For instance, when a central ThyssenKrupp representative discussed climate policy challenges at a conference in 2011, the focus was on the interaction between the ETS and the developing EU energy-efficiency policy (Weddige 2011).

Also more general economic developments can help to explain the similarities in the companies developing climate change measures and positions. The severe economic downturn starting around 2007 began to influence EU industries and production rates in 2008. Steel production decreased and verified emissions fell by about 25 per cent, to a historic low in 2009. This development surely contributed to increasing our companies' abundance of allowances in 2009 and 2010. It is reasonable to assume that the possible incentive effect of the ETS was weakened due to this development, compared to a counterfactual situation where the somewhat tighter allocation which was in fact adopted for Phase II could have induced companies to more active ETS responses. Still, steel production is on the rise again, bringing increased emissions and a need for allowances. In 2010, steel production was up again by about 25 per cent, and this increase continued in 2011 (Point Carbon 2011c, 2011d). Hence, this particular weakening of the ETS' incentive effect may prove only temporary.

Can we then identify other factors that may shed further light on differences in the development of the two companies' climate strategies? We sensed a more reluctant attitude towards the ETS in ThyssenKrupp than in SSAB, with less weight given to setting explicit reductions targets. This may reflect differences in the general institutional fit between the ETS and pre-existing environmental and climate policies in Germany and in Sweden (Skjærseth and Wettestad, 2008). The institutional fit discussion is complex (see Knill and Lenschow 2000, Boerzel 2003), it can roughly be posited that the greater the misfit, the potentially slower and more reluctant the response to the ETS will be among companies.

We know from other studies (e.g. Matthes and Schafhausen 2007, Wurzel 2008) that German societal actors were quite reluctant and negative towards the introduction of emissions trading in Germany. As noted, a main initial policy instrument targeting German industry was the various voluntary agreements (VAs, about 130 by 2001) (Wurzel et al. 2003). Hence, relevant industries had an established tradition of voluntary agreements with the government about emissions reductions and saw no need for a new instrument. A survey conducted among Germany industries in 2005 by consultants EuPD Research concluded that the majority of the respondents perceived emissions trading negatively, and felt that it would lead to higher costs. 'Companies with other installations' (energy-intensive industries) were more negative towards emissions trading than 'energy-related installations' (EuPD Research 2005).

The case of Sweden was not fundamentally different, as voluntary agreements to enhance energy efficiency in major industrial companies had been a main instrument prior to the introduction of emissions trading there as well (Thollander and Dotzauer 2010). Energy-intensive industries had also been exposed to a carbon tax. Still, the tax was quite favourable to Sweden's energy-intensive companies, in order to prevent 'carbon leakage' (Ericsson et al. 2011). Although both Germany and Sweden were perceived as environmental policy frontrunners in the EU at that time, our impression is that Swedish actors were far more positive to the introduction of the ETS. For instance, Sweden was one of the few countries that already stood out as an ETS supporter in connection with the responses to the ETS Green Paper in 2000 (Skjærseth and Wettestad 2008). However, it is interesting to note that the company representatives interviewed for this study tended to downplay this factor. For instance, SSAB emphasized the all-European sector perspective and cross-country contacts and inspirations more than a domestic politics perspective.

Another difference was that ThyssenKrupp gradually became slightly more sceptical towards the CCS option than SSAB. National politics may certainly help in explaining this. The broader political and institutional context surrounding CCS policy development is somewhat more 'benign' in Sweden than in Germany, where public opinion is quite sceptical to CCS storage, with a debate not unlike the debate over nuclear power. There are also important transport challenges (see European Energy Review 2011, ThyssenKrupp interview February 2011). In Sweden these issues seem to create less controversy, although it may in fact have scant options for long-term storage.

We noted also a clear difference in the weight accorded to 'external abatement', with ThyssenKrupp – but not SSAB – involved in several CDM projects. Differences in basic company characteristics might shed some light on this. Handling complex CDM involvement processes requires a certain basic administrative capacity. ThyssenKrupp is clearly a much bigger company, with formidable organizational and administrative resources. The difference between ThyssenKrupp and SSAB also fits with a more general trend for German companies to use far much more CERs (CDM allowances) for ETS compliance purposes than Swedish ones.[13] This

---

13   German installations' offset use so far in the 2008–2012 phase is roughly double that of Sweden (9 per cent versus 4.3 per cent). German installations used 41,100 CERs I in 2011, while Swedish ones used 1,600. See Point Carbon, 2 May 2012.

might indicate a more 'CDM-friendly' culture among German companies than among Swedish ones.[14]

Finally, can other company characteristics – leadership changes in particular – help to explain the striking similarities between the companies? As noted, both companies mentioned 2007/2008 as a turning point in their attention to climate change. That was a time when the EU climate and energy package was taking shape, with a significantly revised ETS post-2012 as a central ingredient. Furthermore, those were days of a serious 'climate hype' in Europe. But was the turning point within the companies also related to internal leadership changes? In the case of SSAB, we know that there was an important organizational change at the time, when a special environmental advisor position was established. That was preceded by a change of CEO in 2006, when Olof Faxander took over from Anders Ullberg. Faxander increased the number of staff at the headquarters and has been described as highly engaged and much more active in the media than Ullberg ever was. However, insiders also emphasize that many other conditions related to the issue of climate change were quite different in Faxander's period (he left in 2010) than in the early days of Ullberg. We are not aware of similar changes in the case of ThyssenKrupp. But it is clear that several relevant organizational changes were carried out in 2008/2009, including the appointment of a Group environment and climate officer.

7. Discussion

Chapter 2 outlined various models and expectations concerning how a regulatory instrument like the EU ETS could affect company climate strategies. According to the first model, companies are likely to be 'reluctant adapters', and the EU ETS will lead to company resistance. This model clearly points to some relevant developments that we have observed in the steel sector and among steel companies. The sector's industry association EUROFER responded rather cautiously and reluctantly to the idea of an ETS. Subsequent developments have not changed this picture radically. Concern among energy-intensive industries over increases in electricity prices and windfall profits reaped by the power producers has been followed by increasing concern over the global competitive effects and 'carbon leakage'. Due to the latter, the energy-intensive industries managed to make continued free allowances and benchmarking a central part of the ETS post-2012. But the steel sector has been deeply dissatisfied with the Commission's development of specific benchmarks; finally, in July 2011 the sector initiated a legal case against the Commission. Furthermore, the steel sector has been a vocal opponent of the EU moving toward a more ambitious 2020 reduction target.

---

14   If so, this is something of a paradox, in light of Germany's earlier scepticism towards the flexible mechanisms in the Kyoto Protocol. See Skjærseth and Wettestad 2008: 92.

Also the steel sector prefers a sectoral or global agreement, holding that a solely European ETS will be flawed.

This model also predicts that choice of ETS compliance response will depend on what pays off in the short term, and that the system will work mainly through an economic incentive mechanism. Similar to other energy-intensive industries, the ETS has not yet been able to provide significant economic incentives for change in the steel industry – in some respects quite the contrary, as brought out in the 'carbon fat cats' concept. Although the ETS probably had a certain initial indirect impact through increased electricity prices, the sector received a surplus of allowances in both the first and second trading period.

However, it is reasonable to assume that the very existence of the ETS and not least the expectation of a stricter system post-2012 have contributed to and stimulated work on improving energy efficiency and making organizational changes, like the appointment of special climate change officers. Such changes have taken place in both ThyssenKrupp and SSAB. The main observation that is seemingly at odds with this model is that the steel industry, with ThyssenKrupp in a leading position, has also done something more than strictly required under the system: here we are thinking of the ULCOS project.

According to the second model, companies will search for new and innovative opportunities as a result of the ETS. As noted above, the ETS has not had any major impact on short-term abatement activities in the steel sector, although there may well have been a more diffuse strengthening effect on ongoing improvements, particularly as regards energy efficiency. Nevertheless, we recall that SSAB announced after 2007 that it aimed to be a climate strategy 'world leader'. The company also announced a target of reducing specific emissions by 2 per cent by 2012 through various measures intended to 'rationalize' the production process.

The main probable impact of the ETS so far is that there has been more attention to the need for more radical technological solutions for the longer term. While the steel sector opposes the ETS and has not been stimulated to benefit from short-term measures (except sales of excess allowances), the sector has begun investing in long-term radical solutions. This makes sense, as the steel industry argues that there are few, if any, possibilities for bringing emissions down with the technology available today.

ULCOS was established at about the same time as the ETS. Indeed, the ETS played a certain initiating role, also interacting with, *inter alia*, a generally increasing concern about climate change in the EU. As stated by one interviewee, 'the ETS was one part of a broader drive' (telephone interviews, April 2012). ULCOS sprang out of a more long-running private–public technology development partnership involving the European Commission and the industry. The revision of the ETS and the development of a more specific CCS policy in 2008 probably contributed significantly to the decision to embark on a second phase in ULCOS. Both ThyssenKrupp and SSAB emphasize the attention/learning dimension of the introduction of the ETS. For instance, ThyssenKrupp representatives talked about a 'mind change' which has occurred in response to the ETS.

This lends further support to other sources and studies. For instance, in an interesting interview in June 2009, chief ETS architect and Commission official Jos Delbeke stated that the most successful element of the ETS was the way it had 'forced company boardroom activity to consider climate change ... Attitudes towards CO_2 and the climate have changed since there has been a price on carbon' (Point Carbon 2009). In a somewhat similar vein, analyst Frank Convery stated that 'carbon emissions trading in Europe has finally lifted environment from the boiler room to the boardroom, and from ministries of environment to ministries of finance. For chief executives of many corporations, the environment has become an omnipresent, if not always welcome, guest at their strategic tables' (Convery 2009: 121).

What then of the third model – has the EU ETS had a positive or negative effect on corporate norms of responsibility and climate-friendliness in the steel industry? This question is by nature tricky and elusive, and data on the steel industry are limited – perhaps because the sector is less keen to 'talk the talk' or that feelings of responsibility have been less developed. The impression is that there have been few concerted or individual CSR programmes beyond ULCOS. Responsibility statements and programmes have emerged mainly after the adoption of the ETS. We find no deep shifts in norms taking place in this sector, but the ETS has drawn attention to the problem and contributed to long-term innovation. And that in turn may contribute to shifting norms in a more 'climate-friendly' direction in the long run.

8. Conclusions and Reflections

As a point of departure we noted the perception of steel companies as 'lazy carbon fat cats'. How meaningful and relevant is such a perception? As to the 'carbon fat cats' part, it is true that companies in the steel sector have accumulated a substantial surplus of allowances. Windfall profits seem not unlikely, and possibly quite high, as these allowances have been distributed for free. But there are some factors that need to be kept in mind. First, ETS companies were not allowed to bank allowances from Phase I (the pilot phase) to subsequent phases, so this initial surplus has been of no use in recent years. Second, the surplus is clearly related to the initial design of the ETS, a decentralized distribution of free allowances, which was a design collectively decided by EU governments. These governments were influenced by commercial interests, but the governments cannot put all the blame on the business sector. Third, the financial crisis has contributed to the surplus of recent years. As dryly noted by a company representative, 'ArcelorMittal would surely prefer to sell steel instead of allowances.'

So what then about the 'laziness' characteristic? In this chapter we have seen that interesting climate strategic developments are underway in the European steel sector. Here it should be noted that the industry appears with a kind of 'Janus face'. In the more specific and high-profile ETS context, the sector has maintained a reluctant and sceptical stance from the very start, as seen not least in the recent years' process of setting industrial benchmarks. These developments may easily overshadow the other, and more technical and low-profile face, where a long-standing technology development process has resulted in the promising ULCOS programme. However, the industry is not doing this alone. ULCOS is a part of a broader public–private technology partnership, with half the initial ULCOS financing contributed by industry and the other half by the European Commission. Hence, the steel industry should share the praise for this initiative with the Commission.

In the companies in focus here, ThyssenKrupp and SSAB, new monitoring and reporting have been introduced, relevant organizational changes have taken place, energy efficiency measures have been carried out, and new attention to the CCS issue can be noted. This supports and complements the findings reported by Ellerman and colleagues, indicating that energy-intensive industries in the ETS have undertaken 'a number of measures to reduce emissions per unit of output in both the short term and the longer term' (Ellerman et al. 2010: 192).

On the whole, our analysis indicates that the causal role of the ETS in the case of the steel sector has been somewhat ambiguous to date. On the one hand, there are several initial ETS design characteristics that have acted to impede the full impact of this instrument so far: the abundance of allowances, distributing them for free, the relatively easy access to external credits – resulting in a volatile and often low carbon price. This has reduced the potential role of the ETS as an 'incentive changer' in the steel sector. The companies have surely requested these design features, but governments have pushed and adopted them for several reasons, not least in order to retain national control.

On the other hand, the ETS is definitely the central causal factor behind new monitoring and reporting of company CO_2 emissions. Furthermore, the substantial changes made to the ETS in 2008 and the revised rules for Phase III have made ETS increasingly important for the companies. Company representatives now emphasize that the ETS and the carbon price as important factors in their thinking about the future and about how to invest.

We have not found many striking differences in climate strategies between the two companies – perhaps not so surprising, given our conclusion that climate strategies have changed only moderately for the sector as such. In seeking to explain the smaller differences in the climate strategies of ThyssenKrupp and SSAB, we have noted the differing environmental policy contexts between Germany and Sweden. For various reasons, central German political and industrial actors have been highly sceptical towards emissions trading, whereas Swedish actors have been generally far more positive. Furthermore, the issue of CCS seems to have developed in a more controversial direction in Germany than in Sweden, which may shed light on why ThyssenKrupp appears to have developed a less enthusiastic CCS position than SSAB. But we should also take differences in company characteristics into consideration, such as the generally higher administrative capacity for handling CDM projects in big ThyssenKrupp than in small SSAB.

Finally, are the sobering findings here as to the role of the ETS really surprising? Is it reasonable to expect the ETS to have delivered more at this early stage? This is a very important but also very challenging discussion: after all, the ETS has been heralded as the 'new grand experiment'. There are many and multi-level political and institutional complexities involved in making such an unprecedented system work really well. Achieving a well-functioning system trusted by all participants is likely to take time and go through phases of learning and corrections. Moreover, considerable uncertainty still remains as to the possibilities of getting a new global climate treaty adopted, levelling the global climate policy playing field. With this in mind, we do not really find it surprising that steel companies have also responded somewhat cautiously to the introduction of the ETS.[15]

List of Interviews and Communications

Alan Haigh, European Commission, DG Research, 27 March 2012 (telephone interview).

Felix Matthes, Öko-Institut, e-mail exchange July 2012.

Hans-Jorn Weddige, ThyssenKrupp, 23 February 2011.

Jean-Paul Birat, ArcelorMittal, 28 March 2012 (telephone interview and e-mail exchange).

Kim Kaersrud, SSAB, 27 August 2010; e-mail exchange July 2011.

Knut Baumann/Marit Foss, Norwegian Federation of Industries, 17 December 2010.

Marco Mensink, CEPI, 28 January 2011.

Tomas Wyns, CAN Europe, 27 January2011; e-mail exchange January 2012.

Yvon Slingenberg, European Commission, DG Clima, 27 January 2011.

References

Birat, J.P. 2011. Challenges & opportunities in the iron & steel industry, powerpoint presentation for ArcelorMittal. IEA-GHG, Dusseldorf, 8–9 November 2011. Available at: http://www.stahl-online.de/deutsch/Dokumente/CCS/03_Birat_ULCOS_IEA-GHG.pdf

Boasson, E.L. and Wettestad, J. 2011. The new drive in EU climate policy: Exploring the role of 'Euroindustry' and institutional entrepreneurs. Paper for ISA Conference in Montreal, 16–19 March 2011.

Boasson, E.L. and Wettestad, J. 2013. *EU Climate Policy: Industry, Policy Interaction and External Environment.* Aldershot: Ashgate.

---

15   On this, see also the discussion in Wettestad, 2011.

Boerzel, T. 2003. *Environmental Leaders and Laggards in Europe. Why There is (Not) A Southern Problem*. Aldershot: Ashgate.

Bloomberg. 2012. EU Court said to dismiss steel producers' suit on carbon, June 7, 2012.

Bruyn, S. de, Markowska, A. and Nelissen, D. 2010. *Will the energy-intensive industry profit from EU ETS under Phase 3? Impacts of EU ETS on profits, competitiveness and innovation*. Report, CE Delft, October 2010.

CAN. 2010. *Horizon 2050 – Steel, cement & paper. Identifying the breakthrough technologies that will lead to dramatic greenhouse gas reductions by 2050*. Brussels, Climate Action Network (CAN) Europe, October 2010.

Carbon Market Data. 2010. Press release: *Carbon Market Data publishes the EU ETS Company Rankings 2009*. 10.06. Available at: http://www.carbonmarketdata. com/cmd/publications/EU%20ETS%202009%20Company%20Rankings%20 -%20June%202010.pdf. Accessed: 15 November 2010

CDP. 2010a. Carbon Disclosure Project Germany 200 Report. London.
_____ 2010b. Carbon Disclosure Project Nordic Report 2010. London.

CE Delft. 2010. *Does the energy intensive industry obtain windfall profits through the EU ETS?* Report no. 10.7005.36, April.

Claes, D.H. and Frisvold, P. 2009. CCS and the European Union: Magic bullet or pure magic? in *Caching The Carbon – The Politics and Policy of Carbon Capture and Storage* (eds), J. Meadowcroft and O. Langhelle. Cheltenham: Edward Elgar, 211–35.

Convery, F. 2009. Reflections – the emerging literature on emissions trading in Europe. *Review of Environmental Economics and Policy*, 3(1), 121–37.

Droege, S. 2009. Carbon leakage and the EU ETS: Sectors at risk and the international context of emissions trading, trade flows and climate policy. *Climate Strategies Issues and Options Brief*, 5 October 2009.

Ecofys. 2004. *Analysis of the National Allocation Plans for the EU Emissions Trading Scheme*. Ecofys, Utrecht, August 2004.
_____ 2009. *Methodology for the free allocation of emission allowances in the EU ETS post 2012 – Sector report for the steel and iron industry, November 2009*. Report together with Fraunhofer and Öko-Institut, Utrecht..

Ecorys 2008. *Study on the Competitiveness of the European Steel Sector*, study by Ecorys, Danish Technological Institute, Cambridge Econometrics, CES Ifo and Idea Consult for DG Enterprise & Industry, Rotterdam, August 2008.

EEA (European Environment Agency). 2009. *Greenhouse gas emissions trends and projections in Europe 2009 – Tracking progress towards Kyoto targets*. EEA report no. 9/2009, Copenhagen.

Ellerman, A.D., Convery, F.J. and Perthuis, C.D. (eds), 2010. *Pricing Carbon – The European Union Emissions Trading Scheme*. Cambridge: Cambridge University Press.

ENDS Daily. 2010a. Industry worried by draft ETS benchmark rules, 9 September 2010.
_____ 2010b. States agree ETS benchmarks for industry, 16 December 2010.

ENDS Report. 2010. EU ETS benchmarks go to the wire, no. 428, September 2010, 53.

_____ 2011. Steel firms pin hopes on low-carbon blast furnace, no. 436, May 2011, 28–30.

Ericsson, K., Nilsson, L.J. and Nilsson, M. 2011. New energy strategies in the Swedish pulp and paper industry – the role of national and EU climate and energy policies. *Energy Policy*, 39(3), 1439–49.

ESTEP. 2011. *European Steel Technology Platform – A Bridge to the Future.* 2010 Activity Report, tp://ftp.cordis.europa.eu/pub/estep/docs/estep-activity-report-2010_en.pdf

EuPD Research. 2005. *Emissions trading in Germany – an abstract.* Available at: http://www.iea.org/work/2005/5ghg/4_Schreiber.pdf. Accessed: 25 October 2010.

Euractiv. 2010. *EU floats method for handing out free CO₂ permits*, 10 September 2010, Foundation Euractiv, Brussels.

Euractiv. 2011. *Steelmakers sue EU over carbon market rules.* 22 July 2011. Foundation Euractiv, Brussels.

EUROFER. 2000a. *The European Steel Industry and Climate Change*, EUROFER, Brussels.

_____ 2000b. *EUROFER view on Emissions Trading – Comments to the Green Paper COM(2000)87*, position paper, 15 September 2000. Brussels: EUROFER.

_____ 2007. *Combating Climate Change – A Global Approach to Foster Growth, Competitivity and Innovation for European Steel.* Brussels: EUROFER.

_____ 2010. *Annual Report.* Brussels: EUROFER.

_____ 2011. Steel industry goes to European Court on EU Emissions Trading Scheme, Press statement, Brussels 21 July 2011.

European Commission. 2000. Green Paper on Greenhouse Gas Emissions Trading within the European Union. COM(2000) 87 final, 8 March.

_____ 2004. EU steel technology platform: A new dynamic for the European steel sector, IP/04/330, Brussels, 12 March 2004.

_____ 2008a. Questions and Answers on the directive on the geological storage of carbon dioxide, 17 December 2008, Memo/08/798, Brussels.

_____ 2008b. Proposal for a Directive of the European Parliament and the Council amending Directive 2003/87/EC so as to improve and extend the greenhouse gas emission allowance trading scheme of the Community. COM(2008) 16 final, 23 January.

European Council. 2009. Directive 2009/29/EC of the European Parliament and the Council of 23 April 2009 amending Directive 2003/87/EC so as to improve and extend the greenhouse gas emission allowance trading scheme of the Community. *Official Journal of the European Union*, 5 June 2009, L140/63.

European Energy Review. 2011. CCS faces mounting obstacles. Available at: http://www.europeanenergyreview.eu/site/pagina.php?id=2739&zoek=ccs%20 faces.

Henningsen, J. 2011. Energy savings and efficiency. In *Towards a Common European Union Energy Policy: Problems, Progress and Prospects* (eds), J.S. Duffield and V.L. Birchfield. New York: Palgrave, 131–45.

IEA. 2007. *Tracking Industrial Energy Efficiency and CO_2 Emissions*. Paris: International Energy Agency (IEA).

Knill, C. and Lenschow, A. (eds) 2000. *Implementing EU environmental policy: New directions and old problems*. Manchester: Manchester University Press.

Kundak, M., Lazic, L. and Crnko, J. 2009. CO_2 emissions in the steel industry. *Metalurgija*, 48(3), 193–97.

Lund, P. 2007. Impacts of EU carbon emission trade directive on energy-intensive industries: Indicative micro-economic analyses, *Ecological Economics*, 63, 799–806.

Matthes, F. and Schafhausen, F. 2007. Germany. In *Allocation in the European Emissions Trading Scheme – Rights, Rents and Fairness* (eds), A.D. Ellerman, B.K. Buchner and C. Carraro. Cambridge: Cambridge University Press, 72–105.

OECD/IEA. 2003. Policies to Reduce Greenhouse Gas Emissions in Industry – Succesful Approaches and Lessons Learned: Workshop Report, OECD and IEA Information Paper, COM/ENV/EPOC/IEA/SLT (2003)2, Paris.

Official Journal of the European Union. 2010. Commission Decision of 24 December 2009 determining, pursuant to Directive 2003/87/EC of the European Parliament and the Council, a list of sectors and subsectors which are deemed to a significant risk of carbon leakage, L 1/10, January 5 2010.

Point Carbon. 2008. *EU ETS Phase II – The Potential and Scale of Windfall Profits in the Power Sector, A report for WWF by Point Carbon Advisory Services*, Oslo, March 2008.

_____ 2009. Jos Delbeke interview, *Trading Carbon*, Oslo, June 2009, 14.

_____ 2010a. Along expectations, *Carbon Market Trader*, Oslo, May 2010, 17.

_____ 2010b. EU steelmaking surges 33% Jan–Sep, Oslo, 21 October 2010.

_____ 2011a. ArcelorMittal books $140m profit from EUA sales, Oslo, 8 February 2011.

_____ 2011b. Steelmakers, utilities scramble for EU CCS funding, Oslo, 9 February 2011.

_____ 2011c. EU steel production rises 7.5%, Oslo, 21 March 2011.

_____ 2011d. EU steel output up 6.9% in Q1, Oslo, 20 April 2011.

Rodgers, I. 2008. Climate change: recent EU developments and steel industry response, Director of UK Steel, Powerpoint presentation made for EUROFER to the OECD Steel Committee, May 2008.

Sandbag. 2010a. *The Carbon Rich List – The companies profiting from the EU Emissions Trading Scheme*. February 2010. London: Sandbag.

_____ 2010b. *Cap or Trap? How the EU ETS risks locking-in carbon emissions*, September 2010. London: Sandbag.

_____ 2011. *Carbon Fat Cats 2011 – The Companies profiting from the EU Emissions Trading Scheme*, June 2011. London: Sandbag.

Sijm, J.P.M., Neuhoff, K. and Chen, Y. 2006. CO_2 Pass-through and windfall profits in the power sector, *Climate Policy* 6, 49–72.

Sijm, J.P.M., Hers, S.J., Lise, W. and Wetzelaer, B.J.H.W. 2008. The impact of the EU ETS on electricity prices – Final report to the European Commission, ECN report, Amsterdam, December 2008.

Skjærseth, J.B. and Wettestad, J. 2008. *EU Emissions Trading: Initiation, Decision-making and Implementation.* Aldershot: Ashgate.

_____ 2010a. The EU Emissions Trading System Directive Revised (Directive 2009/29/EC. In *The New Climate Policies of the European Union: Internal Legislation and Climate Diplomacy* (eds), S. Oberthur and M. Pallemaerts. Brussels: ASP Editions.

_____ 2010b. Fixing the EU Emissions Trading System? Understanding the post-2012 changes, *Global Environmental Politics* 10(4), 101–23.

SSAB. 2009a. *White Paper. SSAB and carbon dioxide emissions.* Available at: http://www.ssab.com/Global/SSAB/Environment/en/001_whitepaper-SSAB%20and%20Carbon%20Dioxide_emissions.pdf. Stockholm. Accessed 10 August 2010.

_____ 2009b. *Sustainability Report 2009.* SSAB: Stockholm.

_____ 2010a. *Sustainability Report 2010.* SSAB: Stockholm.

_____ 2010b. *Investor information, report to Carbon Disclosure Project.*

_____ 2011. *Investor Information, report to Carbon Disclosure Project.*

_____ 2012. Web site information about the company. Available at: http://www.ssab.com/en/Career/About-SSAB/Historical/

Still, G. 2011. Powerpoint presented on the behalf of ThyssenKrupp at CCS Workshop IEA/VDEh, Dusseldorf 8 and 9. November 2011.

Thollander, P. and Dotzauer, E. 2010. An energy efficiency program for Swedish industrial small- and medium-sized enterprises. *Journal of Cleaner Production*, (18)13, 1339–46.

ThyssenKrupp. 2005. ULCOS: Business and universities combine R&D efforts in European Commission sponsored program to further reduce steel industry CO_2 emissions, press release 22 February 2005.

_____ 2009. *Sustainability Report 2009 – Doing the right thing. Right?* Available at: http://www.thyssenkrupp.com/documents/Publikationen/TK_Magazine/sustainability_rreport_steel_doing_the_right_thing.pdf

_____ 2010a. History. Available at: http://www.thyssenkrupp.com/en/konzern/geschichte.html. Accessed 21 October 2010.

_____ 2010b. *Investor information, report to Carbon Disclosure Project.*

_____ 2011a. *Investor Information, report to Carbon Disclosure Project.*

_____ 2011b. *Environmental and climate protection, Management report on the Group, Online Annual Report 09–10.* Available at: http://www.thyssenkrupp.com/financial-reports/09_10/en/environment.html

_____ 2012. Climate protection and energy efficiency, website info about the company. Available at: http://www.thyssenkrupp-steel-europe.com/en/index.jsp Accessed 13 January 2012.

ULCOS. 2011a. Overview presentation. Available at: http://www.ulcos.org/en/about_ulcos/home.php.

_____ 2011b. Where we are today. Available at: http://www.ulcos.org/en/research/where_we_are_today.php

Watanebe, R. 2011. *Climate Policy Changes in Germany and Japan.* London: Taylor and Francis.

Weddige, H.J. 2011. Examining the relation between the EE Directive and the Emissions Trading Scheme (EU ETS), ThyssenKrupp, powerpoint presentation, Energy Efficiency for Asset Intensive Industries, Vienna, 28 November 2011.

Weizsäcker, E.U. von, Hargroves, C., Smith, M.H., et al. 2009. *Factor Five: Transforming the Global Economy Through 80% Improvements in Resource Productivity: A Report to the Club of Rome.* London: Routledge.

Wettestad, J. 2001. The ambiguous prospects for EU climate policy – a summary of options. *Energy and Environment*, 12(2 and 3), 139–67.

_____ 2009. EU Energy-intensive industries and emission trading: Losers becoming winners? *Environmental Policy and Governance*,19(5), 309–20.

_____ 2011. EU emissions trading: Achievements and challenges. In *Towards A Common European Union Energy Policy: Problems, Progress and Prospects* (eds), J.S. Duffield and V.L. Birchfield. New York: Palgrave, 87–113.

Wikipedia. 2012. ThyssenKrupp info. Available at: http://en.wikipedia.org/wiki/ThyssenKrupp. Accessed: 13 January 2012.

Wooders, P., Cook, G., Zakkour, P., Harfoot, P., and Siebert, M. 2009. *The Role of Sectoral Approaches and Agreements: Focus on the Steel Sector in China and India.* Climate Strategies report. Available at: http://www.climatestrategies.org/research/our-reports/category/54/234.html

World Steel Association 2008. *Sectoral approaches to climate change.* Available at: http://www.worldsteel.org/climatechange/

_____ 2012. CO_2 emissions data collection. Available at: http://www.worldsteel.org/steel-by-topic/climate-change/data-collection.html

Wurzel, R.W. 2008. *The Politics of Emissions Trading in Britain and Germany.* Report for the Anglo-German Foundation for the Study of Industrial Society. London: Anglo-German Foundation.

Wurzel, R.K.W., Jordan, A., Zito, A.R. and Brückner, L. 2003. From high regulatory state to social and ecological market economy? 'New' environmental policy instruments in Germany. In *New Instruments of Environmental Governance* (eds), A. Jordan, R. Wurzel and A. Zito. London: Frank Cass, 115–36.

Appendix Table 7.1   Climate strategies of the eight largest steel companies in Europe

Company	Short-term goals and measures	Long-term low-carbon solutions	Acknowledges problem and responsibility for problem-solving?	Stated importance of ETS and EU climate policy
Archlor Mittal (Luxembourg)	Has set a target of reducing emissions by 170 kg per tonne of steel produced by 2020. Such a reduction is equivalent to an 8 per cent reduction in specific emissions	Leading role in the ULCOS-project. Development of new High Strength Steel (HSS) such as HISTAR.	Yes 'ArcelorMittal recognizes that it has a significant responsibility to tackle the global climate change challenge; it takes a leading role in the industry's effort to develop breakthrough steelmaking technologies and is actively researching and developing steel-based technologies and solutions that contribute to combating climate change.' (ArcelorMittal 2008: 1)	Climate change is a global issue and should be addressed through global solutions.

Company	Short-term goals and measures	Long-term low-carbon solutions	Acknowledges problem and responsibility for problem-solving?	Stated importance of ETS and EU climate policy
Corus/ TataSteel (UK/the Netherlands/ India)	According to an agreement with the Dutch government, TataSteel is to work towards an energy efficiency of 2 per cent per annum. In an agreement with the UK the company agreed to work towards reducing the total energy consumption by 15.8 per cent compared to 1997 levels by the end of 2010,	Participating in the ULCOS-project. Corus has worked together with Salzgitter since 2005 on identifying opportunities for the application of High Strength and Ductility (HSD®) steels in selected market sectors.	Yes Corus recognizes that climate change is a global issue, and that the company can contribute to reducing the problem of greenhouse gases.	ETS not mentioned in TataSteel annual reports.
ThyssenKrupp (Germany)	Claim that production processes are already operating at a minimum CO_2 level and that additional reductions are not possible with the technology available today. Have intensified their R&D as a result of this. Do not set Groupwide reduction targets.	Part of the ULCOS-project.	Yes 'At ThyssenKrupp, we regard taking responsibility for society, the environment and the climate as the natural duty of a successful business enterprise' (ThyssenKrupp 2007/2008: 110).	Want global agreement.

Company	Short-term goals and measures	Long-term low-carbon solutions	Acknowledges problem and responsibility for problem-solving?	Stated importance of ETS and EU climate policy
Voestalpine (Austria)	Examples of energy efficiency measures conducted by the company: 'Continual optimization of complete recycling systems for steelmaking top gases to be used in power plants and other industrial furnaces. Substitution of reduction agents of fossil origin (coke, heavy heating oil) in the blast furnace through the use of other process by-products (such as crude tar and coking gas from the coking plant), specially treated residuals (waste oils, plastics, etc.). District heat extraction for communities. CO_2 reductions through optimized logistics (truck to track, optimized transport routes)' (Voestalpine 2010a: 19).	Participate in the ULCOS project.	Yes 'Steel plays a key role in the field of climate protection' (Voestalpine 2010a).	Want a global agreement. Agree with EUROFER in its disagreement with the benchmarking system proposed by the Commission. 'Apart from the fact that the objective of these ambitions is basically a welcome one, from the perspective of the Voestalpine Group, the Roadmap seems, for the most part, to be lacking a holistic approach across all of the impacted value chains' (Voestalpine 2010b: 45).

Company	Short-term goals and measures	Long-term low-carbon solutions	Acknowledges problem and responsibility for problem-solving?	Stated importance of ETS and EU climate policy
Salzgitter (Germany)	Strives to introduce and certify ISO 14001 environmental management systems at its locations. Salzgitter AG, as part of the German steel industry, voluntarily committed itself in May 2001 to reducing specific raw material- and energy-related CO_2 emissions from crude steel production by a total of 22 per cent between 1990 and 2012. Salzgitter AG has made above-average contributions to achieving this target, and has already attained a reduction in specific CO_2 emissions of around 25 per cent in the production areas of the Steel Division. This result will improve further with the 'Power Plant 2010' project for modernizing Salzgitter Flachstahl's power plant.	Salzgitter has been working together with Corus since 2005 on identifying opportunities for the application of High Strength and Ductility (HSD®) steels in selected market sectors.	Yes 'We view the protection of the environment and responsible use of our facilities as an investment in our future' (Salzgitter AG 2010: 74).	In favour of more global regulation. The company has engaged in the political discussions process, arguing in favour of free allocations of CO_2 certificates based on ambitious but technically achievable CO_2 benchmarks (Salzgitter AG 2010).

Company	Short-term goals and measures	Long-term low-carbon solutions	Acknowledges problem and responsibility for problem-solving?	Stated importance of ETS and EU climate policy
US Steel (USA)	All of US Steel's facilities have achieved and maintained ISO 14001 compliance.	The company is committed to investing in technology and research projects to move the steelmaking process in an even more environmentally responsible direction. The goal of these research projects is to reduce, and eventually eliminate, CO_2 emissions from the steel making process (US Steel 2007).	As member of a carbon-intense industry, US Steel is seriously concerned about the amounts of greenhouse gasses produced in the steelmaking process. Even though the steel industry produces less than 3 per cent of the world's CO_2, it is the view of US Steel that this amount can be decreased.	
Dillinger Hütte	Dillinger Hütte has an environmental management system according to DIN EN ISO 14001, introduced in 2004.	Dillinger Hütte GTS is a core member of ULCOS, and is actively supporting the fight against global warming.	Yes The company works to integrate and promote all of its experience, its technical know-how as well as its creativity in order to conserve resources as well as to avoid burdens for humanity and the environment.	It is very important to establish suitable long-term conditions within energy and climate policy, particularly with regard to emissions trading.

Company	Short-term goals and measures	Long-term low-carbon solutions	Acknowledges problem and responsibility for problem-solving?	Stated importance of ETS and EU climate policy
HKM – Hüttenwerke Krupp-Mannesmann	EU Eco auditing scheme: The company has set up its environmental management system according to the requirements of the globally-recognised DIN EN ISO 14001, as well as the EU Eco Auditing Regulation 1836/93.		According to the company website, HKM has protection of the environment and maintaining the natural basis of life as substantial components of its managerial policy.	

Chapter 8
Comparative Analysis

Jon Birger Skjærseth, Per Ove Eikeland, Anne Raaum Christensen,
Lars H. Gulbrandsen, Arild Underdal and Jørgen Wettestad

The previous chapters have analysed ten company cases within five sectors: electric power production, oil, pulp and paper, cement, and steel. This chapter compares these cases to shed light on whether, how and why companies have responded to the world's first international emissions trading system. We commence with the three 'models' to guide the analysis, as set out in Chapter 2: companies as reluctant adapters, as innovators, or as socially responsible. Corporate responses in line with these models may affect pre-existing climate strategies or shape entirely new strategies. The EU ETS co-exists with other EU climate and energy policies; we expect to find that climate strategies are co-produced by company-internal factors as well as wider external contextual factors. Here we will analyse such factors within the framework of the three models, conditioning the effect expected from the ETS. This broader analytical perspective should help in explaining variation in climate strategies and identifying the conditions under which strategies emerge.

In Chapter 1, we outlined various methodological techniques to examine the validity of these models. The first involves comparing expectations with observations. This pattern-matching approach aims to judge the consistency between expectations derived from the models and observed ETS responses and climate strategies. The better match we can observe, the more confidence we will have in the explanatory power of the model. The second type takes into account that our knowledge base sometimes limits our ability to extract empirically assessable expectations or propositions. Under such circumstances, we can evaluate the models in terms of explanation building. This 'narrative' approach is particularly useful for capturing complex relationships between companies and the context within which they operate.

The third methodological technique assesses the heuristic value of our three models. The heuristic value is specifically concerned with the model's assumptions concerning behavioural mechanisms that may link regulation to corporate response. How have the companies studied here actually approached the EU ETS, in terms of our three analytical models? Different empirical tests can be complementary in the sense that a high match between expectations and observations in relation to one model might be explained by causal mechanisms integrated in another model. For example, an outcome apparently caused by the presence of strong social norms of responsibility may actually prove to be the result of a price effect, or vice versa.

Each of the three ways of assessing the models can contribute insights into causal relationships that the others cannot.

The previous chapters have presented in-depth analyses of selected ETS companies and aggregate analyses. The companies in question are mainly from Northern Europe, so the results should be interpreted accordingly. Our aim has been to select two companies within each sector that pursue different climate strategies, in order to shed light on the conditions that trigger different responses. Greater variation could, of course, have been achieved by selecting companies also from other regions of Europe (see Chapter 1 on this point).

The analysis in this chapter is primarily based on observations reported by the case-study authors, supplemented by results reported from other relevant studies. One significant challenge throughout this study has been to 'measure' climate strategies in a precise way to capture what companies actually do. Data have been hard to find for some indicators, such as R&D or patents indicating innovative climate strategies. Where data have been found, they have often been too general to specify expenditures related to intended climate-friendly innovation. For instance, R&D and demonstration of carbon capture in the oil industry has a long history, with CO_2 injected as pressure gas to increase the rate of exploitation from oil reservoirs. Despite these shortcomings, we believe that documentation of change patterns has been sufficiently enabled by data triangulation. The analysis of climate strategies has, to varying degrees, been based on a combination of 'hard data' and qualitative information judged by the authors in each case. Readers should keep in mind, however, that we have not explicitly assessed the 'net impact' of corporate behaviour. This means that, for instance, production growth and related emissions may counterbalance an otherwise proactive climate strategy.

Another significant challenge has been to reduce the risk of exaggerating the effect of the EU ETS as the selected institutional arrangement. We have focused on climate strategies to reduce this problem: if variation in climate strategies cannot be traced back to variation in the EU ETS and other relevant EU climate policies, the case-study authors have considered other factors included in the framework, or exogenous factors such as the financial crisis.[1] In many cases, the EU ETS has been one factor which has co-produced variation in climate strategies together with other drivers. The EU ETS is, for example, one of many factors that have contributed to increase electricity prices in Europe (see Chapter 3). Investments in low-carbon solutions are generally the result of a combination of political, economic, technological and social factors.[2]

With these caveats in mind, we will first assess the three models described in Chapter 2 in light of the observations reported in the individual chapters. The

---

1   The economic crises in Europe since the autumn of 2008 have significantly reduced industrial emissions, making any effort to assess the effect of regulation on emissions extremely difficult.

2   Readers are advised to consult the individual chapters for more nuanced discussion of the specific effects and relationships reported in this chapter.

analysis is based on the extended version of these unitary models, including regulatory risk (carbon intensity and exposure to international competition), managerial capabilities, and national and international context. Second, we discuss various combinations of factors that point towards conditions triggering different ETS responses and climate strategies, and the importance of exogenous factors.

Model I: Companies as Reluctant Adapters

The first and most parsimonious perspective is based on the assumption that rational managers maximize profits and adapt to regulation by minimizing costs. We have explored the following implications of this model: (1) companies will resist the initiation of the ETS; (2) compliance strategies will depend on what pays off in the short term; (3) new large-scale innovations for alleviating climate change will not be initiated. The pure version of this model is concerned only with regulatory pressure and costs to explain differences and change in response. In the following, we apply the extended version (see Chapter 2), including regulatory risk, which is determined by carbon intensity and exposure to international competition.

Initiation of the ETS

How did the companies initially respond to the EU ETS? It is important to underline that the real distribution of costs was largely uncertain when the ETS was adopted. From 2003, it took two years before the ETS became operational. In this period, allocation of allowances was undertaken at the national level through national allocation processes. Companies' initial response strategies were thus shaped by *expectations* of regulatory pressure and carbon prices. Shell, for example, anticipated that a larger share of its installations would be affected by the ETS than later proved to be the case. Steel companies were uncertain whether the ETS would take hold, or would prove to be merely a passing policy fad. And pulp and paper companies expected a more stringent system than later proved to be the case.

At the sector level, we find a marked difference in response between the electric power industry and energy-intensive industries. From the 2000 Green Paper on the ETS, energy-intensive industries were generally sceptical or directly opposed to the introduction of the system. They accepted, more or less reluctantly, the idea of emissions trading, but lobbied actively against a mandatory cap-and-trade system to ease regulatory pressure. Most of these sectors could accept a baseline and credit system built on an energy efficiency baseline, but not a mandatory cap on emissions. Conversely, the electric power industry supported the introduction of the ETS in the form of cap-and-trade, on condition that power companies would receive allowances free of charge.

Plausible reasons for this difference can be found in the combination of (expected) regulatory pressure and regulatory risk. Unlike the energy-intensive industries, the electric power industry expected, and actually became subject to, higher regulatory pressure in the form of shortage of allowances. At the same time, they faced lower regulatory risks due to weaker exposure to international competition. Limitations in transmission capacities mean limited opportunities for electricity trade with non-ETS areas.

Also some same-sector company variation can be explained by this model. While most power companies, including Swedish Vattenfall, accepted the ETS, the German RWE explicitly opposed the system when it was launched. As one of the most carbon-intensive companies in Europe (twice as carbon-intensive as Vattenfall at that point because of the far higher share of coal-based energy generation), RWE was indeed exposed to far higher regulatory risks.

More puzzling, according to this model, are the differences observed between companies within other sectors. The oil industry accepted the idea of emissions trading, but opposed a mandatory cap-and-trade system in Europe. This industry is dominated by major multinational oil companies engaged in exploration, production and refining. Still, the European oil majors Shell (and BP) supported and participated actively in the initiation of the ETS, while US-based ExxonMobil lobbied against any mandatory climate regulation, including the ETS. This difference can only to a very limited extent be traced back to differences in regulatory pressure and risk.

The EU ETS was the first mandatory climate regulation applied to the cement industry in Europe. The cement sector lobbied to reduce the regulatory pressure from the ETS, arguing for voluntary initiatives and international sector-level agreements. Regulatory risk in this sector relates more to carbon intensity than to international trade. Cement companies are global actors, but the actual trade in cement is mostly limited to domestic markets, as transport costs are high compared to the cement price; moreover, raw materials for cement can be found almost everywhere. Still, the cement company Holcim was initially more supportive of the ETS than was HeidelbergCement – mainly because of the technology base, not lower carbon intensity or exposure to international competition. Holcim has in many ways a different understanding of what cement is and how cement should be produced compared to other cement companies, as composite/blended cements are a central part of its business and climate strategies.

The ETS was also the first mandatory climate regulation to target the pulp and paper industry (PPI) in Europe. The PPI sector initially opposed the ETS, arguing that the system would expose the sector in Europe to carbon leakage – relocation of production to regions without a price on carbon. There has been significant focus on the regulatory risk of the system due to carbon intensity and international competition in the PPI. Sweden's Svenske Cellulosa Aktiebolaget (SCA) and Norway's Norske Skog were more supportive of the introduction of the ETS than were the non-Nordic companies. Nordic pulp and paper companies generally have a higher share of renewables in their energy mix than other European pulp and paper companies, which means they face lower regulatory risk.

Also the steel industry had been largely unregulated prior to the ETS. It opposed the new system vigorously – in fact it was the most ETS-resistant among the energy-intensive industries. That the steel industry later chose to challenge the European Commission over benchmark rules that were introduced after the ETS revision in 2009 only reinforces this impression. The strong opposition was justified by the very limited abatement options – claimed by the industry to already be close to the theoretical limit. The ETS would thus, according to the industry, unavoidably mean extra costs that would lead to a loss of competitiveness because of its high exposure to international competition.

These observations show there were considerable differences in initial responses within the five industry sectors. Some but not all of these differences can be explained by 'objective' differences in regulatory pressures and risks.

Compliance Strategies: Short-term Abatement?

What then characterizes the compliance strategies of these companies once the ETS was in place? The EU ETS was adopted in 2003, and trading for the first phase (2005–2007) started in 2005. This was followed by new national allocation plans, adopted mainly in 2006 for the second phase (2008–2012). In December 2008, the EU agreed on a revised ETS for the third phase (2013–2020), introducing significant changes intended to strengthen the system. Our expectation based on Model I is simply that companies would choose compliance strategies to minimize short-term costs. We lack sufficient information to fully rank different response options according to costs, and thus have to rely on some proxies to indicate whether the companies responded as expected under Model I. First, we would expect companies to comply, since penalties for non-compliance significantly surpassed the market price for allowances. Secondly, we would expect continued company lobbying for more lenient regulator pressure when the system came up for revision. Finally, we would expect to find a link between compliance strategies and regulatory pressure. The last indicator is extremely difficult to assess, but we could expect a convergence in strategies within the sectors, as the most reluctant companies would have to step up their activities.

A first observation is that none of the case-study authors identify any instances of deliberate non-compliance as a response strategy – in line with the expectation that profit-maximizing managers would not choose this more costly strategy. Another observation is that none of the case studies report any significant relocation of production to countries without caps on emissions or a price on carbon.[3] However, several energy-intensive industry associations and companies have threatened to relocate, based on expected and actual carbon costs. For instance, the cement industry argued that European cement production would be wiped out if auctioning was applied and the CO_2 price rose above 35 euros/tonne. However, a decision to relocate will be affected by a multitude of other costs and benefits as

---

3   This option has not been specifically analysed in the case studies.

well. In the oil industry, for example, the ETS represents only one of many reasons for the need to reduce refinery production in Europe, such as decreasing demand for petroleum products (see Chapter 4). Carbon leakage, however, was extremely important for companies that argued for lower regulatory pressure once the system was up for revision.

One significant regulatory change made for the third phase was the introduction of new allocation methods. For energy-intensive industries this implied a shift from grandfathering (historical emissions) to benchmarking. They were still allocated free allowances, but had to accept a benchmarking model that would benefit the most energy-efficient installations and companies within specified sectors. Our case studies show that the energy-intensive industry sectors were at least partly successful in lobbying the Commission for as lenient benchmarks as possible, a finding supported also by earlier studies (Skodvin, Gullberg and Aakre 2010). For example, the PPI lobbied successfully for differentiation between a wide range of pulp and paper products in the development of industry benchmarks, getting the number of benchmarks increased from two to eleven. While most industry sectors appeared with unified positions in the negotiations, the cement industry revealed internal conflicts of interests. It was notable that Holcim lobbied for a different benchmark approach from that most advocated by other cement companies.

Despite some variation within the energy-intensive industry lobby, these industries generally lobbied against stricter allocation rules. This appears compatible with Model I – in apparent contrast to the response of the electric power industry. This industry had made profits from free allowances that would be reduced or eliminated by the introduction of auctioning. And yet, the industry accepted (and some companies even supported) auctioning that would reduce their short-term profits. Eurelectric realized that resisting auctioning would be a dead case, given the considerable public criticism concerning the windfall profits reaped by the industry after the system started to operate. This indicates that the power companies refrained from lobbying as a result of their sensitivity to the expectations of others – a mechanism not captured by Model I. An alternative interpretation is that the probability of success in changing the proposed auctioning rules was low.

Likewise, most sectors and companies supported the transition from decentralized allocation of allowances at the national level to a harmonized EU-level system. This marked a change in their positions from when the system was first drafted, as most companies had originally preferred a decentralized ETS. This may indicate change of strategies due to learning and understanding more about the system (see also Skjærseth and Wettestad 2010). Change in strategies due to learning would not be expected under the static Model I.

The different sectors and companies analysed in this study display a clear pattern of convergence in climate strategies over time. Most of the sectors and companies have complied with the ETS in the short term by implementing monitoring, reporting and verification requirements, trading in EAUs, CERs and/ or abatement. With regard to monitoring, reporting and verification, the ETS instructed new procedures that required changes even in companies that had such systems in place before 2005. This applied to varying degrees to the different sectors and companies in this analysis.

The impact of the ETS on abatement has been more significant in the power sector than in the energy-intensive sectors. This finding is compatible with Model I, given variation in regulatory pressure and risks. The 20 main allowance-holders in the power industry cut emissions substantially from 2005 onwards, with the economic crisis accounting for much of the dramatic reductions in 2009. In 2010, emissions were up again because of economic recovery, but final energy supply showed lower CO_2-intensities because of investments in renewable energy and energy efficiency measures. In 2011, the trend of substantial electric power industry emissions cuts continued. We find several indicators of regulatory pressure on individual companies correlating with abatement measures and emissions, particularly fuel switching and energy efficiency improvements, confirmed also by earlier studies (see Chapter 1).

The comparison of Vattenfall and RWE displays convergence towards more proactive and innovative strategies. RWE started out with a more reactive or reluctant approach to the ETS than Vattenfall. Over time, however, these companies responded more similarly, implementing various measures to bring emissions down – which again is compatible with Model I assumptions. Regulatory pressure on both companies increased over time, in line with stricter caps in the second trading period and a stricter cap and full auctioning from 2013 onwards. Another cost-based incentive mechanism linked to the EU ETS started to work for Vattenfall and RWE, related not to higher production costs caused by auctioning but to potentially higher capital costs due to high carbon intensities. The reason was that the global credit rating institution Standard & Poor came to include carbon costs in the evaluations, which gave negative impacts for the two companies regarding their credit worthiness. As such, an entirely new mechanism by which the EU ETS impacted on company climate strategies was established, reinforcing the incentives created for rational cost-minimizing adaptation.

For the energy-intensive industries, ETS regulatory pressure has been low, with allocated allowances generally in surplus of the cap. The oil sector has seen allocation roughly balanced between emissions and caps. The rise in electricity prices, partly as a result of the ETS, has added some pressure. Most major oil companies had their own climate strategies in place before the ETS; typical responses were implementing or strengthening of monitoring, reporting and verification, reinforcement of energy efficiency programmes and allowance trading. The new benchmarks for this sector are expected to stimulate abatement activities further and contribute to putting the least efficient refineries out of business.

This description fits well also with our selected oil companies, Shell and ExxonMobil. The two saw alignment in strategies over time. Exxon had to accept that the ETS was adopted; it then implemented monitoring, reporting and verification, strengthened abatement programmes and generally took a more proactive climate strategy, followed by roughly stable emissions. These changes were partly related to the EU ETS. Shell's compliance strategy has mainly combined strengthening energy efficiency programmes with very active trading in allowances. Still, Shell reported lower energy efficiency and rising emissions, through failure to achieve sufficient effects from its energy efficiency programmes, due to factors independent of the ETS.

The cement industry was allocated abundant free allowances. A key incentive provided by the ETS has been indirect in the form of rising electricity prices, although the cement industry is less dependent on electricity than, for example, the paper industry. Higher electricity prices provide incentives for energy efficiency and use of alternative fuels. CO_2 prices have become an important premise when investment decisions are made. The new benchmarks are expected to contribute to closure of the 'dirtiest' cement plants. Additionally, revenues from selling excess allowances have contributed to the climate strategy, as shown by Holcim's decision to set aside some of the revenues for improving energy efficiency within the company. Convergence in strategies can be observed also in this industry, illustrated by the cases of HeidelbergCement and Holcim. Holcim was an early mover in the cement sector, whereas HeidelbergCement has become more proactive over time. Both companies saw their verified emissions under the ETS increase until the financial crisis turned this trend in 2009. Both companies have reduced specific emissions: less CO_2 is now emitted per ton cement produced. Holcim's strategy has relied mainly on substitution of clinker, whereas HeidelbergCement has focused more on alternative fuels.

The main pressure from the ETS on the PPI thus far has been indirect and caused by an increase in electricity prices, a major concern in this industry. Many companies incorporate carbon costs in their investment planning, but the surplus of allowances has not been able to compensate for the increase in electricity prices. The industry has complied with the ETS through a combination of trading, improving energy efficiency and stepping up the use of biomass and less carbon-intensive fuels. The PPI sector expects stronger regulatory pressure in the third ETS phase. It displays a development towards convergence, as shown in the case studies of SCA and Norske Skog.

Also the steel industry has received allowances in abundance, providing no strong incentives for short-term abatement. Still, both our case-study companies reported up-scaled efforts to improve energy efficiencies. Yet, we also observe significant investments in low-cost external credits for ThyssenKrupp, in line with expectations for a company responding in accordance with this model. SSAB, on the other hand, made less use of this option. The industry association EUROFER, unlike other European-level industry associations, chose to challenge the European Commission legally over the revised EU ETS – another indicator of an industry taking a primarily reactive strategy.

According to Model I, we would not expect to see the initiation of new large-scale innovation in greenhouse gas abatement solutions. However, most sectors and companies have become engaged in low-carbon activities for the longer term. There are significant differences between sectors and companies, but all cases show at least some element of climate-friendly long-term innovation that is not compatible with what could be expected from Model I. We return to these observations in the next section.

Concluding Assessment: Model I

Judged against the consistency between expectations and observations, we conclude that this model can partly explain change and differences in ETS response and climate strategies. The main difference in initial response and compliance with the ETS between the electric power and energy-intensive industries can be understood if this model is extended to include regulatory risk. The essence is that the electric power industry expected, and has been exposed to, stronger regulatory pressure and lower regulatory risk than the energy-intensive industry. Moreover, we have seen a clear tendency towards convergence in strategies corresponding roughly with actual and expected regulatory pressure.

As a heuristic tool, Model I captures well the importance of short-term adaptation by minimization of regulatory costs. The anticipated costs of the ETS and reduced competitiveness have been overriding concerns within industry throughout the evolution of the ETS. Industry associations and companies have lobbied mainly to reduce regulatory pressure and the resultant costs, trying to convince the European Commission that stringent regulation in Europe will lead to a loss of competitiveness and carbon leakage. The electric power industry is a special case in that it has, since the inception of the ETS, expected and actually gained profits from the system (as have some energy-intensive industries after the financial crisis). This observation is not in line with assumption under Model I that environmental regulation will necessarily lead to additional costs. In the electric power industry, we also see an entirely new mechanism by which the EU ETS impact on company climate strategies, involving third-party mediation of incentives through credit rating institutions. This does not conflict with the logic underlying this model, but third-party mediation is not captured in the narrow cost-minimization model.

However, we also note some observations that appear to deviate from the assumption of companies as actors maximizing profits and minimizing costs of regulation – and that limits the ability of Model I to explain company response to the ETS. The main observation that does not fit this model is that all sectors and most companies have engaged in low-carbon activities for the longer term. We find a systematic pattern indicating that companies have not responded exclusively or perhaps not even mainly by ranking different options according to short-term costs and choosing the least-cost option. Other deviating observations are the power industry's acceptance of auctioning, which would reduce their profits; Shell (and BP's) strong support for the EU ETS; and the energy-intensive industries' shift of position on the decentralized approach to allocating allowances. This will hardly be surprising for scholars of environmental strategy; and we should bear in mind that even the extended version of Model I grossly simplifies a complex world of dynamic change in risks and opportunities. The next model takes us closer to this world by emphasizing the limits of making optimal choices and the importance of regulatory design.

Model II: Companies as Innovators

The second 'innovation' model is based on the assumption that managers' efforts to make profits are constrained by bounded rationality (constraints in information and capacities to calculate). Decisions are made by sequential attention, routine, rules-of-thumb or standard operating procedures, all of which act to constrain managers from seeing new and profitable opportunities outside the realm of business-as-usual. Regulation may open a new sequence of attention, enabling managers to discover new and profitable opportunities for innovation. For this to happen, however, regulation must be appropriately designed.

The expectations explored for this model are that: (1) companies will support the ETS by emphasizing new opportunities; (2) companies will comply through implementing outside-business-as-usual incremental innovation (new forms of short-term abatement) beyond what would strictly pay off in the short-term; (3) long-term search activities for new opportunities (R&D) and large-scale innovation project participation will increase; (4) such behaviour is contingent on regulation being appropriate: stringent, harmonized and in line with or slightly ahead of that in other countries. This perspective has also been extended to include company-internal and wider external factors (see Chapter 2).

Initiation of the ETS

The electric power industry supported the introduction of the ETS, whereas the energy-intensive industries generally did not – a pattern that matches well with the expectations under Model I. However, not all electric power companies shared this initial enthusiasm, although they *changed* position in the space of a relatively short period: this applies notably to RWE and other German companies. This pattern points to Model II, with a complementary understanding of initial responses to the ETS that focuses on attention shifts and new perceived revenue opportunities revealed through learning. The simulation exercises on emissions trading initiated by Eurelectric showed that the costs of complying with the system would be outweighed by the extra revenues earned from increases in power tariffs. The repeated rounds of trade simulations also convinced the German companies of this net effect that would accrue if allowances were given out for free. Both RWE and Vattenfall had strong dynamic capabilities prior to the ETS. Similarities in dynamic capabilities cannot explain different initial responses, but can contribute to explaining convergence in strategies over time (see below).

Within the energy-intensive industry, most sectors and companies approached the emergent ETS more in terms of risks than new opportunities. There were some exceptions – Shell, for instance, saw new opportunities for becoming a trade leader for EUAs and CERs within the oil industry. The ETS fed into the proactive climate strategy that Shell had adopted before the ETS. This strategy, which was based on high ambitions in renewables and pioneering implementation of company-internal emissions trading, represented dynamic capabilities built up, positioning the company to benefit from the ETS in a climate-friendly direction. Rather than representing an external 'shock' that changed established practices and routines, the ETS acted to stimulate processes already underway. Further, the national context of Shell's home-base countries is important for understanding the company's proactive strategy compared to the reactive strategy of ExxonMobil. Model II, extended with company-internal and company-external factors, can contribute to explaining variations in initial responses to the ETS.

The cement industry approached the ETS through a two-level strategy. At the sector level, the strategy focused on minimizing regulatory pressure. At the company level, some companies, like Holcim, had supported the ETS from its inception. As in the case of Shell, the ETS fitted well with the strategy that Holcim had adopted before the ETS was launched. Holcim was one of the initiators of the voluntary Cement Sustainability Initiative (CSI) in 2000, and the company aimed to be 'best performer' on specific CO_2 in the sector in the future. The controlling shareholder at the time was a Swiss philanthropist who drew attention within the company to sustainable development. Holcim saw opportunities from the ETS and not only risks, possibly reflecting the company's already established strategy of increasing the supply of blended cement in the market – a product that offers ready opportunities for substitution of clinker as a way of reducing specific emissions.

In the PPI industry, both Norske Skog and SCA supported the idea of a European emissions trading system, although they would have preferred a global system. Compared to the pulp and paper companies in many other European countries that relied on fossil natural gas for much of their electricity and process heat needs, our two case-study companies enjoyed relatively easy access to renewable electricity and CHP based on biofuels. This helps to explain why Norske Skog and SCA were more positive towards the EU ETS than were pulp and paper companies elsewhere in Europe. Compared to Norske Skog, however, SCA appeared more attuned to exploring new opportunities. One explanation is company variation in factors of production (like the location of established mills, availability of natural resources and infrastructures) that constrained or enabled specific innovative and CO_2-lean investment solutions. Illustrative is SCA's extended search for new biomass-based energy solutions to reduce emissions, with investments in the biofuel-based lime kiln at its Östrand mill as a good example (see Chapter 5).

The situation for Norske Skog was somewhat different, in that the company had less need for CO_2-lean innovation for its mills in Norway. They covered the bulk of their electricity needs by hydropower, and only around 1 per cent by fossil fuels. Two additional factors could explain the greater willingness of SCA to invest in low-carbon solutions: availability of human and financial resources, and dynamic capabilities. SCA is not only a far bigger company than Norske Skog, but also one of Europe's largest owners of forests that can be used for innovation and emissions reduction purposes. SCA has a long history of product and process innovation, and ranks among the top three innovators in the industry.

We conclude that the power companies in particular supported the initiation of the EU ETS by emphasising new business opportunities. Examples can also be found in the energy-intensive industries. A combination of a relatively strong technology base for low-carbon solutions, enabling home-base context, and dynamic management capabilities contribute to explain the supportive and opportunity-oriented responses to the initiation of the EU ETS.

Short-term Abatement: Beyond Compliance?

Model II would lead us to expect companies to comply by extensive search for and implementation of more incremental innovation (such as new forms of short-term abatement outside business-as-usual) than strictly required according to what would pay off in the short-term. It is extremely difficult to distinguish between compliance strategies to minimize costs and strategies that represent 'something more' than compliance as managers discover new opportunities influenced by the ETS. With this caveat in mind, we note several observations in the case studies that indicate an opportunity-directed response complementing our understanding of company responses to the ETS.

In the electric power industry, the tightening of the EU ETS from the first to the second trading period was followed by more search activities for new opportunities (growth in annual R&D intensity for many companies). However, major strategic shifts could be observed only after the ETS revision in late 2008 made regulation tighter from 2013 onwards. These shifts included upscaled involvement in business lines outside the core activities of the companies and upscaled investments in options for emissions reductions that would not count against obligations for the companies under the ETS. The former took the shape of greater involvement in carbon-neutral renewable energy projects but also suspension of plans for investments in new, more energy-efficient coal plants, which had been the main option pursued earlier in the decade. The latter shift was observed as stronger involvement in, for instance, the electrification of cars and end-user projects that would reduce energy demand. Expected future carbon costs were an important justification for this upscaled search for new opportunities, and the out-of-coal strategy was additionally justified by the economic downturn in Europe.

We do not see similar clear shifts in the energy-intensive industries. There are, however, some observations indicating activities outside the core business, and targets and measures adopted that appeared beyond what was needed for short-term compliance. In the oil industry, the ETS mainly strengthened and expanded ongoing abatement and trading programmes. In the case of Shell, the company had already introduced carbon pricing in projects in 2000 in anticipation of the ETS and possibly other regulated carbon markets. In addition to seeing trading as a new business opportunity, Shell has been a frontrunner in actively promoting emissions trading around the world, assuming the chairmanship of the International Emissions Trading Association (IETA). With regard to ExxonMobil, the fact that the ETS mandated monitoring, reporting and verification led its Rotterdam refinery to take leadership in establishing monitoring methodologies for the entire refining industry in Europe. Such initiatives point to attention to new opportunities.

In the cement, paper and steel industries, new optional energy efficiency targets have been adopted corresponding to the 2020 deadline for the third phase of the ETS, strengthening efforts already in place on a voluntary basis before the ETS came into being. In the cement industry, additional upscaled activities include investments in alternative fuel use and clinker substitution. In the PPI, the ETS strengthened ongoing engagement in renewable energy, notably in biomass and other less carbon-intensive fuels, contributing also to greater willingness to invest in more far-reaching innovative projects – as exemplified by SCA's decision to replace oil consumption with biofuels (pellets) at its Östrand mill. Importantly, our sector studies show that the effects of the ETS have been conditioned by several factors at the national and regional level, including access to natural resources, electricity supply, and the policy context. For the steel industry, a main impact of the ETS has been greater attention to radical technological solutions needed for the longer term (below).

Thus we note that most sectors and companies had started to exploit resource-saving opportunities on a voluntary basis even before the ETS. In some cases, these activities were instigated in anticipation of the ETS or other regulatory measures. In the power sector, the ETS drew attention to new opportunities, leading to a strategic shift in R&D spending. In the energy-intensive industries, the ETS has mainly strengthened activities implemented before the system entered into operation. There are, however, significant variations among companies, sectors and countries.

Long-term, Low-carbon Solutions

Model II would lead us to expect a search for new opportunities with greater attention not only to short-term opportunities but also to new long-term, low-carbon solutions. We have thus distinguished between short- and long-term strategies, the latter denoting decisions already made to steer the sector or company towards a low-carbon future.[4] Our cases indicate that most sectors and companies have

---

4   This distinction can be difficult to observe empirically since some measures, like those directed at energy-efficiency improvements, may have short- and potentially long-term consequences if major new solutions are discovered. Another challenge of investigating strategic decisions for the long term is that these are reversible, and follow-up decisions still need to be made.

started to engage in activities directed at long-term low-carbon solutions. Whether this development continues will depend, at least partly, on how the ETS develops – a point taken up in Chapter 9.

The most prominent example is found in the European power industry, which is responsible for some 60 per cent of EU ETS emissions. In 2009, the industry association Eurelectric issued a new study on pathways to carbon-neutral electricity in Europe by 2050 that outlined how this ambitious goal could be achieved. It rested on the assumptions of massive energy efficiency efforts combined with electrification and extensive fuel conversion for transport and other sectors of the economy. The report pointed toward new opportunities for the power industry to expand business and profit from higher electricity tariffs as a result of the shrinking amount of allowances. The scenario showed that reduced carbon emissions would save payments for auctioned allowances nearly offsetting the investment costs needed to achieve the goal of carbon neutrality. Recognizing these opportunities outlined by the scenario project, 61 CEOs, representing most of Eurelectric's major member companies, in 2009 signed a declaration committing their companies to achieving carbon-neutral power supply in Europe by 2050.

These activities reinforced the changes already seen in the power industry, that the ETS was not only viewed narrowly as a problematic cost factor, but was also recognized as providing new industrial opportunities. A further indicator of this shift are data that indicate a break with the former trend of falling R&D intensities in many companies, with climate mitigation technologies constituting the main focus of R&D. Reinforcing the impression of a shift is another emerging trend of stronger involvement in joint long-term R&D, involving companies that would otherwise be competitors in the market for electricity.

Perhaps more surprising are some of the activities unleashed in the energy-intensive industries. In relation to the adoption of the revised ETS and the EU climate and energy package in 2008, many major oil companies announced that they would step up or implement new strategies in renewables, biofuels and CCS. The expected increase in regulatory pressure also affected the position of Europia, the European industry association for the oil refining industry, in a more offensive direction, at least temporarily. The European refining industry now paints a dark picture of the future due to decreasing demands for petroleum products, a mismatch between supply and demand for diesel and petrol, and increasing regulatory pressure. From 2013, increased costs from benchmarking and structural challenges are expected to lead to higher environmental performance as well as shutdowns of the least efficient refineries. In the case of Shell, this has led to higher R&D budgets for CCS.[5] In the oil industry, the main importance of the ETS lies in the long-term strategic consequences of a political agreement on carbon pricing, sending a price signal that may be copied in other parts of the world. The main impact on ExxonMobil is that the ETS has contributed in changing Exxon's

---

5   Shell concurrently divested itself of its renewables.

views about the future, with a price on carbon emerging as an important premise of long-term forecasting and investments.

Also the other energy-intensive industries have come up with longer-term initiatives relating mainly to expectations that the ETS will cause a rise in electricity tariffs. In the cement sector, companies expect stronger regulatory pressure from the ETS. This has set the companies into action to explore how they can change a product that has remained roughly unchanged for a long time. Exploration has triggered attention to formerly unattended business opportunities that may represent radical long-term innovations for bringing emissions down in this sector. The most prominent example is HeidelbergCement's attention to CCS. The company has, as the first in this sector, planned to build a test facility for post-combustion CO_2 capture technologies at Brevik in Norway.

The PPI was first among the energy-intensive industries to announce how it would contribute to achieving the EU's goal of decarbonizing Europe by 2050, arguing for a central role in a future low-carbon bio-economy. Key measures highlighted by the industry association CEPI as part of this long-term strategy are fuel shifting and greater use of heat to decarbonize electricity, applying the best available technology, more use of biomass lime kilns, gasification of biomass, using waste and lignin in the fuel mix, and applying as yet undeveloped breakthrough technologies to lower heat demand through more efficient use of water and improved drying methods. SCA's involvement in wind power is an example of new thinking of potential low-carbon solutions for the future. SCA has agreed to let the Norwegian power company Statkraft undertake investments in wind power on SCA forest land area to ensure long-term affordable electricity supply, investigating relatively new modes of siting wind farms in forest areas, less prone to local conflict and thus presumably involving speedier permission processes.

The European steel industry joined forces with the European Commission in a new research programme, ULCOS, aimed at long-term development of breakthrough technologies, including CCS. The first stage of the project (2004–2010) investigated a range of options and identified four main process routes for further development. The project was later extended with a new four-year period 2011–2015, focusing on large-scale demonstration, with a substantially upscaled budget and greater share of total costs covered by the industry. The EU ETS was one factor that impacted on the set-up of the research programme. The extension of ULCOS followed the revision of the ETS and the development of a more specific CCS policy in 2008.

These observations of longer-term initiatives in energy-intensive industries also find support in a major extensive study on the ETS and investment in innovation (Martin, Muûls and Wagner 2011).[6] This study documents that companies

---

6   The study was based on telephone interviews with almost 800 manufacturing firms in Belgium, France, Germany, Hungary, Poland and the UK. It included randomly selected firms that are both ETS and non-ETS companies.

generally expect tighter emissions caps for the third phase and significantly higher carbon prices in the future, averaging 40 euros per tonne in the post-2012 phase. Moreover, 70 per cent of the firms are engaged in R&D on cutting emissions or improving energy efficiency. This 2011 study also finds a significant positive correlation between company expectations about the future stringency of their cap and 'clean' innovation.

A main conclusion from our analysis is that there are many examples of sectors and companies responding according to expectations made under Model II. Responses have involved greater attention and awareness, searching for new opportunities, and stepping up activities to bring emissions down in the short- and long-term. All the same, however, we cannot conclude that there have been any comprehensive, 'deep' shifts toward generally innovative strategies, except for the power sector. Let us ask how appropriate the ETS is, against the backdrop of these observations.

Appropriate ETS Design?

As explained in Chapter 2, Model II is based on the assumption that regulation should be appropriately designed to stimulate innovative behaviour. Properly designed regulation focuses on outcomes, not specific technologies; it encourages continuous improvement and leaves as little room for regulatory uncertainty as possible. Regulation should be stringent rather than lax, and coordinated with other policies to promote radical innovation.

The EU ETS focuses on outcomes, not technologies. The caps and carbon price do not mandate any particular type of technological response. This is in line with Porter and van der Linde's concept of 'appropriate' regulation in the sense that the system does not lock solutions into best-available-technology paths, but continuously encourages innovation in new technologies.

Regulation should be strict to stimulate radical innovation rather than merely incremental adjustment to the system. So far, the regulatory pressure has been low for energy-intensive industries. Some companies have also had large surpluses of allowances that can be transferred to the third phase of the ETS. In addition, companies have had the opportunity to invest in cheaper CDM projects abroad to comply with the ETS, reducing incentives to invest in abatement at home. Combined, these factors have led to less regulatory pressure than intended (and needed), and consequently less exploitation of new and innovative opportunities. This conclusion is also supported in the comprehensive study by Martin, Muûls and Wagner (2011). 'Appropriate' regulatory pressure from the ETS is also related to the criterion that regulation should be in line with or slightly ahead of other countries. Here we note, first, that no major rival or partners of the EU like China, India, Russia or the USA have implemented emissions trading or measures or targets equivalent to the EU's. The USA had planned for a climate and energy package, including cap-and-trade, but these plans stalled in Congress in 2009 (see Skjærseth, Bang and Schreurs forthcoming). Second, the climate negotiations in Copenhagen in 2009 failed to agree on a new, binding and comprehensive agreement to follow up the Kyoto Protocol. The climate negotiations on a new post-2012 agreement in Durban agreed on a road map towards a new comprehensive climate treaty, but where this will lead is highly uncertain. The main point here is that tightening the ETS cap to increase regulatory pressure becomes politically less feasible when other major emitters do not follow and early-mover advantages become more uncertain.

The ETS is based on the idea that a price on carbon provides monetary incentives for continuous improvement. The system has indeed established a carbon price, but it has generally been lower than expected and has varied considerably over time. Most companies refer to volatile carbon prices as a problem that raises the level of uncertainty for investments in future innovation. Oil giant Shell, on the other hand, has made a strategic decision to apply a planning premise for CO_2 at USD 40 per tonne, and this premise has not varied with the fluctuating carbon price.

Finally, the regulatory process should leave as little room for uncertainty as possible at every stage. We have not explored this dimension systematically in the case studies, but unpredictability has been a recurrent problem at the system level. The National Allocation Plan (NAP) processes have been surrounded by significant uncertainty caused by lobbying, administrative inertia and a tug-of-war between the member states and the European Commission (Skjærseth and Wettestad 2008), leading to delays and changes in NAPs. The levels of uncertainty will most likely be reduced in the third phase, with allocations based on harmonized rules and trading periods extended from five to eight years. The quantity of allowances will be reduced in a linear manner, ensuring gradual and predictable reduction in emissions. We should also note that the ETS has no deadline and is intended to continue with a new phase after 2020 in order to contribute to the long-term plans for decarbonizing Europe by 2050. The power, oil and paper industries have responded by developing industry sector plans until 2050. On the other hand, the financial crisis has spurred new attention to the need for political intervention to raise the carbon price, bringing new uncertainty for the post-2012 phase. This is further discussed in Chapter 9.

Coordination between the ETS and other relevant EU policies is particularly important to reduce the uncertainty that can arise when different types of regulation pull in opposite directions. This dimension has become increasingly relevant with the adoption of the EU climate and energy package where the ETS will interact with new policies on renewables, sectors not covered by the ETS, CCS, fuel-quality standards and more stringent vehicle emission standards, affecting industry sectors and companies to varying degrees. Several of our sector studies show that the various policies have pulled or are expected to pull in the same direction. However, the actual complementarity between the various policies adopted is far from perfect, something that could become increasingly evident in the policy implementation stage post-2012. Importantly, when the electric power industry begins to deliver on obligations under the new renewable energy and energy efficiency policies, the need for allowances under the ETS will go down and allowance prices may fall, weakening the incentives for companies to deliver on other climate technology solutions like CCS.

The oil industry will be affected by all parts of the EU climate and energy package, the combined effect being reduced demand for refined oil products. From the perspective of this industry, various parts of the energy and climate package are seen as causing duplication in documentation requirements for biofuels in the renewable and fuel-quality directives, and because of the conflict between the ETS and fuel-quality standards. The former will act to give incentives to energy savings, whereas the latter will entail extra use of energy to produce more environmentally friendly fuels lower in sulphur. This conflict has contributed to reduce energy efficiency at Shell's refineries. In the extended version of Model II, we emphasized the significance of national context, particularly the home-base context. Most cases point to the national context as important for understanding differences in ETS response and climate strategies. In the power industry, the link between public policy and corporate strategies seems particularly strong for companies that are wholly or partly state owned, like Vattenfall and RWE. The analysis shows that national and regional governments can and do seek to influence the strategic direction of companies where they possess ownership shares. Company movements towards more proactive and innovative climate strategies have depended on owners agreeing to ambitious commitments to solving the problem of climate change. The Swedish government held such ambitious commitments even before the EU ETS was adopted, making it an independent force for Vattenfall's strategic work. The municipalities of Germany's North Rhine-Westphalia, owners of RWE, saw local climate policy commitments increase around 2005, justified by adaptation pressure from the EU ETS. Here, state ownership worked more as an important intermediate factor between the EU ETS and strategic changes observed in RWE.

RWE's recently-announced strategy of halting new investments in coal-based power capacity and giving priority to investments in renewables was the result of ETS regulation combined with national and local German energy and climate policies. More restrictive policies on CCS and on nuclear power have made renewables increasingly attractive. With Germany home also to most of Vattenfall's coal-based power plant capacity, its recent similar strategic changes should be understood against the background not only of Swedish but also of German energy and climate policies.

The cement sector study shows that not only policies but also abatement opportunities vary significantly among countries and affect companies differently, depending on where their main activities are located. The Norwegian support scheme for CCS is important in explaining why HeidelbergCement chose a Norwegian installation for exploring this technology. With regard to opportunities, the availability of alternative fuels and alternatives for clinker (furnace slag and fly ash from steel production) varies widely. For example, cement manufacturers in the Netherlands cover 98 per cent of their needs with alternative fuels, whereas in Spain the figure is zero (see Chapter 6).

From the PPI sector we have seen the importance of national and local differences in opportunities for explaining differences in strategies and ETS response. This sector also displays significant differences in how companies exploit existing opportunities. The domestic context matters mainly for the PPI companies when it comes to electricity supply and availability of biomass to replace fossil fuels. This sector study has clearly shown how the effects of the ETS are mediated by factors of production as well as by market factors.

If we view the ETS in light of the EU's long-term climate and energy goals for 2050, we note significant uncertainty. Achieving the EU's long-term target of decarbonizing Europe will require massive investments in new energy infrastructure, particularly grid upgrading and CCS pipelines and storage sites. The need for competence and for public acceptance makes such plans uncertain. New infrastructure is decided at the national level, which has competence over area planning. The European Commission has launched new proposals to involve the EU in streamlining national planning approval of energy infrastructure and financing it. Lack of public acceptance for new energy infrastructure has already affected innovative plans, as we saw in the case of Shell's CCS project for its Pernis refinery in Rotterdam, where plans were blocked by local opposition. The project was technologically feasible and economically profitable, with the expected carbon price that Shell has incorporated to guide its investments.

Other similar examples concern Vattenfall and RWE's CCS storage plans onshore in Germany and in the Netherlands, which have encountered massive local opposition. Germany's federal states decided not to permit any exploration for storage, which led to postponement of the adoption of the German Carbon Storage Law. A draft law passed by the German Parliament in July 2011 would allow the federal states the right to designate/exclude areas for CCS pilot projects, but this was rejected by the Bundesrat, the legal body that represents the German federal states, in September 2011. The draft law postponed the final decision on whether CSS technology should be used as a full-scale climate mitigation option in Germany to 2017.

Concluding Assessment: Model II

Our main conclusion here is that the observations reported in the case studies are partly in line with what we could expect from Model II. We have seen several companies stepping up their short- and long-term innovation, particularly in the electric power industry. Within the energy-intensive industries, various long-term projects have been started up – insufficient, however, to conclude that there has been any fundamental strategic shift giving full priority to the development of long-term low-carbon solutions. This pattern is consistent with the shortcomings of the ETS according to the appropriateness criteria underlying this model. It is impossible to offer any robust conclusions concerning the relative explanatory power of Model I and Model II, as that would require higher-quality data over a longer period. One conclusion that could be made, however, is that the ETS in the first and second trading periods deviated considerably from the idea of 'appropriate' regulation, especially as regards regulatory pressure, and that could explain why we do not find comprehensive changes towards more innovative strategies. After the 2008 revision of the system, it appears that the electric power sector will come closer to such an ideal regulatory situation from 2013 onwards. Hence, the insights provided by Model II serve mainly to complement the analysis based on Model I. Model II explains better why most sectors and companies have responded by increasing their attention to, exploration of, and innovation in low-carbon solutions.

As such, the 'innovation' model complements Model I as a heuristic tool. The model points to other mechanisms – including attention and awareness, learning and future expectations – that resonate with observations in the case studies. Perhaps the most important consequence of the ETS to date has been to draw attention to how a low-carbon future will affect company risks and opportunities. The emphasis on new innovation opportunities is most prominent in the electric power industry, but we can note elements in other industries as well. The cement industry has stressed that incremental innovation is 'good business' and 'everyone is exploring everything' at this stage. The PPI and oil sectors have made an effort to shape their own views on how to survive in a decarbonized Europe by 2050. In the steel industry, the ETS has stepped up its attention to the need for breakthrough technologies. Moreover, companies and sectors have focused intensively on the design of the system. Both rhetoric and actions undertaken show that the companies have approached the ETS more broadly than merely in terms of minimizing regulatory costs.

Still, Model II cannot capture the fact that protection of climate systems is a public good with a normative dimension related to the needs of future generations. The third model takes account of such larger social benefits of climate-friendly innovation. Have companies perhaps been motivated by more than considerations of private risks and opportunities?

Model III: Companies as Socially Responsible

The 'responsibility' model is based on the assumption that company managers have mixed motivations in which social norms of responsibility also figure. Mixed motivations may influence how managers respond to the problem of climate change and regulation. Regulation can both 'crowd-in' and 'crowd-out' social norms of responsibility. A crowd-in effect will take place when regulation is perceived as supporting. This will be strengthened by an inclusive regulatory process based on industry participation. By contrast, a crowd-out effect will occur when regulation is perceived as controlling: this may serve to reduce actor motivation to act voluntarily. The mechanisms here are a shift in the locus of control, which may reduce self-determination and feelings of responsibility.

Emissions trading is expected to crowd-out norms of social responsibility. From a norm-based perspective, cap-and-trade is seen as controlling, indeed even as a potentially 'immoral' market-based instrument. This perspective is far more challenging to assess empirically than the previous models, since corporate norm-guided behaviour is extremely difficult to distinguish from other motivations. In the short-term, norm changes may not necessarily affect outcomes, as a crowding-out effect can be counteracted by a price effect. A change in norms involves a long-term process, and crowding-out can take place only if social norms were strong before the external intervention. With these caveats in mind, we could expect to find that social responsibility for dealing with climate change became reduced after the introduction of the ETS – or we would at least not expect social responsibility to have increased. These expectations are worth exploring, because the EU ETS was the first comprehensive, mandatory climate regulation for most of the sectors and companies assessed in this study, and also because the EU carbon market has experienced various problems that have brought its moral fairness into question. There have been several reported incidents of various types of ETS frauds and corruption with the CDM market and VAT, and hacking into accounts to transfer allowances to other accounts, for resale. The International Emissions Trading Association (IETA) has acknowledged that problems with fraud may threaten to damage the scheme's reputation if not resolved (IETA, 2010). Such incidents may weaken company trust in the EU ETS and its basic legitimacy.

Our case studies have explored whether sectors and companies have acknowledged the problem of climate change as diagnosed by the IPCC and taken responsibility for helping to solve it. We recognize that statements of responsibility often reflect a rhetorical adaptation to outside expectations – companies may simply 'talk the talk'. In order to conclude on social responsibility norms pre-ETS, we have also explored the extent to which companies have engaged in voluntary climate-related initiatives. Moreover, some of the cases have examined the extent of company cooperation with competitors, governments and civil society organizations on low-carbon solutions in the pre- and post-ETS situations. Here the underlying rationale is that genuine social responsibility for collective problem-solving is unlikely to be compatible with inward-looking individual corporate behaviour.

Most of the industry associations and companies stated that they had recognized their social responsibility for dealing with emissions also before the introduction of the ETS. In the cement industry, most of the major companies accepted the problem and took responsibility for problem-solving through voluntary business initiatives initiated before the ETS, including monitoring procedures, exploration of alternative abatement options and the adoption of short-term goals. The European industry association representing the cement industry – Cembureau – has since strengthened its cooperative initiatives with the sector by funding research on, for instance, CCS. There are no signs of reductions in the sector's stated responsibility or voluntary initiatives after the introduction of the ETS.

With regard to the specific company cases, HeidelbergCement acknowledged the problem in 2002, and responsibility statements have been strengthened over time. Holcim also acknowledged the problem prior to the ETS, and responsibility statements have remained firm since then. Both companies still appear serious in their efforts to take responsibility in terms of voluntary action, but this does not seem to represent any 'deep' or genuine change in social norms. Neither company has engaged in significant joint ventures beyond common projects at the sector level: this indicates at most a weak link between the ETS and responsibility statements and voluntary action. Responsibility statements are shaped by many factors, including changing market demand and competition from other building materials. However, the ETS has apparently not had any 'crowding-out' effect, as voluntary initiatives such as the Cement Sustainability Initiative continue to play an important role.

In the PPI, all the largest companies seemed to accept the problem and responsibility for dealing with it, also prior to the launch of the ETS. We have not found indications of a 'deeper' norm shift in the PPI in the form of considerably strengthened collective innovation action to bring about new low-carbon solutions. For instance, no major joint innovation programmes to develop climate-friendly technologies within the sector have been launched. Also the two companies in focus – SCA and Norske Skog – had acknowledged the problem, taken responsibility and adopted voluntary targets and measures before the introduction of the ETS. The companies have later strengthened their targets and measures. They have also shifted to a more positive position on the ETS, as it became clear that the scheme represented less of a regulatory burden than first anticipated. Observations from this sector indicate that the ETS has not 'crowded-out' responsibility as expressed in terms of voluntary action, which may or may not have an element of behaviour guided by social norms.

Our data on the steel industry are limited – perhaps because the sector is less interested in 'talking the talk', or that the sense of responsibility has been less developed. Compared to other energy-intensive industries, steel appears to have had fewer concerted or individual CSR programmes prior to the ETS. Responsibility statements and programmes have emerged mainly afterwards, as with the major voluntary joint long-term innovation programme carried out under the ULCOS framework. No deep shifts in norms were observed for this sector, but the ETS has served to draw attention to the climate problem.

The oil industry was split on climate change. Some European multinationals, like BP and Shell, adopted various voluntary measures shortly after the 1997 Kyoto Protocol was adopted, and lobbied actively in favour of the ETS. Others, mainly US-based companies with ExxonMobil in the lead, denied the problem, did not adopt any voluntary measures, and lobbied actively against the Kyoto Protocol, against US ratification of the Protocol and against the EU ETS. Today, all members of Europia have acknowledged their social responsibility. ExxonMobil shifted its stance cautiously in 2007 and declared that it would stop funding organizations that worked actively to deny the problem. Partly as a result of this, the industry association Europia temporarily expressed more responsibility in relation to the revision of the ETS. Shell and ExxonMobil also report increasing cooperation with the outside world in relation to climate change.

ExxonMobil still opposes emission trading in general and the EU ETS in particular. Interestingly the company substantiates this position by arguing that the allowance market shifts the emphasis away from the aim of reducing carbon emissions. This line of reasoning resembles the logic underlying the 'crowding-out' effect. Turning to the case of Shell, we note the complexity of studying the link between regulation and corporate norms of responsibility. In the late 1990s, Shell aimed to become an energy pioneer by capturing a 10 per cent share of the renewables market before 2005. This ambition formed part of a proactive climate strategy that included the ambition of becoming an allowance trade leader within the oil industry. Since then, Shell has divested itself of major parts of its renewables business, and stated in 2009 that it intended to terminate all new investments in renewable energy. However, this was clearly not linked to the company's perception of the ETS, which it has continued to support strongly. The ETS does not appear to have affected social norms either way, and has been too weak and narrow to create sufficient incentives for countering Shell's renewables divestment. It has, however, provided incentives to step up R&D on CCS.

From this perspective, the most interesting change in the power industry is firmer engagement in long-term collective action aimed at alleviating the problem of climate change. Prior to the EU ETS, Eurelectric surveys had shown cuts in long-term R&D expenditure in the industry – and, more importantly, an evolving disinclination to share business information in collaborative research, brought about by the new competitive conditions after the trade-based reforms in the internal energy market. This indicated that new business norms were evolving that one-sidedly focused on individual short-term profits and the rush for greater market shares, replacing earlier social norms whereby monopoly-protected companies had worked together for solutions to expand the national industry. Now, the new joint industrial initiatives (the common goal set for carbon-neutrality by 2050 and new joint long-term R&D initiatives) indicate a possible break with this trend, with a new industry behavioural pattern compatible with re-internalization of social norms on working together for a common good. This may perhaps mean broader motivation than merely the individual pursuit of short- and long-term profits.

New collective action on climate change coordinated at the level of the industries studied here has been connected to the EU ETS. Eurelectric fronted industry-level ETS allowance trade simulations in the early part of the decade, showing the industry had much to gain if the system did not include auctioning. Concurrent joint scenario projects revealed new opportunities for industrial expansion and profits from committing to long-term carbon-neutral supply, in turn making most companies sign into the agreement for this goal. Once this commitment had been made, the pooling of capital and human resources stood out as a perfectly rational solution, given the enormous capital costs involved in achieving the technological solutions needed to achieve carbon-neutral energy supply.

It is still too early to come to conclusions about the stability of the new behavioural pattern in the longer term, whether it will become internalized as new social norms of industrial collaboration. Companies will still be competitors in the market for electricity, but their new common vision of addressing the climate problem through market extension could well lead to further collaboration to ensure this expansion is a viable strategy for many firms. To the extent that new collective action in the power industry represents new social norms, we can conclude that the ETS has clearly not acted to crowd-out such norms. A more likely interpretation is that the ETS has promoted this development.

All in all, then, we have found no indications that the EU ETS has crowded-out norms of social responsibility. This conclusion should be interpreted with great caution, as responsibility statements, voluntary CSR activities and more widespread cooperation on low-carbon solutions may, or may not, be rooted in an element of social norms. Still, the industries have clearly been working to reduce emissions, also those not related to their obligations under the ETS. Such activities may be purely strategic, aimed at reducing reputational costs, avoiding further state regulation or lowering R&D costs. Alternatively, such abatement activities may be an expression of genuine norm-guided behaviour. A modified conclusion is that the EU ETS has not reduced voluntary social responsibility activities independent of their motivation. On the contrary, the power sector in particular shows that such activities have actually increased after the EU ETS was put into place.

As noted in Chapter 2, the crowding-out effect may be counterbalanced by a crowding-in effect: new norms of social responsibility created in the process of making the system. According to theory, such crowding-in would be stimulated by mutual acknowledgement of responsibilities, communication and participation in the regulatory process. With regard to responsibilities, the idea is that public authorities fix the cap and the rules of the game, and companies are free to trade within those restrictions. The power industry has accepted this approach, whereas the energy-intensive industries dislike the cap aspect of the ETS. However, the development of the system has been based on open communication and participation by industry. Since the first Green Paper on the ETS in 2000, industry has been involved in formal and informal consultation with the European Commission through the European Climate Change Programme and other channels (Skjærseth 2010). Our case studies have not systematically analysed the legitimacy of the decision-making process, but this aspect is noted by several of the authors. Hence, we have for example seen that the electric power industry was particularly active when the system was established, as noted above, and cooperation between representatives from the Commission and the power industry on trade simulations facilitated mutual support through communication.

Participation and communication in the regulatory process have probably contributed to reduce perceptions of control from 'above', particularly within the power industry, which has supported the system from its inception. We cannot rule out the possibility that a strategic response to a cap-and-trade system conforming to the 'crowding-in' regulatory process criteria, regulatory pressure and new IPCC knowledge may actually induce a corporate sense of responsibility in the longer term.

Concluding Assessment: Model III

Our main conclusion based on Model III and the case studies is that the EU ETS has not served to reduce voluntary social responsibility activities or collective action on low-carbon solutions. Observations from the energy-intensive industries, the electric power industry in particular, indicate that such activities have actually increased since the EU ETS was introduced. With the power industry, the main reason appears to be that the ETS has increasingly been perceived of as more supporting than controlling. The power sector has participated actively in the development of the system and has gradually seen new market opportunities in the electrification of Europe. However, we cannot say to what extent this represents a shift in genuine norms of climate responsibility.

Nevertheless, this perspective adds to our understanding of climate strategies and responses to regulation. By focusing on the social benefits of regulation, the responsibility perspective brings in how companies relate to each other and their social context. Climate change has a value component related to the needs of future generations. Most sectors and companies now accept this, and acknowledge their responsibility with respect to reducing greenhouse gas emissions. The extent to which such statements actually affect business decisions in a climate-friendly direction is extremely difficult to judge at this stage. On the other hand, it is also hard to envisage how the IPCC recommendations for the industrialized countries (80–95 per cent reductions by 2050) can be achieved without some elements of behaviour that is driven by norms of social responsibility.

Conditions for Reactive and Innovative Responses to Regulation

In this final analysis, we attempt to paint a picture of the factors that characterize typical 'reactive' and 'innovative' companies, based on observations from our study.[7] This may enable us to identify combinations of factors that work to co-produce different types of responses to the EU ETS.

What is it, then, that characterizes a company with an innovative response to the EU ETS? A first observation is that companies with strong dynamic capabilities are most likely to respond by innovation. 'Dynamic capabilities' refer to company management routines aimed at keeping an eye out for the opportunities emanating

---

7   We do not have sufficient information to provide a meaningful picture of socially responsible companies. However, it is reasonable to assume that such companies will pull in the direction of innovative strategies.

from changes in the firm's surroundings and environment, and then amassing and realigning resources, particularly significant in-house R&D capacity, to act on these opportunities. Some companies have stronger dynamic capabilities than others. The study of the power industry has shown that shifts in top leadership can be a necessary factor for re-orienting internal resources towards low-carbon solutions. Dynamic capabilities may also be a factor at the sector level, where such capabilities could be hosted by European industry associations, in turn linked to various organizations that conduct scientific and technical research. In sectors characterized by smaller companies, small size can to some extent be compensated by strong European industry associations, often linked to organizations for scientific and technical research, such as the European Cement Research Academy (ECRA). Eurelectric in the power industry stood out as the industry association that was by far the most active in participating in the development of the ETS at an early stage. A flipside is that such engagement may also be used to weaken the regulatory pressure.

Even though managers have flexibility to do more than merely administer given resources within a locked-in business strategy and resource base, our empirical assessment indicates that 'technology base' should be included as a factor independent of dynamic capabilities. General business strategy and technological lock-ins thus constitute a second factor. Some companies, independent of any dynamic capabilities, may have inherited business strategies and core technologies that make innovation towards low-carbon solutions easier than is the case for other companies within and across sectors. Cases in point include oil companies engaged in CCS for industrial purposes, cement companies relying heavily on clinker substitution, power companies with a high share of renewables in their energy mix or paper companies that own large tracts of forest. Conversely, many steel companies argue that their current process technologies have reached the theoretical limit for further reductions in emissions. The upshot is that companies based on core technologies that may be developed further into low-carbon solutions are likely to respond by innovation in the short-term. On the other hand, companies facing strong constraints with their current core technologies may actually have extra incentives for developing breakthrough technologies in the longer term.

Third, we see that regulatory risk is important, linked to inherited business strategies and core technologies. Low-carbon-intensive companies are less likely to respond with resistance than are companies within the same sector with high carbon intensity. However, this factor should not be exaggerated – our case studies show a systematic pattern of convergence within sectors, whereby laggard companies catch up with the leaders. This is particularly evident for power and oil companies. Moreover, high carbon intensity can serve to make incremental adjustment less likely and radical breakthrough innovation more likely. The results from the case studies are inconclusive as to the relationship between high carbon intensity and innovation.

Fourth, exposure to international competition appears to be a more important regulatory risk than carbon intensity. Companies in sectors exposed to limited competition from outside Europe are significantly more likely to respond innovatively. We found a clear difference in response between European power companies and companies in energy-intensive sectors. Finally, companies with headquarters and significant activities in home-base countries that are 'enabling' in terms of a stringent and supportive climate and R&D policies are more likely to respond with innovative strategies than are those operating under less-enabling national conditions.

Table 8.1   Combination of factors shaping 'reactive' or 'innovative' companies

	Companies tend to adopt reactive strategies	Companies tend to adopt innovative strategies
Dynamic capabilities for low-carbon solutions	Weak	Strong
Technology base for low-carbon solutions	Weak	Strong: promote innovation also in the short term Weak: additional incentives for breakthrough technologies in the long term
Carbon intensity	High carbon intensity will lead to resistance to regulation	Indeterminate
Exposure to international competition	High	Low
National home-base context	Constraining	Enabling

A company that responds to regulation such as the ETS by innovation could display one or more of the following features (see Table 8.1): strong dynamic capabilities, a technology base suitable for low-carbon solutions, and exposed to limited international competition within an enabling national context. Conversely, a reactive company will typically have low dynamic capabilities, a technology base unsuitable for low-carbon solutions, and be exposed to significant international competition within a constraining national home-base context.

Whether companies respond in either direction will depend on the institutional and regulatory context. With regard to cap-and-trade, we note that corporate involvement in creating the system, strong regulatory pressure, coordination with other relevant regulation and predictability will increase the likelihood of innovative responses. The wider institutional context may also affect corporate strategies. Harmonized international regulation at sector or country level will reduce the regulatory risk in terms of competitive disadvantage, increase the advantages for early movers and promote enabling national contexts.

Finally, we have seen that exogenous factors are extremely important in explaining variation in corporate climate strategies. Two such factors in particular have worked against innovative strategies in the context of the EU ETS and EU climate policies. The first is the succession of the economic crises in Europe since 2008. These crises have reduced the regulatory pressure of the ETS, directed attention away from climate change mitigation and drained financial resources from investments – including low-carbon solutions. A second factor is public acceptance of new infrastructure or CCS. In the power and oil sectors, we have seen how solutions that are technologically feasible and economically profitable have not materialized, due to local opposition.

Conclusions

This chapter has offered a cross-cutting assessment of our ten company cases within five sectors in light of three models designed to explain whether, how and why companies have responded to the EU's Emissions Trading System. We have analysed the EU ETS in the context of other relevant EU policies and other company-internal and wider external drivers likely to co-produce variation in climate strategies. Exogenous factors have also been taken into account.

Our first conclusion is that all companies and sectors have, to varying degrees, shifted their short- and long-term climate strategies in a more 'innovative' direction since the introduction (or anticipation) of the EU ETS. The most significant changes can be observed in the electric power industry. Interesting changes have also taken place in the energy-intensive industries, but these changes have been less substantial and less widespread. Moreover, we note that inter-company collaboration on low-carbon solutions has increased.

Second, that particularly the two first models can explain these changes. The 'reluctant adaptation' model captures well the main difference in response between the electric power industry and the energy-intensive industries. The electric power industry has been exposed to significantly higher regulatory pressure and lower regulatory risk in terms of competitive disadvantages than the energy-intensive industries – and the impact of the EU ETS on climate strategies has been significantly higher in the electric power industry. The 'innovation' model, based on the Porter Hypothesis, helps to explain variation in corporate response within the sectors, regarding long-term low-carbon activities in particular. This perspective shows that regulation in the form of the EU ETS has triggered attention and stimulates exploration, experimenting, learning and also investment into low-carbon solutions that may benefit companies. It also contributes to explaining variation between the electric power industry and energy-intensive industries by highlighting the importance of 'appropriate' regulation extending beyond regulatory pressure. The relatively limited but varying shifts in the climate strategies of energy-intensive companies can be explained by a lack of comparable climate policy in major trade partners of the EU, by low and unstable carbon prices, by unpredictability, and to some extent by poor horizontal and vertical coordination between the ETS and other relevant policies within Europe.

The 'responsibility' model has contributed less in explaining varying climate strategies in the sense that there are apparently no specific observations reported in the case studies which cannot be explained without reference to this model. The reason is not necessarily that social norms of responsibility are unimportant since they are hard to observe and different models can explain the same patterns with reference to different behavioural mechanisms. This model might still be fruitful as it directs our attention to the social benefits of low-carbon solutions and the complex relationships between different types of regulation and norms. The analysis has not supported the theory-based expectation that emissions trading will 'crowd-out' voluntary individual or collective corporate action on climate change: such activities have actually increased after the EU ETS. Admittedly, we do not know the extent to which the voluntary actions reported in place before and after the ETS reflect changes in norms or cost–benefit calculations.

The third important observation is that the impact of the EU ETS has been conditioned by company-internal factors and the wider political, economic and social context in which the companies operate. Climate strategies have also been affected by exogenous factors, such as the financial crisis and public acceptance of low-carbon solutions, with the financial crisis and low acceptance constraining innovation as part of climate strategies. Against this backdrop, we have sought to explore the conditions for responses to regulation by painting a 'portrait' of company characteristics conducive to reactive or to innovative responses to the EU ETS.

The results reported in this study have been based on data collected mainly before the European crisis fully unfolded and the accompanying drop in attention and political energy invested in climate-change mitigation. Chapter 9 concludes this study with a discussion of the road ahead.

References

IETA. 2010. EU Emissions Trading System: Closing the Door to Fraud. International Emissions Trading Association, Press Release 11 February 2010.

Martin, R., Muûls, M. and Wagner, U. 2011. *Climate Change, Investment and Carbon Markets and Prices – Evidence from Manager Interviews*. Carbon Pricing for Low-Carbon Investment Project, Climate Strategies, Climate Policy Initiative, Berlin, January 2011.

Skjærseth, J.B. 2010. EU Emissions trading: Legitimacy and stringency. *Environmental Policy and Governance*, 20, 295–308.

Skjærseth, J.B., Bang, G. and Schreurs, M. Forthcoming 2013. Explaining growing climate policy differences in the European Union and the United States. *Global Environmental Change*.

Skjærseth, J.B., and Wettestad, J. 2008. Implementing EU Emissions Trading: Success or failure? *International Environmental Agreements*, 8(3), 275–90.

Skjærseth, J.B., and Wettestad, J. 2010. Fixing the EU Emissions Trading System? Understanding the post-2012 changes. *Global Environmental Politics*, 10(4), 101–23.

Skodvin, T., Gullberg A.T. and Aakre S. 2010. Target-group influence and political feasibility: The case of climate policy design in Europe. *Journal of European Public Policy*, 17(6), 854–73.

Chapter 9

Concluding Remarks and the Road Ahead

Per Ove Eikeland and Jon Birger Skjærseth

The EU ETS is the EU's climate policy 'flagship', intended to steer Europe towards a low-carbon economy in the short- and long-term. The EU has invested significant political capital in making the system work, and researchers have invested significant intellectual capital in explaining how it works. This study seeks to contribute to this pool of knowledge by asking whether, how and why *companies* have responded to the system to date.

The previous chapter analysed our main empirical findings concerning the electric power, oil, cement, steel and pulp and paper sectors. These findings can be summarized as follows. First, all companies and sectors studied have, to differing degrees, adopted more proactive strategies with 'innovative' elements since the introduction or anticipation of the ETS. Second, cooperation between companies on low-carbon innovation has increased. Third, the most significant aggregate changes can be observed in the electric power industry, which has embarked on a strategy to decarbonize power supply in Europe by 2050. The changes observed in the energy-intensive industries appear less significant but are interesting, as there have been few studies of company responses in these sectors. Fourth, the ETS has affected these changes significantly, particularly in the electric power sector, which has faced the strongest direct regulatory pressure from the system. Still, the strength of the causal relationship between the ETS and corporate climate strategies is extremely hard to determine precisely, because of other international, EU and national policies that have co-evolved with the ETS. The key to explaining changes in corporate climate strategies from a regulatory perspective is found in how the ETS works together with other policies.

A fifth finding relates to the question of *how* the ETS has affected company climate strategies. Our study has found evidence of ETS impacts through a wide range of mechanisms – notably through regulatory pressures that create incentives for cost-cutting in periods when the ETS price was expected to increase, but also through triggering attention, experimentation, learning and investment with respect to low-carbon solutions outside business-as-usual for the companies. We have seen that corporate responses take place mainly at the intersection between these mechanisms.

Variations in response to the EU ETS across sectors and companies were found to be linked to differences in regulatory risks and opportunities. These were in turn tied to company-internal factors such as the inherited resource base and managerial capabilities, as well as external factors outside the ETS. The

ETS has clearly worked in combination with company-internal factors and the wider political, economic and social context in which the companies operate. As could be expected, we also noted factors unrelated to the ETS and other drivers incorporated in the analytical framework: these factors include the finding that corporate climate strategies are influenced by public acceptance of low-carbon solutions and the financial crisis.

A tentative conclusion is that innovative responses seem most likely in a company that is exposed to limited international competition. Moreover, companies that have inherited a technology base suitable for low-carbon solutions will have greater opportunities for responding innovatively, even though they may face low regulatory risks. On the other hand, resources and technologies do not determine response strategies. Companies with strong dynamic mangement capabilities will have greater opportunities for innovative strategic responses. Finally, companies that enjoy supportive and enabling national contexts are more likely to respond with innovative short- and long-term strategies.

Three alternative 'behavioural' models of corporate response to regulation were developed to assess the empirical material. These depict companies as reluctant adapters, as innovators and as socially responsible. We believe that this choice has shown the fruitfulness of dealing with a richer range of potential mechanisms for ETS impact on company strategies than what would have been possible with a single-model study. There is also another advantage: the ETS would probably demonstrate its greatest total effect on company climate strategies if various mechanisms were working in a coherent direction. Thus our study has focused not only on the ETS as a producer of cost-incentives and an attention-steerer, but on also its relationship with norms of social responsibility. We have, however, found no strong evidence of a causal link between the ETS and norms of responsibility – but this may change in the future.

We have also noted some specific mechanisms within the behavioural models not well documented in the literature. The study of the cement industry (Chapter 6) showed that the 'windfall profits' gained from the ETS, which have been criticized as indicating a malfunctioning system, in fact appear to have been used to scale up climate change mitigation activities. 'Income effects' in addition to 'price effects' from the ETS stand out as interesting research topics for future studies. Such income effects could work in either direction: by enabling funding and thus stimulating greater efforts at developing long-term low-carbon solutions; or by levelling out the incentive effect provided by carbon prices, thus leading firms to reduce their long-term efforts.

An interesting observation from the electric power industry study concerns the linkage established between the ETS and company credit-worthiness evaluations, in turn affecting company climate strategies. When the global credit rating institution Standard & Poor began including carbon costs in its evaluations, this gave some power companies a negative connotation to their credit worthiness, thereby reinforcing incentives created by the ETS. This observation points to the need to include not only production but also capital costs.

Another strength of our research approach, we believe, is our use of systematic comparison of sectors and companies across explanatory factors and different points in time. The comparative approach has enabled a systematic search for similarities and differences in climate strategies and the potential explanatory factors included in the analytical framework. We believe our analytical framework has merits for the study of regulation or the effect of policy instruments more generally, by showing one way of studying how different mechanisms may stimulate desired changes in industry. This approach is in principle applicable to all issue-areas characterized by efforts to govern large corporations as the causes of environmental problems.

Despite the already broad explanatory focus, our empirical findings point to the existence of a more extended or sophisticated set of explanatory factors and causal mechanisms than those covered by the upfront analytical framework. These could well gain greater clarity through research employing alternative or complementary analytical lenses. A widespread empirical pattern observed is that corporate climate strategies have apparently evolved in the direction of *convergence* among the companies in the industry sectors explored in this study. Institutional theory – notably the theory of institutional isomorphism, with its focus on convergent organizational strategies and structures – could shed light on a greater variety of mechanisms at play (see DiMaggio and Powell 1991). The theory is based on the concept of 'organizational fields' as a separate analytical level describing a network of companies and related actors bound together by frequent interaction. According to this theory, organizations adopt practices not primarily in order to achieve greater efficiency, but to adapt to changes in such a greater organizational field that are considered proper, natural or legitimate (DiMaggio and Powell 1991, Galbreath 2009, Gunningham 2009).[1]

In fact we have found empirical evidence pointing towards the EU ETS as a factor that has triggered more frequent interaction among agents, indicating that 'the organizational fields' could be relevant as analytical concept. Discussion of the ETS was a key part of the European Climate Change Programme established by the European Commission in 2000 (see Chapter 1). This programme led to new institutional arenas where industry, green groups and the European Commission exchanged ideas, shared information and clarified their positions. This interactive process was in turn linked to various European technology initiatives and platforms that stimulated companies to collaborate more closely within industry associations, and stimulated these associations to interact more closely with policy-makers and civil society actors. One key for understanding the changes that have taken place in corporate climate strategies lies in the combined influence of such processes, the ETS and other relevant EU policies on climate and energy.

---

1   Institutional isomorphism theory deals with three mechanisms for the convergence of practices, strategies, and structures within organizational fields: coercive, mimetic and normative isomorphism (DiMaggio and Powell 1991).

Yet another framework that could fill in nuances in the convergence of climate strategies is Schelling's 'tipping model' (Schelling 1978, Skjærseth and Skodvin 2003: 210). This model sees group dynamics as 'critical mass phenomena', and can be applied generally to identify critical masses and aggregated outcomes. The model would predict that change in strategy by one important company, or a critical number of companies, could trigger large-scale changes within and possibly across industry sectors. An underlying mechanism is 'the more the merrier': the more companies that join, the stronger will be the incentives for others to join. This could, under certain conditions, lead to a 'race to the top' where early-mover advantages serve as important motivation.

An analytical extension could also be recommended, based on the observations of changes in company *beliefs* about the future – in the form of expectations about future EU allowance prices but also the expected future composition of energy systems, as indicated by the many scenarios presented by the various companies and industry associations. Such manifestations of beliefs constitute important premises for how ETS responses could be expected to evolve. The shaping of beliefs concerning future ETS prices should be seen within the context of the totality of EU (and national and international) climate and energy policies, not least the EU ambition to decarbonize Europe by 2050. The point to note here is that changes in beliefs appear to have had the most significant impact on 'laggards' or companies that behave most in accordance with the 'reluctant adaptation' model (see Chapter 2). We find the most concrete examples in the electric power and oil industries. In the oil industry, change in the strategy of one of the major 'laggards' influenced the position of the European industry association in a more proactive direction, at least temporarily. Companies may oppose changes, but at a certain point they will feel isolated. While early-mover advantages can explain why some companies take the lead, isolation may be important in explaining why some companies catch up, and why aggregate changes in the positions of industry associations occur.

Finally, we acknowledge in retrospect the need for a more sophisticated approach to how, and with what consequences, the EU ETS interacts with other climate, energy and technology policies at national and EU levels. This will become increasingly important when the 2008 EU climate and energy package is fully implemented for the post 2012 period. Some companies and sectors, like the oil industry, are affected by most policies on renewables, energy efficiency, CCS, fuel quality and vehicle emissions, in addition to the ETS.

In addition to the need for analytical extensions and refinement, we should note that our research approach and methods have had some limitations. The first limitation is related to the importance of more thoroughly investigating the role of norms of responsibility and potential changes in norms associated with the introduction of the ETS. Future work should make use of a greater range of field-research methods than this study – not least, more intensive face-to-face interviewing and direct observation of the objects studied. A second limitation relates to our investigation of management capabilities for change and innovation as a potentially fundamental company-internal factor that influences the choice of climate strategies, such as exploration of new business lines and engagement in collaboration on long-term R&D. Although sparse, the empirical findings have still given fruitful inputs to our analysis of dynamic capabilities in interaction with the ETS.

Finally, the selection of objects, a relatively small set of companies mainly based in Northern Europe, has limited the full potential for this research approach to provide empirically generalizable conclusions. The generalization aimed at here is mainly analytical, based on a systematic assessment of a limited number of cases in light of the expectations derived from the three models. However, we have also sought to capture more general empirical patterns by exploring aggregate changes at sector levels. Future comparative studies should expand the focus to include companies from a broader range of national contexts – particularly to shed more light on the conditions under which different corporate responses to the EU ETS are likely to occur. Likewise, future comparative studies should include the new sectors added to the system after the 2008 revision, like aluminium and aviation.

The Road Ahead

This study has shown that companies have responded to the EU ETS by adopting more proactive climate strategies with elements of low-carbon innovation. The issue now is increasingly becoming whether the ETS is appropriately designed to drive industry responses further toward a low-carbon economy when significant changes in economic circumstances occur. This challenge raises questions not only about political feasibility, but also about the choice of policy direction: should the EU ETS merely meet reduction targets set for 2020 or impose a sufficient carbon price to drive investments in low-carbon technologies to meet 2050 targets? Is the objective to drive certain technologies (e.g. renewables), and if so, how can we avoid those technologies displacing the development and diffusion of other low-carbon technologies (e.g. CCS)? These are questions about governance where responsibility rests heavily with the regulator.

The first question is essentially about the design and stringency of EU ETS. If the system is to drive investments into future low-carbon solutions, stringency would be required quite independently of what behavioural model would best describe company responses to the ETS. Companies that adapt to the system with a purely short-term cost-minimization focus would in fact need a very stringent ETS in order to respond with innovative strategies. Companies that have already adapted to the ETS by attention to new opportunities will still need regulatory stringency to uphold their attention and search behaviour for new opportunities. Finally, companies may also be expected to respond positively in terms of social responsibility if future efforts to tighten the system are perceived as legitimate and are based on industry involvement. In essence, very much will depend on how the system in place from 2013 will evolve as to regulatory stringency and expected carbon prices.

Predicting future carbon prices is no easy endeavour, given the multiple economic, political and technological forces at play impacting on the demand and supply of allowances. The basic driver of higher prices is fairly straightforward, however: scarcity in allowances – a situation where demand for allowances exceeds supply, clearing the market at a higher price level. Demand for allowances is essentially decided by the level of economic activity (production of goods and services needing allowances) and the carbon intensity of producing goods and services.

The supply of allowances is basically fixed by the cap set for the short term (decreasing by 1.74 per cent annually in the period 2013 to 2020). Still, various policy interventions have recently been proposed to lower the supply of allowances and, more directly, to intervene in the price-setting mechanism. The reason such extra-market options have attracted renewed attention is the recent collapse of the ETS spot-market price. The average carbon price recorded in the final quarter of 2011 was €9 per ton CO_2 (Spencer and Guérin 2012). The market simply sees little scarcity, mainly because of economic output and energy demand figures pointing downwards in Europe and thus also the demand for allowances. As expected carbon prices differ fundamentally from those that this study has documented as being used to guide investments – notably the $40/tonne flagged by Shell – we cannot expect the ETS to provide appropriate signals in the short run for realizing the massive energy infrastructure investments currently planned in Europe.

Fundamental economic driving forces notwithstanding, allowing the EU ETS to produce higher carbon prices actually boils down to a question of politics – the political willingness to make changes in the design of the ETS or impose additional policy measures to support higher carbon price levels. Such political willingness was shown when European leaders in 2008 adopted the revisions of the system that extended it in the long term and standardized the allocation of allowances to make it more robust to strategic adaptation by member states to create undue competitive advantages for national companies.

The quest for new revisions has resulted in a range of policy proposals aimed at creating higher and less volatile carbon prices. The European Commission's DG Climate Action has proposed raising the unilateral EU emissions reduction goal from 20 per cent to 25 per cent for 2020, with a corresponding reduction in the annual ETS cap, thus increasing prices through reduction in supply of allowances. The proposal enjoyed the strong support of various member states but was not accepted by the Polish presidency.

Another proposal aimed at limiting supply is to permanently set aside a quantity of emissions allowances from auctioning during the third phase to eliminate the current oversupply. Other proposals more directly aimed at carbon price regulation include imposing a price floor to the ETS. Yet a final proposal, championed by French President Nicolas Sarkozy, was to shelter European industries exposed to competition by enforcing a carbon price on international competitors in the form of a border tax on products from non-regulated industries abroad.

Reaching agreement on such measures does not seem politically feasible at present, when political attention is heavily focused on the series of credit and economic crises that have continued to strike Europe. In fact, these economic troubles may have provided new ammunition to industries that wish to see the stringency of the system lowered. As we have seen, many energy-intensive companies were of this opinion when revisions of the ETS were debated in 2008. The industries exposed to international competition were given considerable concessions through the decision to continue handing out parts of their allowances for free, also for the third phase. Their cautious support of the system moreover appeared conditional on it being only the pilot for an envisaged broader international system.

When the Copenhagen climate summit in 2009 failed to provide for more ambitious international commitments and the proposed federal US cap-and-trade system became stalled in Congress, industrial expectations of a future global system faded. And, after the Durban Summit in December 2011 postponed a new international climate treaty to 2015, to take effect in 2020 at the earliest, the prospects of getting immediate broad industrial support for measures to produce substantially higher carbon prices may be even bleaker.

Higher carbon prices still have many supporters, among them the electric power industry and its industry association Eurelectric. Eurelectric wants policymakers to accept that the EU ETS, in contrast to renewables and energy-efficiency policies currently executed through national approaches, should drive the decarbonization of Europe.

This directs our attention to the second question, which is essentially about conflicting instruments that push certain technologies and works to press down carbon prices in Europe: EU and nationally mandated investments in renewable energy. Most notably, the electricity supply industry has blamed both EU renewables and energy efficiency policies for exaggerating the downward trend of demand and allowance prices, and has urged the EU to drop its 20 per cent renewables targets for 2020. The underlying premise for this argument is that the industry wants no top-down political selection of low-carbon solutions, which is held to impose additional costs on achieving targets set for emissions reduction. Dropping the specific goals as to renewable energy would, however, not be seen as a proper policy option by a great many policy actors, and certainly not in the fast-growing renewable energy supply industry. Member-state governments were united in setting binding goals for renewables at the EU level, not least since renewables are seen as serving also other energy policy goals than emission reductions, notably by reducing imports of fossil fuels, thereby strengthening security of energy supply as well as creating new industrial growth and employment.

There are clear signs of differing opinions between the two Commission services – DG Energy and DG Climate Action – as to the role the ETS should play as compared to other policy measures in guiding the EU towards its long-term goal of 80–95 per cent reductions in GHG emissions by 2050. DG Climate

Action has been fighting for the ETS to provide sufficient prices signals to drive low-carbon investments in all sectors covered by the system. DG Energy, in its Energy Roadmap 2050, underscores that a higher carbon price would not only create incentives for investment in low-carbon technologies but also increase the risk of carbon leakage. This Roadmap thus supports the EU ETS in combination with instruments designed to achieve specific energy-policy objectives – notably, research and innovation, promotion of energy efficiency and development of renewables.

DG Climate Action has recently signalled that the first annual report on the ETS, scheduled for 2013, would be hastened to include a review of the auction time-profile for the third trading period – which is equivalent to a temporary set-aside of carbon allowances. This option seems attractive, because it will not restrict further intervention options and can be adopted by a qualified majority in the Climate Change Committee. Yet, it still remains to be seen whether the planned hold-back of allowances to be auctioned can achieve the needed support, whether it will create sufficient scarcity to raise short-term prices and whether it will be viewed by the market as sufficiently credible to create stable higher prices. What seems obvious today is that, with the current price expectations, the EU ETS is unlikely to drive new innovative climate strategies in European industries. That role now appears to be held by other policies at the EU and (particularly) the national levels. Renewable energy policies will certainly induce innovation in energy supply. In fact, they may also create incentives for innovation in energy-intensive industries, should energy prices rise because of the costs associated with developing and deploying new renewable energy. Low carbon prices could, on the other hand, deter innovation in technologies not supported by binding targets or feed-in tariffs, such as CCS.

What the longer-term future will look like is veiled in greater uncertainties – when and how the EU will recover from its deep economic crisis, whether and how the EU will continue to play a leadership role in international climate policies, and indeed, whether the ETS will be accorded new political priority as climate policy instrument after 2020. The 'flagship' of EU climate policy is still afloat, but it is headed into extremely troubled waters.

References

DiMaggio, P.J. and Powell, W.W. 1991. The iron cage revisited: Institutional isomorphism and collective rationality in organizational fields. In *The New Institutionalism in Organizational Analysis* (eds), W.W. Powell and P.J. DiMaggio. Chicago, IL: University of Chicago Press.

Galbreath, J. 2009. Corporate governance practices that address climate change: An exploratory study. *Business Strategy and the Environment*, 19, 335–50.

Gunningham, N. 2009. Shaping corporate environmental performance: A review. *Environmental Policy and Governance*, 19, 215–31.

Schelling. T.C. 1978. *Micromotives and Macrobehaviour*. New York and London: W.W. Norton.

Skjærseth, J.B. and Skodvin,T. 2003. *Climate Change and the Oil Industry: Common Problem, Varying Strategies*. Manchester: Manchester University Press.

Spencer, T. and Guérin, E. 2012. Time to reform the EU Emission Trading Scheme. *European Energy Review*, 23 January.

Index

Page numbers in *italics* refer to figures and tables.

GLOBAL ENVIRONMENTAL GOVERNANCE SERIES

Full series list

Renewable Energy Policy
Convergence in the EU
The Evolution of Feed-in Tariffs in
Germany, Spain and France
David Jacobs

The EU as International
Environmental Negotiator
Tom Delreux

Global Energy Governance in a
Multipolar World
*Dries Lesage, Thijs Van de Graaf
and Kirsten Westphal*

Innovation in Global Health Governance
Critical Cases
*Edited by Andrew F. Cooper
and John J. Kirton*

Environmental Skepticism
Ecology, Power and Public Life
Peter J. Jacques

Transatlantic Environment and
Energy Politics
Comparative and International Perspectives
*Edited by Miranda A. Schreurs, Henrik Selin,
and Stacy D. VanDeveer*

The Legitimacy of International Regimes
Helmut Breitmeier

Governing Agrobiodiversity
Plant Genetics and Developing Countries
Regine Andersen

The Social Construction of Climate Change
Power, Knowledge, Norms, Discourses
Edited by Mary E. Pettenger

Governing Global Health
Challenge, Response, Innovation
*Edited by Andrew Cooper, John Kirton
and Ted Schrecker*

Participation for Sustainability in Trade
*Edited by Sophie Thoyer and
Benoît Martimort-Asso*

Bilateral Ecopolitics
Continuity and Change in Canadian-
American Environmental Relations
*Edited by Philippe Le Prestre
and Peter Stoett*

Governing Global Desertification
Linking Environmental Degradation,
Poverty and Participation
*Edited by Pierre Marc Johnson,
Karel Mayrand and Marc Paquin*

Sustainability, Civil Society and
International Governance
Local, North American and
Global Contributions
Edited by John J. Kirton and Peter I. Hajnal

A World Environment Organization
Solution or Threat for Effective International
Environmental Governance?
Edited by Frank Biermann and Steffen Bauer

Hard Choices, Soft Law
Voluntary Standards in Global Trade,
Environment and Social Governance
*Edited by John J. Kirton and
Michael J. Trebilcock*

The Politics of Irrigation Reform
Contested Policy Formulation and
Implementation in Asia, Africa and
Latin America
*Edited by Peter P. Mollinga
and Alex Bolding*